First International Interdisciplinary Scientific and Practical Conference Man in the Arctic (IIRPCMIA 2021)

E3S Web of Conferences Volume 378 (2023)

St. Petersburg, Russia
17-19 November 2021

Editors:

Aleksei Makhovikov
Ekaterina Samylovskaya

ISBN: 978-1-7138-7241-2

Printed from e-media with permission by:

Curran Associates, Inc.
57 Morehouse Lane
Red Hook, NY 12571

Some format issues inherent in the e-media version may also appear in this print version.

For additional information, please contact EDP Sciences – Web of Conferences
at the address below.

EDP Sciences – Web of Conferences
17, Avenue du Hoggar
Parc d'Activité de Courtabœuf
BP 112
F-91944 Les Ulis Cedex A
France

Phone: +33 (0) 1 69 18 75 75
Fax: +33 (0) 1 69 28 84 91

contact-edps@webofconferences.org

Additional copies of this publication are available from:

Curran Associates, Inc.
57 Morehouse Lane
Red Hook, NY 12571 USA
Phone: 845-758-0400
Fax: 845-758-2633
Email: curran@proceedings.com
Web: www.proceedings.com

TABLE OF CONTENTS

The Arc-Of-Meridian Expedition to the Spitsbergen Archipelago 1899-1901 as an Example of International Scientific Collaboration: Tasks and Results 1
Kyzyurova Nadezhda

The Extreme Anthropogen of the Arctic: The Formation of the Great Glacier, the Emergence a Man and the Arctic Ocean in the Early Holocene 7
Zhirnov Anatoly

Study of Static Charge Accumulation in HDPE Gas Pipelines 12
Pshenin Vladimir, Menshikov Stepan, Komarovskiy Maksim

Rational Methods of Geodetic Control of Technogenic Consequences of the Development of Oil and Gas Fields 16
Volkov Victor, Volkov Nikita, Volkova Tatiana

Small-Sized Magnetolevitation System of the Trestle Type for the Arctic 21
Sundukov Eugene, Shifrin Boris, Tarabukina Nadezhda, Sundukova Veronica

North Pole Ice-Resistant Self-Propelled Platform as an Innovative Complex for Research in the Arctic 26
Likhomanov Vladimir Alekseevich, Ugrumov Yuri, Savitskaya Anna, Patseva Polina, Krupina Nina, Chernov Alexei, Svistunov Ivan, Likhomanov Vladimir Antonovich

Reducing Harmful Air Emissions from Vehicles, Ships and Stationary Facilities Operating in the Arctic 31
Antipov Boris, Shotidi Konstantin, Novikov Denis

Implementation of an Electrical Network Monitoring System Based on a Non-Contact Temperature Sensor and a Current Sensor Based on the Hall Effect 35
Burylov Dmitriy, Smirnov Vladimir, Dvoynikov Vladimir, Shpenst Vadim

Study of Water Purification Reactor Operating Modes 39
Chuikov Alexander, Volkov Andrey, Barskii Igor, Shabanov Dmitrii

Investigation of the Effect of Algorithms of Positioning Robots on Their Power and Energy Consumption 45
Volkov Andrey, Kolesnikova Anna, Kornilova Anastasia, Matsko Olga

Thermodynamic Modeling of the Reaction of Simultaneous Interaction of Nickel (II), Cobalt (II) and Iron (III) Oxides with Carbon 50
Kolesnikova Olga, Nurbaeva Farida, Sabyrbayeva Guizhan, Nurshakhanova Lazzat

Quality Analysis of Voxel Models Obtained with Remote Sensing 55
Vystrchil Mikhail, Sukhov Arseniy, Rybakov Anton, Chura Mikhail, Artemova Galina

Peculiarities of Climate Change in the Arctic and Implications for Environmental Management 60
Drachkova Lyudmila, Kondratov Nikolai, Trofimova Anna

Variability of the Ice Area and the Possibility of Achieving an Ice-Free Regime in the Barents Sea 65
Istomin Eugene, Petrov Yaroslav, Martyn Irma, Abramov Valery, Yagotinceva Natalia

Assessment of Hunting Pressure on Arctic-Nesting Shorebirds: First Results from the Northeast of Russia 69
Klokov Konstantin, Gerasimov Yuri, Syroechkovskiy Eugene

Indigenous Hunting in the Russian Arctic: Toward Sustainable Wildlife Resource Management 75
Klokov Konstantin

Efficiency of Sorption-Biological Treatment of Oil-Polluted Coastal Areas in the Euro-Arctic Region 81
Myazin Vladimir

Problems of Environmental and Technogenic Safety Regulation in the Arctic Region: Oil Spills 85
Yakovlev Sergey, Shemyakin Alexey

Oil Spill Response Technology for Ice-Covered Waters in the Arctic 90
Mingazhev Askar, Shchipachev Andrey, Mingazheva Alisa

Permafrost Degradation and Devastating Floods of the 20th – 21st Centuries for the Livelihoods of Yakut Arctic Villages 95
Suleymanov Alexander

Effectiveness of the Right to Health and Favorable Environment in the Arctic Regions 102
Markus Jean-Paul, Borisov Uolan

Problems of Entrepreneurship Development in the Arctic: Russian and Foreign Experience of Observing the Principles of Green Economy 107
Kuchumov Artur, Pecheritsa Elena, Blazhenkova Natalia

Problems and Prospects of Economic Growth in the Context of the Environmental Component of Sustainable Development in the Arctic Zone of the Russian Federation 113
Kuchumov Artur, Pecheritsa Elena, Blazhenkova Natalia, Chaikovskaya Alexandra

A Paradigm Shift in the Agricultural System in the Arctic Zone of the Russian Federation 119
Ingiri Andrey, Ageeva Polina, Ikko Natalia, Yashkina Anna

Prospective Industrial Complexes in the Russian Arctic: Focus on Rare-Earth Metals 124
Solovyova Victoria, Cherepovitsyna Alina

Associated Petroleum Gas Utilization: New Opportunities for the Oil and Gas Complex in a Circular Economy 132
Riadinskaia Arina, Cherepovitsyna Alina

Global Challenges and Opportunities for Arctic Oil and Gas Projects 138
Cherepovitsyn Alexey, Rutenko Evgeniya, Yudin Sergey

Arctic Climate Policy Development: Taking Public Beliefs into Account 144
Klyuchnikova Elena, Klyuchnikov Gleb

The Karelians: Securing the Right to Preserve Traditional Habitats (experience of the Republic of Karelia) 148
Yalovitsyna Svetlana

Influence of the Arctic Climatic Conditions of the Kola Peninsula on the Reactivity of the Sympathetic-Adrenal and Cardiovascular Systems of Junior High School Children 152
Scherbina Fyodor, Shelkov Mikhail, Trotsenko Alla, Scherbina Anatoly, Volkova Tatyana, Zvyagintsev Peter

Transformation of Daily Food Practices of Indigenous Peoples of Taimyr Through the Prism of Environmental Change 156
Bobrova Vasilisa

Paleoreconstruction of Natural Conditions as the Basis for the Rational Use of Natural Resource Management on the Territory of the Neva Lowland........ 160
Paranin Roman, Paranina Alina, Grigoryev Alexey, Larchenko Lyubov, Kiseleva Margarita

Prospects for the Use of Arctic Routes in the Context of Climate Change and the Implementation of the Concept of Sustainable Development of the Arctic Zone........ 166
Gogoleva Valentina, Kirsanova Natalia, Zyabkina Tatyana, Dolgopolova Tatyana

Analysis of the Hydrodynamic Modeling Market Under the Environmental Conservation Concept 171
Ivanov Artem, Kruk Marina, Litvin Vladimir

Financial Potential as a Basis for Ensuring Environmental and Economic Security of the Arctic Regions........ 178
Ulchenko Mikhail, Ponomarev Nikita

Strategies for Large Oil and Gas Companies Operating to Support Sustainable Development and Environmental Safety in the Russian Arctic........ 182
Bianco Ishel, Fadeev Alexei, Illinsky Alexander

Environmental Tourism Potential of the Arctic Regions of Russia 189
Roslyakova Natalia, Dorofeyeva Ludmila

Development of Tourism in the Arctic Regions of the Russian Federation: The Environmental Aspect........ 195
Kuchumov Artur, Karpova Galina, Pecheritsa Elena, Voloshinova Marina

Northern Labyrinths: Potential for Interdisciplinary Research and a Resource for International Tourism........ 200
Paranina Alina, Paranin Roman, Grigoryev Alexey, Larchenko Lyubov, Kiseleva Margarita

Opportunities and Limiting Factors for Tourism Development in the Khibiny........ 206
Vorobievskaya Elena, Kirillov Sergey, Sedova Natalia, Slipenchuk Michael

Author Index

MAN

IN THE ARCTIC

Preface

The First International Interdisciplinary Scientific and Practical Conference Man in the Arctic (IIRPCMIA 2021) was successfully held during November 17th to 19th, 2021 in Saint Petersburg Mining University (Petersbourg, Russia). The conference is an international forum for scientists, engineers, and practitioners to present their latest research results and inspirations in all areas of Arctic research.

The purpose of the conference is to foster international networking and partnerships in interdisciplinary research and to provide researchers from Russia and other countries with a venue for sharing information and experience regarding the elaboration of an integrated interdisciplinary approach to solving problems associated with Arctic exploration and development.

The scope of the conference covered all areas of Arctic research, including: History of Arctic exploration and development, Infrastructural support for the Arctic, Sustainable development of the Arctic regions, Economic and industrial systems in the Arctic: changes in management approaches in the context of the emerging green economy,Tourism development in the Arctic, Arctic development and climate policy, Indigenous peoples of the Arctic, Human capital in the Arctic, Innovative technologies for developing Arctic resources, Mining in the Arctic, Geology, tectonics, and geodynamics in the Arctic, Development of the energy sector in the Arctic, Occupational and environmental safety issues in the Arctic, Digital technologies in the Arctic, Issues associated with mine surveying, geodetic surveying, geomechanics, cadastre, and construction in the Arctic

The conference brought together 175 participants, primarily from Russia, but also from France, Serbia and Kazakhstan.

After a rigorous selection process, the Organizing Committee decided to publish 42 papers in the E3S Web of Conferences, an open-access proceeding in environment, energy, and earth sciences.

We would like to thank all the participants in the First International Interdisciplinary Scientific and Practical Conference Man in the Arctic. We would also like to thank the Organizing Committee and all the Committee members, for the work they put in to organize this event. Finally, our special thanks go to the Earth and Environmental Science, for the endeavor of all its colleagues in publishing this paper volume. We deeply appreciate their efforts in making the conference successful.

The Committee of IIRPCMIA 2021.

ORGANIZING COMMITTEE

Chairman:	**M.A. Pashkevich,** Doctor of Technical Sciences, Professor, Head of Geo-ecology Department, Director of the Scientific Center "Ecosystems' Technogenic Transformation Assessment", Mining University (Russia)
Deputy Chairmen:	**A.B. Makhovikov,** Candidate of Technical Sciences, Associate Professor, Dean of the Fundamental and Humanitarian Disciplines Department, Mining University (Russia) **M.V. Dvoinikov,** Doctor of Technical Sciences, Professor, Head of the Well Drilling Department, Director of "Arctic" Competence Center, Mining University (Russia) **S.N. Rudnik,** Candidate of Historical Sciences, Head of the History Department, Mining University (Russia)
Organizing Committee Secretary	**E.A. Samylovskaya**, Candidate of Historical Sciences, Associate Professor of the History Department, Saint Petersburg Mining University (Russia)
Organizing Committee Members:	**I.M. Avlasenko,** Candidate of Technical Sciences, Associate Professor of Diplomatic and Consular Service Department, Belarusian State University (Belarus) **V.T. Borzenkov,** Candidate of Technical Sciences, Vice-Rector for International Affairs, Saint Petersburg Mining University (Russia) **Hanna Wlodarkiewicz-Klimek,** Doctor of Economics, Professor, Dean of Management Engineering Department, Poznan University of Technology (Poland) **A.V. Golovnev,** Doctor of History, Corresponding Member of the Russian Academy of Sciences, Professor, Anthropology and Ethnography Museum Director named to honour Peter the Great (Kunstkamera) of the Russian Academy of Sciences (Russia) **Zaya Yan,** Doctor, Professor, Head of Drilling and Geoengineering Department, Krakow Mining and Metallurgical Academy (Poland) **Zafar Javed,** PhD, Scientific Coordinator in Study and Research Centre, Jamia Millia Islamia (India) **Capilupi Stefano Maria,** PhD, Director in Italian Language and Culture Institute, Russian Christian Humanitarian Academy; Scientific Consultant of Political Philosophy Department, University of Rome "La Sapienza" (Russia, Italy)

Statement of Peer review

In submitting conference proceedings to *Web of Conferences*, the editors of the proceedings certify to the Publisher that

1. They adhere to its **Policy on Publishing Integrity** in order to safeguard good scientific practice in publishing.
2. All articles have been subjected to peer review administered by the proceedings editors.
3. Reviews have been conducted by expert referees, who have been requested to provide unbiased and constructive comments aimed, whenever possible, at improving the work.
4. Proceedings editors have taken all reasonable steps to ensure the quality of the materials they publish and their decision to accept or reject a paper for publication has been based only on the merits of the work and the relevance to the journal.

Title, date and place of the conference

Title: The First International Interdisciplinary Scientific and Practical Conference Man in the Arctic (IIRPCMIA 2021)
Date: November 17th to 19th, 2021
Place: Saint Petersburg Mining University (Petersbourg, Russia)

Proceedings editor(s):

Aleksei Makhovikov, Candidate of Technical Sciences, Associate Professor, Dean of the Fundamental and Humanitarian Disciplines Department, Mining University (Russia)
Ekaterina Samylovskaya, Candidate of Historical Sciences, Associate Professor of the History Department, Saint Petersburg Mining University (Russia)

Date and editor's signature

Aleksei Makhovikov **Ekaterina Samylovskaya**

21.03.2023

E3S Web of Conferences 378, 01001 (2023) https://doi.org/10.1051/e3sconf/202337801001
IIRPCMIA 2021

The Arc-of-Meridian Expedition to the Spitsbergen Archipelago 1899-1901 as an example of international scientific collaboration: tasks and results

Nadezhda Kyzyurova[1,*]

[1]A PhD student, Department of Interdisciplinary Humanitarian Research, Komi Scientific Center, Ural Branch, Russian Academy of Sciences, 167982, Syktyvkar, Russian Federation

Abstract. As one of the most striking indicators of environmental change,, the Arctic is a region of bifurcated interests of various states. An important role in the process of cognition and development of the Arctic is assigned to the scientific community. The article discusses the experience and scientific tasks of the Swedish-Russian expedition on arc-of-meridian measurement in 1899-1901. The main results of the expedition and its contribution to the further development of research in the Arctic are discussed. The influence on the international scientific community is considered.

1 Introduction

One of the Sustainable Development Goals developed by the UN General Assembly in 2015 is to take urgent action to combat climate change and its consequences. At the end of the last century, discussions on climate change acquired a particular relevance at the global level. The Earth's climate is changing; atmospheric and ocean temperatures have risen, sea levels are rising, and the Arctic is one of the most obvious indicators of climate change. This has opened up new opportunities for consideration of exploration of natural resources that used to be covered by ice or permafrost. Even nowadays, oil plays a dominant role among other energy sources [1]. Climate is a long-term observation of weather that can be evaluated, so the study of these issues requires long-term monitoring. That is why it is necessary to look back at the experience of scientific expeditions studying the environment in the Arctic. One of such expeditions was the Arc-of-Meridian expedition to the Spitsbergen archipelago in 1899-1901, the scientific tasks and results of which are discussed in this article.

Since the middle of the 20th century, scientific and technological revolution began, which contributed to the creation of unique equipment for working on the Arctic shelf [2], improving transport, which made remote areas of the Arctic Ocean accessible, which led to the emergence of the problem of saving ecosystem as one of the key elements of sustainable development. At the present stage, the development of science allows to argue the claims for an increase in the continental shelf, but the issue of expanding the shelf in the Arctic became relevant at the turn of the 19th - 20th centuries. There are also international disputes about the ownership of certain territories in the Arctic. Therefore, the accumulated experience of research and promotion of scientific interests in geopolitically attractive territories should be taken into account at the present stage in order to avoid mistakes in the allocation of zones of influence.

In this regard, it is important to analyze the experience of the Spitsbergen Expedition of 1899-1901 as a scientific international enterprise. Understanding that this event is one of the first experiences of such cooperation, it is necessary to analyze the prehistory of the measurement of the meridian arc, to consider the course of the expedition; to evaluate the results and significance of the expedition for the scientific community.

2 Main body

Before the beginning of the 20th century, there were no permanent residents on the archipelago. Since the 16th century, it was visited by Norwegian and Russian sailors, whalers and hunters. In the 19th century, the era of colonialism was not over, and not all places in the world were developed. Among them were the northern territories and waters of the Arctic. Redistribution, establishment of control over the no-man's land and ownerless territories and sea-fishing in the Arctic began in the 19th century. This led to a number of political, economic and legal international disputes. The attention of the scientific community was drawn to the Spitsbergen since the 1870s, when the Geographical Society in Paris proposed to establish a meteorological observatory on the Spitsbergen archipelago, and the Swedish polar explorer A.E. Nordenskjold proposed to the Swedish government at the same period to consider the status of the archipelago [1]. Moreover, by the beginning of the 20th century, Spitsbergen faced a new global process of industrialization; this way, active exploration and mining of coal began. The availability of valuable raw materials attracted even more attention from various states. The

* Corresponding author: kamasheva.nad@gmail.com

uncertain political status made it impossible to carry out industrial activities at the proper level.

The issue of the discovery of the archipelago is still the subject of debate. There are several versions of who pioneered the discovery of Spitsbergen. The Norwegian historian T.B. Arlov highlights the following theories: the "Stone Age" hypothesis; the "Viking hypothesis" (12th century); the expedition of V. Barents (1596–1597); the Pomors hypothesis (16th century) [4]. Researcher I. Baranovsky emphasizes that historically Russia and the archipelago have been linked by a series of events. One example is the wintering of the Pomor Ivan Starostin, who visited Spitsbergen 17 times and stayed for 15 years on the archipelago until his death in 1826 [5]. Thus, despite the fact that until the second half of the 19th century Spitsbergen had been attracting mainly not explorers but those who had trapping and fishing purposes, the accumulation of knowledge about the archipelago begun. Moreover, the experience of sailing in harsh weather conditions gained by the natives and travelers contributed to the subsequent scientific exploration of the archipelago.

More international attention was paid to the archipelago in 1870. The French Geographical Society proposed to establish a meteorological observatory on the archipelago to study the climate indicators in the Arctic. Another proposal was made by the Swedish polar explorer A.E. Nordenskjold to the Swedish Foreign Office of the United Kingdoms. He proposed establishing sovereignty over the Spitsbergen Islands before any foreign power took over them and potentially created difficulties for further exploration [3]. Next year, an attempt was made by the Kingdom of Sweden to claim sovereign rights to the Spitsbergen Archipelago. The government appealed to Russia, Germany, England and Holland to submit objections to the inclusion of Spitsbergen as part of Sweden. However, the Russian government, due to the fact that the issue of historical rights to the archipelago remained disputed, could not agree to such inclusion [6]. The Russian Empire opposed the consolidation of exclusive influence of any European power in Spitsbergen because of the geopolitically significant territory due to its location at the outlet to the Atlantic Ocean.

In 1898, the Imperial Academy of Sciences (St. Petersburg) received a proposal to organize a joint scientific expedition to carry out degree measurements on Spitsbergen submitted by the Swedish Royal Academy of Sciences The Russian side agreed with the Swedish proposal. This would allow polar research to move to a new level. Moreover, the Swedish side undertook part of the funding [6]. The letter from the Minister of Foreign Affairs of the Russian Empire M.N. Muravyov to the Minister of Finance S.Yu. Witte of March 21, 1898, emphasized how important it was to accept the Swedish proposal, otherwise, rejection would mean recognition that Spitsbergen is outside Russian interests. The proposal was positively met by S.Yu. Witte, as this allowed not only to establish positive international relations but also to attract Swedish finances. He emphasized that "Regardless of this, one cannot but recognize it is as extremely desirable for our expedition to join the enterprise of the Stockholm Academy and begin the difficult and complex work with united forces and at half the cost"[7].

Since the geopolitical status of the Spitsbergen archipelago had not been determined by the second half of the 19th century, the issue began to appear more often on the political agenda. This way, international studies were important not only from a scientific but also from a diplomatic point of view. At the same time, the issue of studying the environment in polar regions appeared on the scientific agenda.

The change in the situation with equipment suitable for expeditions in the harsh weather conditions, made it safer and more productive to conduct expeditions in the polar regions. Moreover, scientific research in other parts of our planet, such as measuring the meridian arc or magnetic research, required a logical continuation at high latitudes in order to build up a holistic picture on a planetary scale. One such question that required further disclosure of data from northern latitudes was the question of the shape and size of Earth. A similar expedition was undertaken by the French Academy of Sciences in 1735-1745 sending one expedition closer to the equator, to Peru, and the second one to Lapland, closer to the poles. It had already been recognized that the Earth had some ellipsoidal shape rather than a true sphere, but scientists disagreed on where this ellipsoid was flattened - at the poles or at the equator [8]. In the 19th century the concept of geoid was introduced, which led to the idea that to determine Earth shape and size it is necessary to obtain accurate data on the linear magnitude of the meridian arc at different latitudes.

Technologies such as triangulation, a method of determining the size of Earth by covering its surface with a network of adjacent triangles, have allowed researchers to build an accurate model of the shape of our planet. However, this required measurements to be made in different parts of Earth. This method was based on the idea that the meridians have different curvatures. One successful project focused on degree measurement was conducted by Russian and Scandinavian scientists from the Danube to the Arctic Ocean from 1816 to 1855. However, this enterprise lacked data obtained at high latitudes where the difference in the value of degrees was more drastic. The project of such an expedition was proposed by the Englishman E. Sabin, a polar explorer, who conducted a series of experiments with a pendulum to determine the shape of Earth. He was convinced of the possibility of carrying out a triangulation network on the Spitsbergen islands [9]. Thus, the research carried out at the beginning of the 19th century by determining the length of the meridian arc, predetermined the further line of research in the polar latitudes, namely, taking degree measurements in the northern latitudes.

In 1899, E. Sabin's plan was implemented in high latitudes. The tasks of the expedition were outlined in detail by F.N. Chernyshev in his work "On the course of the expedition for degree measurements on the Spitsbergen islands in 1899-1900". He noted that the program of the expedition was developed at joint meetings of the Russian and Swedish Commissions and included a number of research works, not only geodetic and astronomical work, but also observations of the northern lights, meteorological and magnetic observations throughout the year. Moreover, a network of triangulation points was developed, consisting of 22 triangles distributed between Ross Island and Mount Keilhau. The joint geodetic work included a meridian measurement , made to measure the size and shape of our planet, "... covering 4 degrees 19 minutes from the southern tip of Spitsbergen, at 76o30' to the northernmost of the Seven Islands, at 80o49'" [9].

As the head of the Swedish commission, which was entrusted with the development of the expedition itself, King Oscar appointed Prince Gustav. Among representatives of the Royal Swedish Academy of Sciences were academician A.E. Nordenskjold, mathematician M.G. Mittag Leffler, professor A.E. Ederin. On the part of the Imperial St. Petersburg Academy of Sciences, a special commission was formed including members of the Academy and representatives of the military and naval departments. His Imperial Highness the Grand Duke Konstantin K. Romanov was appointed chairman of the Russian commission for the preparation of this expedition. The Spitsbergen Commission included the Chief of the Military-Topographic Department of the General Staff Lieutenant General O.E. von-Stubendorf, the Chief of the Main Hydrographic Department of the Naval Ministry Lieutenant General K.I. Mikhailov and the Major General A.R. Bonsdorf. The development of the expedition program was entrusted to O.A. Backlund (chairman), F.B. Schmidt, A.P. Karpinsky, M.A. Rykachev, B.B. Golitsyn and F.N. Chernyshev [6].

The location of the expedition was determined "along the eastern coast of Spitsbergen from the South Cape along Sturfjord Bay and the Hinlopen Strait to the northernmost island in the archipelago of seven islands". The Swedish expedition determined that there were no serious climatic difficulties in the practical implementation of the project. The location of the expedition was not chosen by chance. The large length from south to north made it possible to carry out more complex measurements at a distance of more than 4 ¼ degrees, and the presence of high peaks made it possible to establish signals for astronomical and geodetic observations [6]. For wintering, research programs for magnetic and meteorological observations were developed. It was necessary to set geodetic signals in the south of the island at the selected observation points. The work on physical geography and topography was to be carried out by the reconnaissance party, and the geodetic party was supposed to facilitate this process by surveying the immediate vicinity of the signals. The work in geology was headed by F.N. Chernyshev. Thus, the expedition was entrusted not only with the task of measuring the length of the meridian arc, but also with multifaceted studies, since the selected area had been little studied before, and for some areas there was no data at the time of the study.

The reconnaissance work, the initial survey of the terrain, and the installation of the triangles were distributed between the Swedish and Russian parts of the expedition. The reconnaissance was carried out in order to inspect the territory and select reference points for further observations. During the preparatory expedition in 1898, the Swedes completed work only in the northern part. In 1899, the Russians had to conduct a reconnaissance in Sturfjord Bay, but due to the fact that the bay was later freed from ice and its shores were still poorly studied, these works seemed very complicated, which was difficult to accomplish in one summer [9].

The environment of the archipelago had not been properly studied yet, that is why the Imperial Academy of Sciences (St. Petersburg), despite the degree measurement of the meridian, had astronomical, geodetic, meteorological, physical and geological observations on the agenda. "In particular, to determine the specific gravity of the rocks composing it [the archipelago], which may be presented for studying their influence on the deviation of the plumb line and for understanding the distribution of gravity"[6]. It was planned to conduct familiarization work with the mineral and fish resources of the archipelago, in anticipation of practical results for the development of industries and trade, for raising the welfare of the north of Russia. In the 19th century, the absolute determination of the force of gravity could be made more accurately than before. In 1881, the Austrian scientist R. Sternek invented a pendulum, the measurements of which better corresponded to the necessary parameters [10]. And one of the items on the expedition's agenda was gravity measurements.

As for the financing of the expedition, the situation was as follows. At the beginning, the Russian government planned and allocated 100,000 rubles t for the expedition for one year from 1899 to 1900. The budget of the expedition was estimated at 200,000 rubles, half of which was covered by the Swedish side [6]. However, there were unforeseen expenses associated, first of all, with the weather conditions, because of which it was necessary to extend the expedition, due to which it was necessary to make adjustments to the estimate.

From the letters of the Chairman of the Russian Spitsbergen Commission and President of the Imperial St. Petersburg Academy of Sciences, Grand Duke Konstantin K. Romanov to the Minister of Finance S.Yu. Witte, one can judge about the budget of the expedition. It was already determined before the start of the expedition that the costs would exceed the primary ones. The letters of 1898 mentioned the expenses associated with wintering - the construction of houses and observatories, the purchase of all the necessary tools and food supplies for a year and a half, as well as the sending of reinforced scientific personnel for simultaneous work in the northern and southern parts of East Spitsbergen.

In 1899, Konstantin K.Romanov repeatedly referred to the fact that additional funds were needed for the expedition, as a large amount of work had been completed, but the goal had not been achieved, and to stop the expedition with the results obtained without giving them a logical conclusion would have been considered a failure. Additional funding for the expedition was also necessary to purchase the necessary equipment, the construction of an observatory and houses, covering costs connected with staff. It was not foreseen that a larger number of workers would be needed than initially determined, or that a special steamer would have to be hired due to the difficulty of delivering all the necessary materials [7]. It should be noted that even projects implemented nowadays in the Arctic require significant funding and remain unprofitable [11].

Despite the lack of funds, archive documents note that the Russian expedition was organized much cheaper and, at the same time, more scientifically complete than the Swedish one. A letter to the Minister of Finance in January 1901 from the Grand Duke Konstantin Romanov notes that, thanks to the good spirits of domestic sailors and scientists, it was possible to perform the most difficult work, but since the work was not completed, it was not worth giving up without completing the tasks, since the completion of the work would be given to the Swedes, thereby the achievements of Russian researchers would fade into the background. At the same time, it is noted that the Swedes were not so successful in fulfilling the assigned tasks, and only completed one-fourth of what the Russians had accomplished [7].

The cost of the expedition went beyond the original budget, however the Russian government was aware of the importance of the research being carried out. Thus, the requests for an increase in funds were considered justified, since the scope of work was expanding as the expedition proceeded, and besides, the approval of the estimate was made at the highest level due to the political importance of this enterprise. It was important for Russia to gain a foothold in the archipelago. Moreover, it was extremely important not to concede all the achievements of Sweden at the final stages due to the only reason - lack of funds. In total, the Russian side spent 160,000 on the expedition, that exceeded the original limit on 60,000 rubles. And for 1903, 1904, 24,000 rubles were allocated to process the collected data. The Swedes also did not meet the limit. According to some researchers, they exceeded the limit by more than 2.5 times [12].

The expedition's results can be traced year by year. In the joint work plan, it was calculated that the end of the expedition would be in 1900. However, F.N. Chernyshev stressed that the ice conditions in which their Swedish colleagues found themselves were especially unfavorable. But the members of the Russian expedition also faced severe hardships due to climatic conditions, therefore he highlighted the selfless courage of the participants of the Russian expedition that allowed them to obtain great results [9]. In the summer of 1899, the work was completed on the placement of the signal network in Sturfjord Bay. In 1899, geodetic and astronomical observations in the most inaccessible research points were completed. Equipment was installed and ready for measurements in the magnetic and meteorological observatory. For the first time, a photographic magnetograph was put into operation, and geological, topographic and hydrographic information about Sturfjord Bay, which had previously been obtained during the Nordenskjöld and Duner expedition in 1864, became more reliable [9].

In 1900 almost the entire research program was fulfilled due to the increase in the volume of work carried out. Three observation points - Cape Agarda, Cape Misunderstanding, and the Helwald plateau - remained outside the scope of observation. However, due to climatic conditions, the Swedish side was unable to fulfill the planned program. The members of the expedition encountered unfavorable ice distribution conditions around the northern shores of the archipelago, which significantly slowed down the progress of research. Therefore, the research continued already in 1901. In the summer of 1901, triangulation work in the southern part of the archipelago was not completed, and the Swedish party was forced to go there during the summer of 1902 to finish its work [3].

A total of 3.5° meridian arc was measured over the three years of the expedition, most of the measurements of which were made by Russian researchers (arc of 2.5°). The Swedes were in charge of the northern part of the triangulation chain in the Spitsbergen archipelago. Their main station was at Cape Crozierpinten in Sorgfjord. There they established a baseline as a reference point for the northern part of the triangulation chain. Most of the research was carried out during the wintering of 1899-1900, which included geodetic work, magnetic and meteorological observations. A zonal survey from Hornsund Bay through the entire Sturfjord Bay to Cape Wallace Point was carried out, that was unique in its geographic location and scope. As a result, the topographic survey covered over 10,000 km2. Hydrographic data were obtained through continuous measurements of the depth of Sturfjord Bay. Crossing the routes of Western Spitsbergen in three places made it possible to obtain hydrological, biological and glaciological information about the inner area of the archipelago, which had not previously been studied [5].

The results of glaciological observations are used nowadays to monitor climate change. For example, researchers A. Pälli, J.C. Moore, J. Jania, and P. Glowacki compared the current state of some glaciers in the south of the archipelago, among them Hornbreen, Hambergbreen, with their state in 1899-1901 during the expedition under consideration. According to their study, the area of the glaciers has decreased by 37-50% [13]. That raises further discussions about climate change.

Observation of aurora borealis during the expedition held by J. Sykora and J. Westman allowed determining the height of the northern lights and its shape and speed. As it was the first study conducted using modern instruments, the obtained data was the beginning of a new era in the understanding the phenomenon [14]. Moreover, the topography of the centre, north, and northeast of Spitsbergen was introduced by G. De Geer and his team during the expedition [15]. Moreover, the data he obtained also allow researchers to analyze the state of glaciers, and the recession of glaciers indicates the ice melting.

For the first time, the results of these studies were presented at the International Congress of Surveyors and Astronomers in Paris in 1900. In October 1903 they were discussed at the 14th General Meeting of the International Geophysical Union and were met with great interest in the scientific community. During the expedition, a tremendous amount of information was obtained in various fields of science, particularly, in geodesy.

Therefore, the question arose of processing data and continuing research, which also required additional funds. During the 1900s, the question of financing the proceeding of the data obtained was on the agenda of the Ministry of Finance several times [6]. The results of the expedition, formalized and compiled in a general way, were eagerly awaited in the international scientific community. But, as D.P. Belyaev noted, having begun in 1904, the publication of the results ended only in 1919 [12].
Despite the fact that the completion of the expedition was delayed, and the research was carried out for three years, not least due to the strong personality of F.N. Chernyshev, the goals of the expedition were fulfilled. The information obtained during the expedition not only contributed to the advancement in the issue of determining the shape of Earth, but also significantly expanded the knowledge of the "terra incognita" – the Spitsbergen archipelago.

3 Conclusion

At the end of the last century discussions about climate change acquired particular urgency at the global level. This opened up new possibilities for consideration of exploration of natural resources that have previously been covered by ice or permafrost. The Arctic is a region with fragile environment but it is also a bifurcation point where the interests of different states intersect. The combination of these factors makes us pay attention to the need to study the Arctic in a historical context. The scientific community has an important role to play in the process of cognition and development of the Arctic, without whose active assistance, if we look back at the historical past, the modern development of the Arctic expanses is impossible. The Russian-Swedish expedition to the Spitsbergen archipelago to measure the meridian arc was a major international undertaking. It was the first successful research expedition in the Arctic latitudes. In 1899-1901, members of the expedition managed to penetrate into the previously unexplored central regions of the archipelago and obtain material in the field of geodesy, astronomy, hydrography, zoology, botany, geology, meteorology, physics and other areas of natural sciences. The obtained material became of interest to the international scientific community.
The main research task, which the expedition was to solve – the measurement of the meridian arc - was successfully accomplished. Thanks to the persistence of F.N. Chernyshev, the research was brought to its logical conclusion and the Russian side succeeded in fulfilling the task. These studies made it possible to clarify the shape and size of our planet, the magnitude of the compression of Earth's ellipsoid. This logistically and scientifically difficult task was successfully accomplished through the cooperation of the two countries. Both the purely scientific motivation of this expedition, as well as the international cooperation and its successful completion without human casualties make the expedition one of the most successful projects. The expedition participants not only made a contribution to the development of polar science, but also to world science as a whole.
The data obtained during the expedition set new goals for further research. The series of topographic maps made during the expedition were supplemented with data with greater positional accuracy over the next century. New equipment and instruments enabled the research to advance polar science. The results obtained are used to compare the current state of the environment with its state a century ago. The ongoing processes in the environment demand our attention and can be analyzed more properly relying on past experience.
The expedition can rightfully be considered a major scientific enterprise. It is important that decisions on the financing of the expedition were made at the highest level, since in the 19th century, when the states nearest to the archipelago began to claim their rights to Spitsbergen. As a result, Russia's attention also shifted to the Arctic territories, as it gradually realized the importance of polar waters as potential transportation arteries, "storehouses" of natural resources, as well as an access to the Atlantic Ocean. And already in 1900, a Russian polar expedition was sent in search of Sannikov Land with the aim to further develop the region.
Analysing the current situation, the Arctic states include the issue of global climate change in their Arctic strategies. On the one hand, new opportunities appear for the development of natural resources and sea routes, but on the other, new risks of environmental security arise. And these issues already encompass the entire Arctic region, and therefore moving away from the practice of joint decision-making can affect the fragile Arctic environment and, in general, the stability of the Arctic development. It is important that environmental policies contribute to the prevention of man-made disasters and their consequences, as well as to the response to climate change. Even planning the budget of expeditions requires data on the conditions of the areas to avoid unforeseen costs associated with natural conditions, as it happened during the expedition. That is why it is necessary to take the previous experience of scientific research into account.

References

1. P.P. Tsyglianu, et al. Engineering projects in the russian fuel and energy complex: actual problems, factors and recommendations for development, Ugol', 3, pp. 45-51, doi: 10.18796/0041-5790-2023-3-45-51 (2023) (In Russ.)

2. S.I. Podolsky, I.A. Tropov, Dmitry A. Kazakovsky and development of mine surveying in the USSR, Gornyi Zhurnal, 2, pp. 97–100 (2022) (In Russ.)

3. R. Berd, From Spitsbergen" to "Svalbard". Norwegianization in Norway and in the "Norwegian Sea", 1820-1925, *Acta Borealia*, **30**, pp. 154-173 (2013)

4. T.B. Arlov, The Discovery and Early Exploitation of Svalbard. Some Historiographical Notes, *Acta Borealia*, **22**, pp. 3-19 (2005)

5. I. Baranovskiy, *Shpitsbergenskaya konferentsiya [Spitsbergen Conference]* (Sankt-Peterburgskie vedomosti, 1910) p. 166 (In Russ.)

6. Rossiyskiy gosudarstvennyy istoricheskiy arkhiv [Russian State Historical Archive], Fund 1278, Register 2, № 196 (In Russ.)

7. Rossiyskiy gosudarstvennyy istoricheskiy arkhiv [Russian State Historical Archive], Fund 565, Register 8, № 29377 (In Russ.)

8. J.R. Smith, The Meridian Arc Measurement in Peru 1735-1745, *FIG XXII International Congress. Washington, D.C. USA*, **12** (2002)

9. F.N. Chernyshev, O khode rabot ekspeditsii po gradusnym izmereniyam na ostrovakh Shpitsbergena v 1899—1900 gg [On the progress of the expedition on degree measurements on the Spitsbergen islands in 1899-1900], *Izvestiya IAN*, **14**, pp. 255-280 (1901)

10. D.D. Sergiyevskiy, Doklad v Komissii Imperatorskogo Russkogo Geograficheskogo obshchestva po issledovaniyu sily tyazhesti v Rossii [Report to the Commission of the Imperial Russian Geographical Society for the study of gravity in Russia], *Izvestia RGO*, **39**, pp. 508-542 (1903)

11. A.E. Cherepovitsyn, et al., Critical analysis of methodological approaches to assessing sustainability of arctic oil and gas projects, *Journal of Mining Institute*, **249**, pp. 463-478 (2021)

12. D.P. Belyaev, Nauka v interesakh geopolitiki: rossiysko-shvedskaya ekspeditsiya na arkhipelag Shpitsbergen [Science for Geopolitics: Russian-Swedish Expedition to the Spitsbergen Archipelago], *Vestnik MSTU*, **16**, pp. 267-270 (2013)

13. A. Pälli, J.C. Moore, J. Jania, P. Glowacki, Glacier changes in southern Spitsbergen, Svalbard, 1901–2000, *Annals of Glaciology*, **37**, pp. 219-225 (2003)

14. S. Chernouss and I. Sandahl, Comparison and significance of auroral studies during the Swedish and Russian bilateral expedition to Spitsbergen in 1899–1900, *Ann. Geophys.*, **26**, pp. 1127–1140 (2008)

15. O. Orheim, A. Hoel, The place names of Svalbard (Oslo : Norsk Polarinstitutt, 2003) p. 537

The extreme Anthropogen of the Arctic: the formation of the Great Glacier, the emergence a man and the Arctic Ocean in the Early Holocene

Anatoly Zhirnov[1,*]

[1]Doctor of Geological and Mineralogical Sciences, Leading Researcher, Laboratory of geological and environmental Studies, Institute For Complex Analysis of Regional Problems, Far Eastern Branch, Russian Academy of Sciences, 679016, Birobidzhan, Russian Federation

Abstract. The relevance of the article stems from the need to consider recent events in the Arctic in recent time, which are largely contradictory. The aim of the study is to show extreme environmental changes in the Arctic of anthropogenic nature, including the emergence of man and the Arctic Ocean in the Early Holocene. The territory of the Arctic was a landmass with high mountains, in the period preceding the Holocene, on which a large glacier formed 30-18 thousand years ago (Late Pleistocene). The glacier slid irrepressibly southward over great distances, where it covered vast areas of Europe and North America. In the years that followed, a dramatic warming and active melting of the glacier began. By the beginning of the Holocene, the glacier had melted, and a narrow strait of the Atlantic appeared at the foot of Greenland. Coniferous-deciduous forests, numerous rivers and the first human settlements appeared on the flat territory of the Arctic, and various animals – mammoths, horses, bulls - spread out. However, a major catastrophic event occurred by the middle of the Holocene (6.0 thousand years ago), after active volcanism in the Arctic: the collapse of the central part of the Arctic to a depth of about 5 km and the formation of the Central Arctic graben, associated with the appearance of a huge amount of endogenous water. There began a rapid movement of water on the flat parts of the Arctic and the formation of the modern huge, shallow (50-100 m) shelf - the Arctic Ocean. Many human settlements were flooded, animals escaped, in part, on high uplands. For example, huge animal cemeteries were preserved on the Novosibirsk Islands. A new cooling of the climate occurred 4 thousand years ago, and an ice sheet formed on the surface of the ocean, which led to the name of the North Glacial Ocean in Russia. Modern man began to explore the coastal territories of the Arctic shelf since the mid - Holocene, but active industrial development of the Arctic began in the 21st century.

1 Introduction

The Anthropogen (Quaternary) is the newest and very short stage in the geological history of the Earth - 1.6 million years (and its end, the Holocene - only 11 thousand years), but it is saturated with major climatic and oceanic logical events in the Arctic and human life history, including dramatic ones [10, 11, 13, 19]. However, it has been little discussed in science.

The modern literature focuses on various aspects of the movement of individual blocks of the Arctic territory in the Mesozoic and Early Cenozoic [23]. At the same time many researchers have missed the newest stage of the World Ocean's advance on the continents [25]. It should be added to this that the entire Anthropogenic period with its major climatic and ocean logical events has also fallen out of consideration and human exploration of the Arctic in the modern time [33].

The issues of changes in the Arctic environment during the Anthropogenic, the emergence of man and the formation of the Arctic Ocean in the Early Holocene, as well as the modern industrial development of the Arctic are briefly discussed in this report.

2 Materials and methods

The materials of the study were the latest data on the evolution of the Arctic in the Anthropogenic, which bring some clarity on this issue [3, 8, 10–12, 15–21, 23–26]. The main method of research is the generalization of established modern and previous facts.

3 Results

3.1 Extreme changes in the Arctic environment at the end of the Pleistocene - Early Holocene and the emergence of man

* Corresponding author: zhantmich@yandex.ru

A long Paleozoic-Mesozoic stage of marine sedimentation is distinguished in the geological history of the Central Arctic. However, in the early Cenozoic, the territory rises and by the Neocene (25 million years ago) becomes completely dry, with the high Gakkel and Lomonosov ridges to the north [22, 26], adjacent to the highland ridges of Greenland.

The Quaternary (Anthropogenic) period was characterized by abrupt climatic changes. First of all, there were several epochs of sharp cooling and formation of large glaciers – 0.5 million years ago, 0.3 million years ago, and 70-18 thousand years ago. The location of glaciers is assumed to be in Canada, Scandinavia and on the islands of the Arctic [13]. However, a special study of this issue has shown that the place of formation of the giant glacier was the high mountain ranges of the Arctic (on Greenland and on the Central basin's territory – Gakkel and Lomonosov ranges), from where it slid over great distances to Europe and North America [33].

The final stage of the Pleistocene, the Valdai glaciations epoch (Worm – in Western Europe, Wisconsin - in North America), about 20-18 thousand years ago, was characterized by extremely harsh conditions when the global temperature dropped by 4-5°C [13].

However, later, during the period 13-11 thousand years ago, rapid climate warming began, active melting of glaciers and formation of river valleys that flowed from the north to the south [10]. In particular: "in the first half of the Holocene, between 9 to 7 thousand years AGO (by14 C), the average July temperature exceeded the modern temperature by 6-7°C [14]. The coniferous-deciduous forests grew in a warm climate, large herds of mammoths, horses, bulls, and sheep appeared, and there were settlements of Mesolithic people on some islands (Svalbard, etc.) [11]. There were also large human settlements on the east coast of Greenland [28].

The extensions of many modern rivers are established on the land between Iceland and Scotland, and in the North Sea, east of Great Britain (Figure 1).

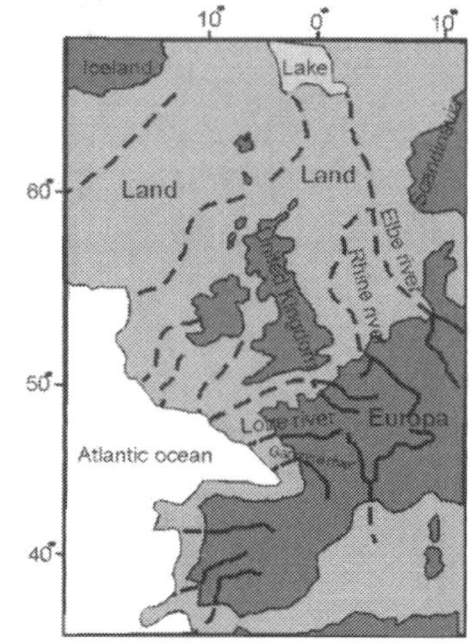

Fig. 1. Scheme of river network's position (dashed lines) on land between Western Europe and Iceland at the beginning of the Holocene [11].

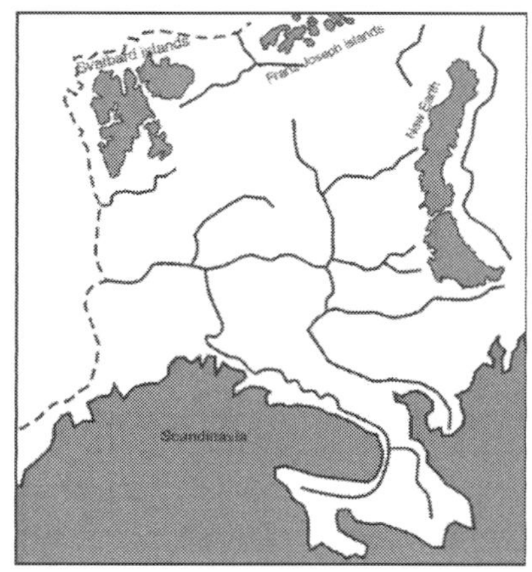

Fig. 2. Scheme of rivers on the Barents Sea's territory in the Early Holocene [11].

3.2 Tectonic collapse of the Central Arctic Arch

A sharp intensification of volcanism in the Arctic began in the first half of the Holocene, due to a decrease in the stability of the Earth's crust and upper mantle, after the repeated formation of giant upland glaciers in the Arctic. At this time, the largest volcanic eruptions took place over a period of 10-6 thousand years ago (Fig. 3). Large outbursts of magmatic material during volcanic eruptions led to disruption of isostatic equilibrium in the Earth's crust and collapse of large blocks of the Earth's crust to a depth of up to 5 km in the center of the Arctic, along vertical faults (Figure 4). Thus, deep grabens were formed – the Central Arctic, the Greenland, and the Norwegian graben [25, 26].

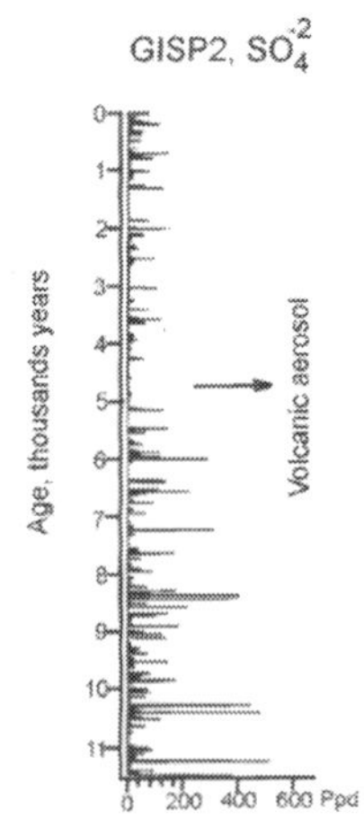

Fig. 3. Large volcanic eruptions in the Arctic at the beginning of the Holocene; the last major eruption occurred 6 thousand years ago [2].

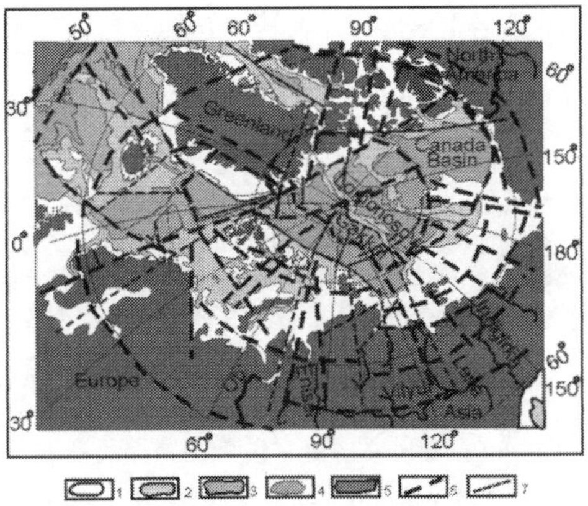

Fig. 4. Scheme of the Arctic's water basins, islands, underwater ranges and faults in the Late Holocene [3, 9, 25, 26].

Legend: 1 – Shallow shelf (50-100 m); 2 – Deep-sea shelf (200-600 m); 3 – Continental slopes of underwater ridges (1-2 km); 4 – Central Arctic deep-water (5 km) graben; 5 – Eastern edge of the Atlantic Ocean (in the north-western part of the figure); 6 – Major faults; 7 – Subordinate faults.

3.3 Catastrophic flooding of the Arctic plains and continental margins – the "Great Flood"

The formation of large grabens at the bottom of the World Ocean in the Mesozoic-Cenozoic is associated, as a rule, with the simultaneous arrival of endogenous waters and the beginning of sedimentation in them [6, 10]. The deep collapse of the southern part of the Arctic Arch, along rectilinear subvertical faults is the most large-scale geological event of the Holocene. The emerging Central Arctic, Greenland and Norwegian deep grabens, quickly filled with endogenous water, divided the formerly single megacontinent into two separate geographical continents – the American and the Eurasian-African (Fig. 4).

As a result of the rise of enormous amounts of water, rapid flooding of the vast plains of the Arctic and continental margins began. The water level in the World Ocean rose by 50-120 m [13, 25, and 30]. According to different data [7, 24 and 28], this event occurred in the period 6.5-5.5 thousand years ago, and has been preserved in the memory of the peoples of many countries as the "Great Flood".

Many millions of people died then [21]. For example, a layer of rocks 6-5 thousand years old with numerous remains of human skeletons was discovered on the east coast of Greenland [28].

In particular, the Old Testament of the Christian religion says: "and the Earth opened its mouth and swallowed them, and all the people of Karee, and all their possessions" [19]. For example, a layer of rocks 6-5 thousand years old with numerous human skeletal remains was discovered on the east coast of Greenland [28].

The great French paleontologist J. Cuvier considered the "Great Flood" as the last major catastrophe of the Earth, which occurred no earlier than 6 thousand years ago and caused the death of people in many countries [11].

The rapid rise of water in the Arctic caused animals to flee in masses to higher elevated areas, where they subsequently perished, when the higher elevations were surrounded by water on all sides and turned into islands. An example is the Novosibirsk Islands near the southern edge of the modern Central Arctic graben, where a giant cemetery of bone remains of mammoths, horses, bulls, and sheep was discovered [11].

Modern man began to inhabit the coastal parts of the emerging Arctic Ocean from the mid–Holocene.

4 Industrial development of the Arctic in the modern time

The active industrial development of the Arctic by man began in the 21st century. This is due to two main reasons. Firstly, the Arctic has become the most important sea route for cargo deliveries to the northern regions of the country, to Chukotka and Kamchatka. Secondly, many deposits of important minerals have been identified in the Arctic and coastal areas. In particular, projected hydrocarbon resources are estimated at 250 billion tons of oil equivalent, including ~43 billion tons of oil and condensate and 206 trillion m^3 of natural gas in the western regions of the Arctic, in the Pechora-Barents Sea and others [5]. Oil and gas production is now the backbone of the country's economy. Deposits of gold, silver and other metals are being developed in the coastal areas [1, 20]. But some problems have appeared in the development of the Arctic.

Many coastal islands are sandy, which makes it very difficult to build production facilities on them. Therefore, a progressive method of preliminary creation of modular piles is proposed, which greatly facilitates the development of such areas [17]. During the development of ore deposits, rivers and lakes are polluted with toxic elements, such as manganese, which is dangerous for the population [20]. A special approach based on innovation is needed for the sustainable development of the Arctic [4]. To increase production efficiency, it is recommended to actively introduce digital technologies which are still underused [29]. The most important issues are environmental protection, ethics, and safety of work. Therefore, increased requirements for higher qualifications of engineers and management of mining companies are needed [18].

5 Discussions

The location of the great glacier, on the high ridges of the Arctic, was not known before the new work [33]. There is a lot of research on the warm climate of the Early Holocene and the emergence of humans and many animals [7, 10, 11, 13, 28]. However, there is very little information about the location of human settlements at this time, since during the catastrophic flooding of the territory 6.0 thousand years ago, they were mostly destroyed.

There is still little research on the grandiose collapse of the Central Arctic and the appearance of a huge Central Arctic deep-sea graben, about 6 thousand years ago [15, 16, 27]. This problem demands further comprehensive research by many scientists.

Human settlements are best preserved in the south of Europe, where many coastal cities and settlements of the Mediterranean and Black Seas are now at a depth of 10-20 m or more [10].

Whereas many settlements of Early Holocene in the Arctic are now at a depth of 5 km, in the central part of it, and at a depth of 50-100 m in a shallow shelf. The task of researching them is a matter of the future. The Arctic has become the most important industrial region of the country in modern times, as a source of oil, gas and solid minerals. But some problems have appeared during the development of the region, for the elimination of which various measures are proposed.

6 Conclusions

A very short Pleistocene-Holocene geological history of the Earth turned out to be exceptionally rich in geological, anthropological and ocean logical events.

Firstly, giant glaciers formed three times on the high-altitude arch of the Arctic in the second half of the Pleistocene, which uncontrollably slid southward, over huge distances - up to 3 thousand km, to the territory of almost all of Europe and North America [13, 33].

Secondary, a very warm climate came in the Early Holocene after the melting of the last glacier, and human settlements appeared on the plains of the Arctic, near rivers, and numerous herds of various animals.

Thirdly, the level of the World Ocean rose sharply (by 50-120 m), after the collapse of the southern part of the Arctic arch, at the end of the Early Holocene (6.0 thousand years ago) and the rapid rise of a large volume of endogenous water. As a result, oceanic waters flooded many marginal, flat parts of the Arctic and continents, creating a widespread shallow shelf with a depth of 30-50 to 200 m or more [8, 10, 15, 25].

And, finally, the Arctic Ocean, which appeared around the middle of the Holocene, divided the single megacontinent of the Earth into two large parts of the geographical lands – the American and Eurasian-African continents [26].

The author is very grateful to E.A. Samylovskaya and other reviewers for serious comments on the original manuscript, which made it possible to significantly improve its content and quality.

References

1. T.N. Aleksandrova, et al., Gold–sulphide deposits of the Russian Arctic zone: Mineralogical features and prospects of ore benefication, *Chemie der Erde*, **00092819(80)**, doi: 10.1016/j.chemer.2019.04.006 (2020)
2. O.K. Borisova, Landscape and climatic changes in the Holocene, *Izvestiya RAS, series geograf.*, **2**, pp. 5-20 (2014).)
3. K.L. Buchan and R.E. Ernst, A giant circumferential dyke swarm associated with the High Arctic Large Igneous Province (HALIP) Gondwana, *Research*, **58** pp. 39–57, doi:10.1016/j.gr.2018.02.006 (2018)
4. G. Gerasimova, I.S. Oblova, E.I. Golovina, The Demographic Factor Impact on the Economics of the Arctic Region, Resources, 10, 117. doi: 10.3390/ resources10110117 (2021)
5. A.S. Egorov, et al. Deep structure, tectonics and petroleum potential of the western sector of the Russian Arctic, *Journal of Marine Science and Engineering*, 21100830140(9), pp. 1-26, doi: 10.3390/jmse9030258 (2021)
6. E.M. Emelyanov, et al., *Paleoceanology of the Atlantic Ocean* (Leningrad, Nedra, 1989) p. 247
7. A. K. Hufthammer, et al., Animals and humans in the European Russian Arctic towards the end of the last ice age and during the mid-Holocene time, *Boreas*, **48(2)**, 397-406, doi: 10.1111/bor.12343 (2019)
8. P.A. Kaplin, *Modern history of the coasts of the World Ocean* (Moscow, Publish. House of Moscow Univ., 1973) p. 265
9. S.N. Kashubin, et al. Map of the crustal thickness of the Circumpolar Arctic, *Regional geology and metallogeny*, **46**, pp. 5-13 (2011)
10. A.M. Kondratov, *Traces on the shelf* (Leningrad: Gidrometeoizdat, 1981) p, 152
11. A.M. Kondratov, *The Great Flood: Myths and Reality* (Leningrad: Gidrometeoizdat, 1982) p. 151
12. T.C. Kluvitkina and E.I. Polyakova, Changes paleogidrology conditions in the Laptev Sea in the Late Pleistocene and Holocene based on the materials of the study of aquatic palinomorf, *Oceanol.*, **53(3)**, pp. 433-448 (2019)
13. N.V. Koronovsky, N.A. Yasamanov, *Geology* (Moscow, Publish. House "Academy", 2014) p. 448
14. V.M. Kotlyakov and A.A. Velichko, Cryosphere of the Arctic: past and present, *Scientific and technology problems of the Arctic development* (Moscow, Nauka, 2014). pp. 74-81
15. R.B. Krapivner, The last planetary transgression of the World Ocean: glacial-melting or tectonics?, *Geotectonic*, pp. 427-244, doi: 10.31857/S0016853X20040050 (2020)
16. A.A. Kremenetsky, New model for the formation of the Atlantic-Arctic heterochronous rift system: concept and basic provisions, *Native Geology*, **3-4**, pp. 18-27, doi: 10.47765/0869-7175-2021-10018 (2021).
17. A. Cherepovitsyn, A. Lebedev Drill Cuttings Disposal Efficiency in Offshore Oil Drilling, Journal

of Marine Science and Engineering, 11(2), p. 317, doi: 10.3390/jmse11020317 (2023)

18. E. Samylovskaya, et al, Transformation of the personnel training system for oil and gas projects in the Russian Arctic, *Resources*, **21100808642 (9)**, pp. 1-20, doi: 10.3390/resources9110137 (2020)
19. V.B. Lyatsky, Earthquakes and geotectonic, *Leningrad. Geograf. Soc.*, 2, p. 22 (1967)
20. V.A. Matveeva, et al., Manganese Pollution in Mining-Influenced Rivers and Lakes: Current State and Forecast under Climate Change in the Russian Arctic, *Water (Switzerland)*, **21100255400(14)**, doi: 10.3390/w14071091 (2022)
21. O.V. Mikhailov, *Half-forgotten secrets of Space, Earth, Sea, Technology and History* (Moscow, Lenand, 2017) p. 304
22. D.V. Nalivkin, *Essays on the geology of the USSR* (Leningrad, Nedra, 1980), p. 158
23. A.M. Nikishin, et al., The main problems of the structure and history of the geological development of the Arctic Ocean, *Herald of the Russian Academy of Sciences,* **90(3)**, pp. 345–356, doi: 10.1134/S101933162003003X (2020)
24. S.B. Nikolaeva, et al. Catastrophic Holocene event in the bottom sediments of the Kola region (Northeast of the Scandinavian Shield), *Doklady RAS*, **473(1)**, pp. 88-92 (2017)
25. B.B. Orlenok, *Global volcanism and oceanization of the Earth and planets* (Kaliningra, Publishing house of the Russian State. Immanuel Kant University, 2010) p. 196
26. A.D. Pavlenkin and L.G. Poselova, Neotectonic collapse of the Arctic vault. Geol.-geof. characteristics of the Arctic region's lithosphere, *Sb. Tr. NIIGA-Vniioceanology*, 210(6), pp. 194-200 (2006)
27. N.I. Pavlenkova, et al., New trends in global oceanic crust tectonics, *Herald of Russ. Academy of Science*, **88(1)**, pp. 88-92 (2018)
28. L. Polyak, et al., History of sea ice in the Arctic, *Quaternary Science. Rev.*, 29, pp. 1757–1778 (2010)
29. E. Samylovskaya, A. Makhovikov, A. Lutonin, et al., Digital Technologies in Arctic Oil and Gas Resources Extraction: Global Trends and Russian Experience, *Resources*, **211 00808642(11)** doi: 10.3390/resources11030029 (2022)
30. F.P. Shepard, *Marine geology* (Leningrad, Nedra, 1976) p. 488
31. N. Syring, et al. Holocene interactions between glacier retreat, sea ice formation, and Atlantic water advection at the inner Northeast Greenland continental shelf, *Paleoceanog. and Paleoclimat*, **35**, e2020PA004019, doi:10.1029/2020PA004019 (2020)
32. G.B. Udintsev, *Relief and structure of the ocean floor* (Moscow, Nedra, 1987) 239 p.
33. A.M. Zhirnov, Arctic mountain arch as a place of great glaciers slipping in the Late Pleistocene, *Discovery*, **42(194)**, pp. 132-143 (2015)

Study of static charge accumulation in HDPE gas pipelines

Vladimir Pshenin[1,*], *Stepan* Menshikov[2,*] and *Maksim* Komarovskiy[3]

[1]Candidate of Technical Sciences, Associate Professor, Department of Transport and Storage of Oil and Gas, 199106, St. Petersburg Mining University, Saint Petersburg, Russian Federation
[2]Postgraduate Student, 199106, St. Petersburg Mining University, Saint Petersburg, Russian Federation
[3]Student, 199106, St. Petersburg Mining University, Saint Petersburg, Russian Federation

Abstract. The accumulation of static charge in polyethylene pipes of gas network systems is a familiar process, which is paid attention to mainly to prevent accidents on pipelines. incidents related to static electricity can occur both during assembly works (coil tapping) and during the operation of gas pipelines (gas venting, etc.). Despite the fact that repeated attempts to study this process have been made by major operating organizations, today we can state that these regularities have not been studied in full. In this work we have made an attempt to describe theoretically the process of static charge accumulation on a pipe body, as well as to evaluate experimentally the adequacy of the proposed models.

1 Introduction

At the present time, the modern trend in the development of gas distribution networks is the widespread use of plastic materials, including the introduction of polyethylene pipelines wherever it is permissible from environmental conditions [3, 14, 16]. Earlier the temperature range of operating was considered as one of the main limiting factors, but at the present time the accumulated experience proves the possibility of PE pipelines application even at negative temperatures [1, 15].

One of the important peculiarities of PE pipelines is the accumulation of a static charge in the process of operation. Serious research of such processes, however, was carried out only for insulation of high-voltage wires (the occurrence of delamination and breakdown in the vapor-gas cavity was studied). The process of static charge accumulation has both potentially positive features (pitting detection detection on pipelines) and negative ones, which can manifest themselves in the phenomenon of pipe body breakdown. In addition, static electricity can cause accidents during some technological operations, such as blowdown, pipeline sectioning and so on [2, 19].

For a detailed study of the process of static charge accumulation, a laboratory setup was developed and a number of experiments were conducted, based on the results of which experimental values of the ignition threshold of independent discharge were obtained.

2 Theoretical investigation

Since 1993, there have been a number of micro-hole (5 to 50 μm diameter) incidents on gas distribution pipelines in the United States [13]. Most of these have been identified during accident investigations, while other cases have been accidental. In terms of pipeline geometry, the identified cases involved pipelines in the range from 1/2" to 4" and SDR from 10 to 11.5. The greatest number of incidents were observed during blowdown and filling operations of the pipeline [12, 17]. Thus, in a number of cases, polyethylene is capable of accumulating a charge on the inner surface, which can lead to breakdown and local melting of the pipe body.

3 Conducting research

3.1. Experimental studies

The experimental setup consists of a section of pipeline with a high-voltage electrode ("ionizer module") at one end in the form of a ring with four needles. An image of the experimental setup is shown in Figure 1.

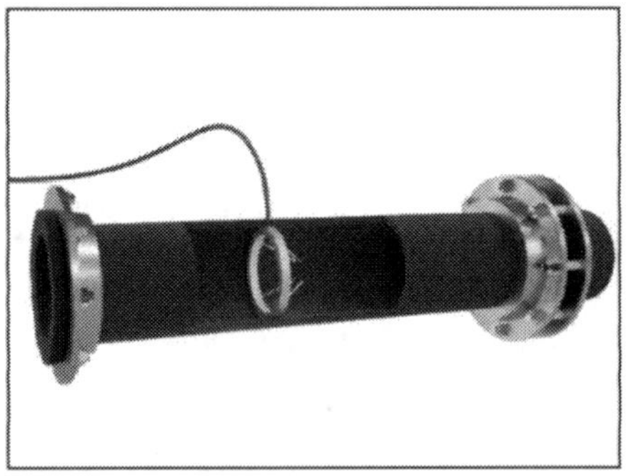

Fig. 1. 3-d model of the ionizer experimental setup

Four steel needles are arranged symmetrically around the perimeter of the ring. A more dense

* Corresponding authors: Pshenin_VV@pers.spmi.ru, stepan47@bk.ru

arrangement of the needles should not increase the discharge, because there is an effect of "mutual" quenching of discharges on neighboring needles - even when the needles are arranged symmetrically, when the tension on all needles is the same, the discharge is not ignited on all needles if the needles are too close to each other [2]. A needle is the most convenient form for igniting an electrical discharge; a metal ring is used to embed the needles in the structure of a sealed pipeline. The beads on the edges of the ring, facing into the air, serve to reduce the electric field in the air near the ring and prevent the ring from being coronated toward the air [4]. Figure 2 shows how the electrodes are positioned relative to each other and the pipe. The grounded electrode is shown in blue and the high-voltage electrode is shown in red. The black lines show the contour of the polyethylene pipe.

Fig. 2. Electrode - insertion into a polyethylene pipe

Measurement of the electrostatic field is carried out by the measuring transducer PZ-80 (Figure 3), which contains a converting element in the form of a mechanical modulator. The electric signal comes from the converting elements through the signal matching unit to the analog-digital converter and signal processor, which calculates all the measured quantities. Technical characteristics of the digital converter PZ-80 are presented in Table 1.

Table 1. Technical characteristics of the digital transducer PZ-80

Parameter	Unit of measurement	Meaning
Measurement range of electrostatic field intensity (ESI)	kV/m	0.3 – 200
Permissible error limits of measurement of the electrostatic field intensity ESI	%	±15
Weight	gram	250

Figure 3 shows the measuring device PZ-80 and the display unit Ecophysics-110A.

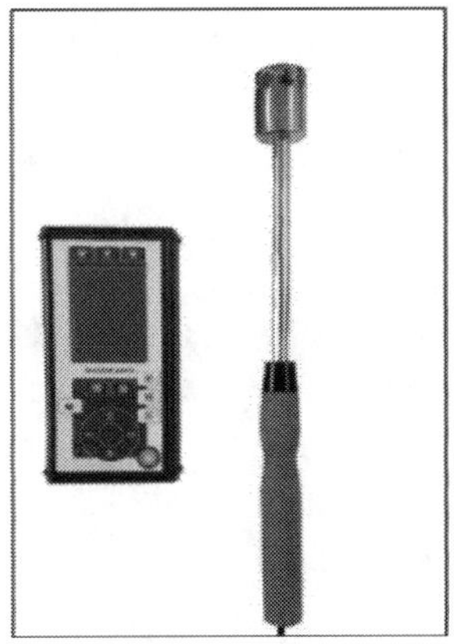

Fig. 3. Measuring device PZ-80 and the display unit Ecophysics-110A

Operating principle of PZ-80 electric and magnetic field intensity meters is based on conversion of field intensity into electric voltage signal, which then goes to analog-digital converter [5]. After filtering and detection operations the measurement of field intensity and its representation on the indicator is performed. Transformation of the alternating electric field strength is performed by means of a dipole measuring transducer, formed by two plane-parallel plates. The test side for static electricity accumulation by polyethylene pipelines of gas distribution systems is shown in Figure 4.

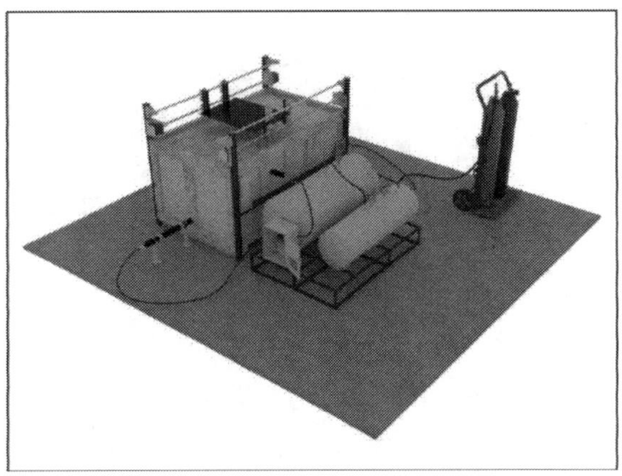

Fig. 4. Test side for static electricity accumulation in polyethylene pipelines

3.2 Theoretical investigation

In the process of experimental work, we measured the current thatthe "ionizer" gives at different lengths of the section of the tube between the "ionizer" and the ion collector grid (starting with a small distance, 10 centimeters, 20 cm, 50 cm, etc., all the way up to the length of the box) [6]. Of the measurements in this case, only the measurement of the electric current is involved. Calculation of threshold voltage, at which an independent electric discharge occurs in the given geometry, is performed by integration of ionization coefficient, which is a function of field strength specific for the given type of gas [7].

$$\int_A^B \alpha(E)ds = M_{cr} (1)$$

Integration is performed along the electric field line AB, the beginning and the end correspond to the surface of the electrode or solid insulator (depending on where this field line goes from the gas region).

The method consists in finding the voltage when the value of integral (1) at least on one power line in this area reaches the critical Mcr number which value depends on the type of discharge which can be implemented in this system: for corona or barrier discharges Mcr is in the range of 6...15, for streamer discharge - in the range of 15...20. In this case we take Mcr = 15 - since corona discharge should occur in this system, we will get an estimation for the threshold voltage from above. In addition to the threshold voltage itself, the method makes it possible to determine the area in which a discharge can occur at a given voltage [8]. The dependence of the CH4 ionization coefficient on the field strength is shown in Figure 5.

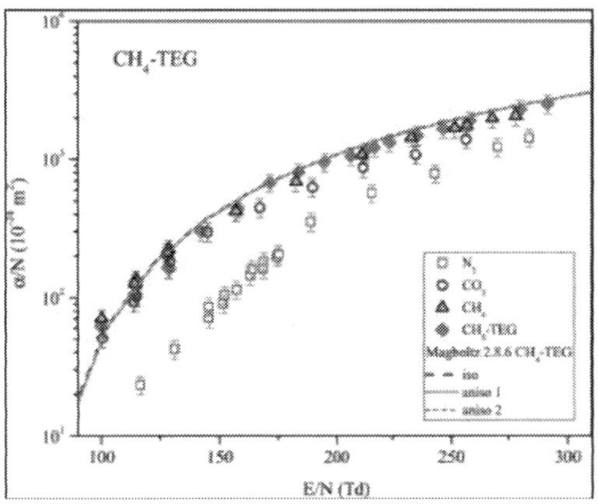

Fig.5. dependence of the CH4 ionization coefficient on the electric field strength

The electric field is calculated in the quasi-static approximation, the Laplace equation is solved by formula (2) (without taking into account the volume and the surface charge, because the method considers the field distribution immediately before the ignition of the discharge)

$$-\nabla \cdot \varepsilon \nabla \varphi = 0 \, . \qquad (2)$$

In the volume the dielectric permittivity of polyethylene (ε=2.35), gases (ε=1) are set. On the surfaces of electrodes, the boundary conditions of fixed potential (equal to voltage or zero for grounded elements) are set, on the outer surface in the air zero potential is set (imitation of grounded room walls), on the outer "ends" of the pipe the condition of power lines parallel to the boundary is set.

The key parameters that affect the discharge ignition threshold are needle length and the needle radius [9]. With the parameters listed below in the description of the system components, an ignition threshold of 18 kV was achieved. This is a relatively high value compared to the ignition threshold value typical for corona discharge in a single-needle system - 4...8 kV.

Consider the question of whether the electric field will prevent the ions from drifting outside the electrode system through the tube. As shown in the graph (Figure 6), in the region where the field is rotated toward the grounded electrode (and is thus rotated relative to the direction of gas pumping), the drift velocity of the ions is below 10 m/s, i.e., a gas flow with a speed near the tube axis of about 20 m/s should effectively carry the ions outside the electrode system.

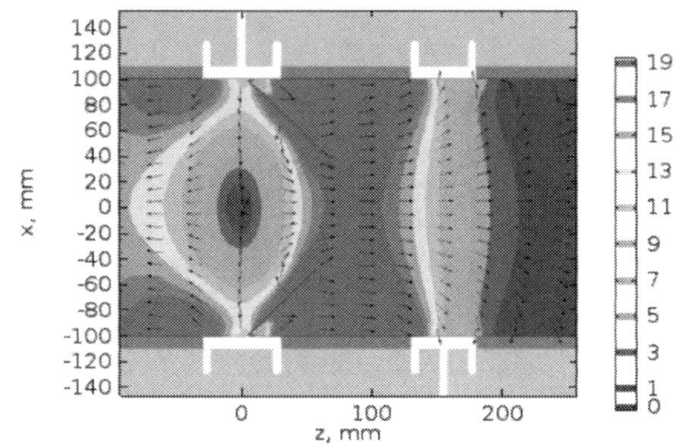

Fig. 6. Ion flux drift velocity under the action of electric field at 18 kV

In the course of preliminary tests of the developed unit, for the first time in the practice of industrial application on a real section of PE pipe, important results were obtained: it was established that polyethylene gas pipelines accumulate charge most quickly when the gas contains hot moisture, and also that the charge tends to accumulate when there in the presence of suspended mechanical particles in the gas. The opposite is also true, which in its turn leads researchers to look for effective methods to reduce the risks of static electricity [10, 11].

4 Conclusion

As a result of the studies, the peculiarities of the phenomenon of static charge accumulation in polyethylene gas pipelines have been established. Theoretically predicted and the first results in terms of gas flow ionization have been obtained. For further consideration, the introduction of antistatic additives (aerosols) into the gas flow before carrying out technological operations seems to have a certain potential as a method of minimizing the negative effects [18].

References

1. V.S. Litvinenko, The Role of Hydrocarbons in the Global Energy Agenda: The Focus on Liquefied Natural Gas, *Resources*, **5**, pp. 59-81 (2020)
2. N.J. Cassidy, Electrical and Magnetic Properties of Rocks, Soils and Fluids Ground Penetrating Radar Theory and Applications ed H M Jol (New York: Elsevier, 2009) pp. 41-72.
3. Y. Zhukovskiy, P. Tsvetkov, A. Buldysko, et al., Scenario modeling of sustainable development of energy supply in the Arctic, *Resources,* **10(12)**, pp. 124 (2021)
4. J. De Urquijo et al., Drift and proton transfer reactions of positive ions in methane parent gas J. Phys. B: At. Mol. Opt. Phys., **30**, 4395 (1997)
5. S.S. Nunayon et al., Experimental evaluation of positive and negative air ions disinfection efficacy under different ventilation duct conditions Build. Environ, **158**, pp. 295–301 (2019)
6. J. Sunget al., Improvement of particle clean air delivery rate (CADR) of ion spray electrostatic air cleaner with zero ozone emission on diffusion charging in indoor space, *Proc. IEEE Ind. Appl. Soc. Annu. Meet.* (Baltimore: IEEE), 1-7 (2019)
7. P.S. Zhidkova and A.V. Samusenko, A Computer Model of the Ionic Wind in the Unipolar Approximation with the Boundary Condition on the Ion Flow Variation Rate, *Surf. Eng. Appl. Electrochem*, **52(4)**, pp. 370–379 (2016)
8. Yu.P. Raiser, Physics of Gas Discharge (Moscow: Intellect , 2009)
9. A.R. Petri et al., Measurement of the first Townsend ionization coefficient in a methane-based tissue-equivalent gas, *Nucl. Instrum. Methods Phys. Res.,* Sect. A, **849**, pp. 31-40 (2017)
10. T.N. Zubkov et al., Simulation of the Seat Form of a Negative Corona Discharge in a Point–Plane Electrode System, *Surf. Eng. Appl. Electrochem.*, **49(6)**, pp. 474–479 (2013)
11. Sung J. H. et al. Ultrafine particle cleaning performance of an ion spray electro-static air cleaner emitting zero ozone with diffusion charging by carbon fiber Build. Environ., **166(2)**, 106422 (2019)
12. Cherepovitsyn, A.; Lebedev, A. Drill Cuttings Disposal Efficiency in Offshore Oil Drilling. J. Mar. Sci. Eng., 11, 317 (2023).
13. Zemenkova, M. Y., Chizhevskaya, E. L., & Zemenkov, Y. D. Intelligent monitoring of the condition of hydrocarbon pipeline transport facilities using neural network technologies, *J. Of Min. Inst.*, **258**, 933-944 (2022)
14. Tsyglianu P.P., Romasheva N.V., Fadeeva M.I. & Petrov I.V. Engineering projects in the russian fuel and energy complex: actual problems, factors and recommendations for development. *Ugol'*, **3**, pp. 45-51 (2023)
15. Shammazov, A. M., Aminev, A. N., Pirogov, A. N., Shammazov, I. A., Pirogov, N. E., Petrenko, S. V., and K. E. Denisov. Solving problems of reconstruction and development optimization of a pipeline system (Russian), *OIJ,* **8**, pp. 80-83 (2018)
16. Zhdaneev, O.V. Technological sovereignty of the Russian Federation fuel and energy complex. J. Of Min. Inst., **258**, pp. 1061-1070 (2022)
17. Romasheva N. V., Babenko M. A., Nikolaichuk L.A. Sustainable development of the Russian Arctic region: environmental problems and ways to solve them., *MIAB*., **10(2)**, pp. 78—87 (2022)
18. Razmanova, S. V., & Andrukhova, O. V. Oilfield service companies as part of economy digitalization: assessment of the prospects for innovative development. Journal of Mining Institute, 244, 482-492 (2020).
19. Litvinenko, V. S., Petrov, E. I., Vasilevskaya, D. V., Yakovenko, A. V., Naumov, I. A., & Ratnikov, M. A. Assessment of the role of the state in the management of mineral resources. Journal of Mining Institute, 259, 95-111 (2023).

Rational methods of geodetic control of technogenic consequences of the development of oil and gas fields

Victor Volkov[1], *Nikita* Volkov[2,*] and *Tatiana* Volkova[3]

[1] Associate Professor, Department of Geodesy, Land Management and Cadastre, 190005, Saint Petersburg State University of Architecture and Civil Engineering, Saint Petersburg, Russian Federation
[2] Doctor of Technical Sciences, Professor, Department of Engineering Ecology and Municipal Economy, 190005, Saint Petersburg State University of Architecture and Civil Engineering, Saint Petersburg, Russian Federation
[3] Candidate of Technical Sciences, Associate Professor, Department of Engineering Ecology and Urban Economy, 190005, Saint Petersburg State University of Architecture and Civil Engineering, Saint Petersburg, Russian Federation

Abstract. The methods of controlling the consequences of the influence of manmade geomechanical processes caused by the extraction of hydrocarbons from the subsurface in the fields under development are considered. The possibilities of methods from the standpoint of separate determination of spatial and temporal quantitative parameters characterizing the deformed state of the skeletons of reservoirs and host rocks, the entire thickness of the rock mass above the deposit in an uneven field of compressive stresses and deformations of the earth's surface are shown. The inconsistency of methods of classical repeated leveling, geodetic positioning and traditional radar interferometry in connection with the low level of representativeness of their results is noted. scientific novelty lies in the development of ways to improve the effectiveness of the method of radar satellite interferometry. As a result, the conditions for the effective use of radar satellite interferometry on the basis of a network of stationary points which coincide with the points of the leveling network method of spot geodetic sounding of deformation processes in the oil and gas fields under development have been determined and characterized.

1 Introduction

Oil and gas fields in the Russian Federation are mostly located in the territories of the Arctic Zone of the Russian Federation. Special attention is paid to the study of natural and manmade deformation processes developed in the near-surface layers of the earth crust in connection with the extraction of hydrocarbons from the subsurface. This study is implemented in the form of monitoring deformations of the skeletons of reservoirs, their host rocks and the entire thickness of the rock mass, lying over the fields and manifested in deformations (shifts) of the earth's surface and activation of tectonic disturbances. These deformations may lead to the disruption of technological processes of hydrocarbons extraction from the subsurface and destruction of elements of field development. The priority in organizing monitoring is to obtain results that separately characterize the spatial and temporal parameters of natural and manmade deformation processes that have developed in the developed near-surface layers of the earth crust, which, together with exogenous natural deformations, manifest themselves in deformations of the physical (daytime) earth's surface. An important component of the vertical displacements of the physical earth's surface recorded during monitoring in the natural conditions of the Russian Arctic, is numerous and diverse deformations of the near-surface layers of the earth's crust generated by natural exogenous geomechanical processes [1, 2]. Such processes are noise in the monitoring of technogenic deformation processes.

Currently, several methods of studying deformations of the physical earth's surface and the near-surface layers of the earth crust have been proposed for monitoring technogenic deformation processes [2, 3, 4, 5]. At the same time, the design of geodetic observations, does not consider the issue of achieving the goals of monitoring and the possibility of practical implementation of the recommended methods in the prevailing natural and man-made conditions. The purpose of the article is to establish the factors determining the possibility of effective application in the conditions of the Russian Arctic of common and little-known methods for studying the deformations of the earth crust, its near-surface layers and the Earth's surface for monitoring man-made deformations of reservoir skeletons, their host rocks, displacements of the Earth's surface and activation of man-made disturbances in the territories of oil and gas fields being under development

2 Methods

At the present stage, the main method of monitoring of technogenic consequences of the development of oil and gas fields located in the Russian Arctic is the geodetic

* Corresponding author: volkov.nikita@yahoo.com

method of re-leveling the control leveling points of profile lines. At the same time, in the territories of the developed oil and gas fields located in the Russian Arctic, traditionally the structural solution of observation systems is reduced to the construction of mutually intersecting re-leveling lines with a length from several tens to hundreds of kilometers. leveling lines (profile lines) are fixed standard ground and rock (in the presence of rocks) leveling points without assessing their resistance to repeated leveling of profile lines is carried out by the Class II program without scientific substantiation of the accuracy of leveling, their periodicity and requirements for the stability of leveling points [2, 4, 5, 6].

Using the example of the Bovanenkovskoye field, it looks as follows: at the predicted values of rates of technogenic subsidence of the earth's surface vpredicted = 9.2 mm/year and length of leveling lines L ≈ 130 km, the standard error in determining rates of vertical displacements of the earth's surface by results of repeated leveling of class II ($\eta = 2\sqrt{2}$ mm/km) is

$$m_{v\,(predicted)} = \frac{\eta}{T}\sqrt{2L} \geq 32$$ mm/year

(η, mm/km is the mean square error of leveling for 1 km of travel; T, year – frequency of leveling) with the unaccounted forecast instability of ground leveling indicates the influence of exogenous processes [5, 6]. in the unaccounted forecast instability of ground leveling indicates the influence of exogenous processes

3 Discussion

Exogenous natural factors and the geomechanical processes generated by them influence the results of repeated leveling and the physical earth's surface through exogenous near-surface movements of the Earth's crust. The dynamics of natural exogenous geomechanical processes and the displacements of leveling points and the physical surface generated by them is complicated by the diversity and variability in space and time of moisture and temperature fields in the near-surface layers of the Earth's crust, which are heterogeneous in composition and depositional conditions. Natural exogenous deformations are superimposed on the studied technogenic deformations of reservoirs and rock massifs and manifest themselves in complexly differentiated displacements of leveling points and the physical earth's surface, which are mistakenly attributed to the parameters of the studied deformation processes, namely:

- Seasonal thawing and freezing of dispersed soils in the conditions of the Russian Arctic is accompanied by their frost heave, generating vertical irreversible displacements of standard leveling points at a rate of up to 250 mm/year and fluctuations of the Earth's surface ± (4÷300 mm/year) [2, 7, 8].
- Thermal and moisture exchange in the near-surface layers of the earth's crust, composed of dispersed rocks, generates in them volume deformations (swelling and shrinkage), jointly manifested in poorly studied oscillatory and short-period vertical displacements of leveling points and heterogeneous in space and time hydrothermal movements of the physical earth's surface with an amplitude of several millimeters to meters [1, 2, 4].
- Periodic changes in the geothermal field of rocks and their temperature deformations cause inhomogeneous in space and time vertical displacements of ground and rock leveling points and the physical earth's surface with amplitudes from 0.2 mm to 39 mm [1, 2]. Consequently, the lack of consideration of significant effects of exogenous factors on the results of traditionally performed re-leveling, as well as other known disadvantages [2, 4, 5, 8], inherent in repeated leveling, exclude the possibility of its use for monitoring of man-made deformations at oil and gas fields developed in the Russian Arctic.

3.1. Method of spot geodetic sounding of deformation processes in oil and gas fields. Developed by the authors of the article and tested at the Yamburgskoye oil and gas condensate field [4], method of point geodetic observation of deformation processes in oil and gas fields allows us to separately monito the deformations of the reservoir skeleton, its host rocks and rock masses lying above the reservoir, as well as their joint manifestation in the movements of the Earth's surface.

At the same time, a geodetic network of observation stations is being created on the territory of oil and gas fields, consisting of reference (initial) and control (deformation) leveling points (LPs). Within the boundaries of the predicted manifestation of dangerous technogenic processes outside the integral contour of hydrocarbon production vertically or laterally, reference LP are equipped on the basis of mothballed exploration and observation wells. To control the development of deformations of reservoirs, rocks and the earth's surface at the lateral distance

$$S_{km} \leq \frac{3 \cdot 10^{-2} v_{predicted}^{2} T^{2}}{M^{2}}$$

($v_{predicted}$, mm/year – the predicted values of deformation rates; M, mm/km - the design accuracy of leveling) from the reference LPs on the basis of preserved exploration or observation wells, control deep LPs are equipped in pairs with control ground LPs. The base of the reference LPs is located above the developed deposit , and the bearing pipe of the reference soil LPs is embedded outside the zone of seasonal thawing (freezing) of soils in permafrost rocks to the depth

$$z \geq f\left(1 + 1,4\frac{P_{Spec\ Heaving}}{P_{Spec\ Freezing}}\right)$$

at the thickness f of seasonal thawing and freezing of soils with specific values of soil heaving ($P_{Spec\ Heaving}$) and freezing of the bearing pipe with soil ($P_{Spec\ Freezing}$), in which the laying is carried out. Control LPs (based on wells) record reservoir deformations, while surface ones record total deformations of both collectors and rock massifs overlying the deposit, manifested in displacements of the earth's surface. The permissible influence of exogenous factors (interference of repeated leveling) on t LPs is set by an inequality $m_{interf} \leq 3 \cdot 10^{-2} v_{predicted} \cdot T$, and the distance of the reference LPs from the reference LPs of observation stations is designed as S < 1 km. This design of measurements makes it possible to register the deformations of reservoirs, rock massifs located above them, as well as the earth's surface throughout the year with an average square error of less than 1 mm.

Approbation of this method [4] confirmed the high accuracy, significance, and representativeness of the results of repeated leveling pf points performed as part of monitoring of technogenic deformations of the skeleton of reservoirs, rock massifs overlying them, and activation of tectonic disturbances in the territories of oil and gas fields located in the Russian Arctic. At the same time, it should be noted that the detail control of technogenic deformation processes on the territory of oil and gas fields by this method depends on the presence of preserved wells used as reference and control leveling points and the density of their location within the mining allotment.

3.2. Method for monitoring deformations of the Earth's surface based on global positioning stations.

The comparatively low accuracy of the results of determining the vertical component of the deformations of the Earth's surface 4-10 mm [3] and the level of their representativeness at high cost of staging and complexity of organizing satellite observations in the far North do not allow using this method when monitoring deformation processes in oil and gas fields located in the territories of the Russian Arctic.

3.3. Method for monitoring deformations of the Earth's surface based on the traditional radar interferometry techniques.

At the same time, the rates of such vertical displacements, as shown by studies [1], change in the natural conditions of the Russian Arctic during a day, month and year in the range from millimeter to 250 mm/year, with man-made shifts of the Earth's surface characterized by rates of 5-10 mm/year. Vertical displacements of the physical surface generated by exogenous geomechanical processes are difficult to differentiate in space and time due to the heterogeneity of the composition of the soils of the surface layers of the Earth's crust, their physical properties and natural and man-made conditions of occurrence [2, 4]. At the same time, the rates of such vertical displacements change in the natural conditions of the Russian Arctic during the day, month and year in the range from a millimeter to 250 mm/year, and manmade shifts of the Earth's surface are characterized by speeds of 5-10 mm/ year.

The determination of the deformations of the Earth's surface using the results of interferometric radar surveys carried out from spacecrafts is based on comparing the phase of radar signals reflected at points on the physical surface at certain time intervals. The choice of intervals between observations is not scientifically justified. In this case, the distances Di (i is the number of repeated interferometric radar survey) traveled by the radar signals in the direction from the spacecraft to j = 1, ... , m point of the physical surface and back are determined. By the sign ("+" and "-") and the magnitude of changes in the distances, the character (rise and fall) and the scale of deformations of the Earth's surface are judged. The development of deformations is caused by extraction of hydrocarbons from the subsurface, whereas the change in distances is the result of the combined impact on the physical surface of the Earth of natural exogenous geomechanical processes and deformations of mountain massifs and reservoirs caused by extraction of hydrocarbons [9, 10, 11].

It is obvious that exogenous geomechanical deformations of the physical surface of the Earth are a significant hindrance in determining the deformations of the Earth's surface, making it difficult or impossible to interpret the results of repeated radar surveys. The impossibility of interpreting the results of repeated radar surveys is also explained by the presence of snow cover on the reflecting surface during the autumn and winter periods and herbaceous vegetation in the summer period.

Consequently, the use of the interferometric radar survey method to monitoring the technogenic consequences of of hydrocarbon deposit development in the territories of the Russian Arctic is impractical because there is no possibility of obtaining reliable and representative data on the deformations of the Earth's surface. The obtained values of deformations indicate the stress-strain state of reservoirs and the overlying rock massifs.

3.4. Method for monitoring deformations of the Earth's surface using interferometric radar based on stationary points (InSAR).

The InSAR (Interferometric synthetic aperture radar) method is similar in principle to the method of monitoring the physical surface of the earth based on interferometric radar imagery from spacecraft. The difference can be attributed to the use of stationary reflectors, which are triangular prisms with a right reflecting inner angle. The levels of reflected signals from such reflectors are significantly higher than the levels of signals reflected from the physical surface of the Earth. The reflectors installed on the bases form a network of stationary InSAR points, which also includes roofs of buildings, large metal and other natural and man-made objects, which have reflectivity in relation to radar signals and do not experience vertical displacements due to deformations of their bases and foundations.

Studies of this method show that the accuracy, representativeness of its results and the effectiveness of their use in controlling the technogenic consequences of the development of hydrocarbon deposits in the Russian Arctic depends on the implementation of a number of conditions [9, 10, 11, 12, 13], namely:

- Combination of radar interferometric survey sessions with geotechnical monitoring cycles, which makes it possible to exclude from the survey results vertical displacements of real estate objects caused by deformations of their foundations and settlements (or elevations) of foundations.

- The size of the edges of the reflective prisms should provide a sufficient level of reflected radar signal and should be 1 meter or more.

- The design of the bases of reflective prisms should ensure their resistance to the effects of natural exogenous geomechanical processes within $\Delta_{destabilization} \leq 0{,}18 \cdot v_{predicted} \cdot T$ ($v_{predicted}$ - forecast rates of controlled deformations mm/year, T - frequency of repeated radar interferometric survey) [2].

- The use for the analysis of radar images obtained from the maximum possible number of spacecraft with various geometric groupings of spacecraft.

- The frequency and number of radar survey sessions should exclude the influence of the inhomogeneous ionosphere and atmosphere, that is, the requirements for the minimum number of freely distributed radar images (at least 12 pieces) must be met [9, 10].

- The value of the controlled vertical displacements based on the results of two radar surveys should be no more than 25% of the wavelength of the radar signal [10, 11].

- Inclusion in the network of stationary InSAR points, equipped on the basis of moth exploration and observation wells, located outside the integral contour of hydrocarbon production (reference points), as well as on wells, the bases (bottoms) of which are located above the developed deposit and on stable tubular ground leveling points [2, 4, 5].

- Mandatory verification of the results obtained by the InSAR method, based on their comparison with the results of repeated deformation processes obtained by the geodesic method of point probing.

- The use of radar images obtained from spacecraft in order to significantly reduce the influence of tropospheric noise (interference) [12, 13].

- Radar survey should be carried out in the absence of snow cover on the reflective elements of the terrain and prisms.

4 Conclusion

1. In the practice of monitoring to control the technogenic consequences of the development of oil and gas fields located in the territories of the Russian Arctic, over the last century and currently, there is essentially no even a formulation of research problems, as a consequence, there are no necessary solutions to the problems related to the choice of method for monitoring technogenic deformation processes, design of repeated instrumental and remote repeated observations, their implementation, as well as the description, analysis and interpretation of their results.
2. At the present stage [14, 15], the control of technogenic consequences of the development of oil and gas fields is possible with the joint application of methods of point re-leveling and interferometric radar survey as part of monitoring observations, given that the results of radar interoferometry allow minimizing the interpolation errors of the deformation characteristics point-set by repeated leveling over all areas of mining allotment.

References

1. V.I. Volkov, Modern formulation of the problem of the influence of non-tectonic factors on the results of geodetic studies of modern vertical movements of the earth's crust, *Izvestia vuzov «Geodesy and aerophotosurveying»*, **3**, pp. 17-25 (1998)
2. N.V. Volkov, Location and selection of construction and depth of laying of leveling points on the geodynamic polygons of oil and gas fields, *Izvestia vuzov «Geodesy and aerophotosurveying»*, **5**, pp. 54-59 (2017).
3. S.S. Arsenyev-Obraztsov and A.P. Pozdnyakov, Application of interferometric synthetic aperture radar (inSAR) technique to address the challenges of petroleum field geology and oil and gas field development, *Gas Industry Journal*, **798(3)**, pp. 38-44 (2020)
4. V.I. Volkov et al., New concept of usage of mine surveying and geodetic observations for monitoring the technogenic consequences of the development of oil and gas fields, *Mine Surveying Bulletin*, **3**, pp. 45-50 (2018).
5. V.I. Volkov and N.V. Volkov, Use of the program and goal-oriented approach to observe the vertical displacements of the earth's surface in Russia *E3S Web of Conferences,* **91**, 07023 (2019)
6. V.I. Volkov and V. Smirnov, Program- and Goal-Oriented Approach to Organization of Monitoring Defo rmations of Buildings and Structures Applied Mechanics and Materials, 725-726, 118-123 (2015)
7. G. Bondarik, Foundations of the theory of variability of engineering-geological properties of rocks (Moscow, Nedra, 1971), p. 271.
8. Y.A. Kashnikov and S.G. Ashihmin, Rock mechanics in the development of hydrocarbon

deposits (Moscow, Nedra-Businesscenter, 2007), p. 467.

9. H. Klemm et al., Monitoring horizontal and vertical surface deformation over a hydrocarbon reservoir by PSInSAR First break, **28(5)**, pp. 29-37 (2010)
10. D.W. Vasco et al., Reservoir monitoring and characterization using satellite geodetic data: Interferometric Synthetic Aperture Radar observations from the Krechba field, Algeria, Geophysics, **73(6)** (2008)
11. A. Rucci et al., Monitoring the geologic storage of carbon dioxide using multicomponent SAR interferometry, *Geophysical Journal International,* **193(1)**, pp. 197-208 (2013)
12. A. Rucci et al., InSAR for Ground Displacement Continous Monitoring, doi: 10.3997/2214-4609.201701936 (2017)
13. G. Bitelli et al., Updating the subsidence map of Emilia-Romagna region (Italy) by integration of SAR interferometry and GNSS time series: the 2011–2016 period, *Proceedings of the International Association of Hydrological Sciences*, 382, pp. 39-44 (2020)
14. A. Kopteva, L. Kalimullin, P. Tcvetkov, A. Soares, Prospects and Obstacles for Green Hydrogen Production in Russia, *Energies*, **14(3)**, p. 718, doi: 10.3390/en14030718 (2021)
15. E. Samylovskaya, A. Makhovikov, A. Lutonin, D. Medvedev, R.-E. Kudryavtseva, Digital Technologies in Arctic Oil and Gas Resources Extraction: Global Trends and Russian Experience, *Resources*, **11(3)**, pp. 29, doi: 10.3390/resources11030029 (2022)

Small-sized magnetolevitation system of the trestle type for the Arctic

Eugene Sundukov[1,*], *Boris* Shifrin[2], *Nadezhda* Tarabukina[3] and *Veronica* Sundukova[4]

[1]Ph.D. in Economics, Senior Researcher, Federal Research Centre «Komi Scientific Centre of Ural Branch of Russian Academy of Sciences», 167982, Syktyvkar, Russian Federation
[2]Ph.D. in Technology, Associate Professor, Department of Mathematical Methods in Management, Saint Petersbourg State Forest Technical University, 194021, Saint Petersbourg, Russian Federation
[3]Senior engineer, Federal Research Centre «Komi Scientific Centre of Ural Branch of Russian Academy of Sciences», 167982, Syktyvkar, Russian Federation
[4]Bachelor, Syktyvkar Forest Institute, 167982, Syktyvkar, Komi republic, Russian Federation

Abstract. The construction of large technical systems (industrial and transport) poses a threat to the environmental situation in the Arctic. On the other hand, the operation of such systems in harsh Arctic conditions is also unsafe, especially in permafrost conditions. We offer a small-sized maglev system of the trestle type with "movers" and "fellow travelers" type suspension. In this system, the track magnetic field is created by a moving sequence of interacting magnetic field sources –movers in a special control channel (beam, pipe), which also interact with the magnetic field sources of transport modules –fellow travelers. Fellow travelers are suspended in controlled channels due to interaction with permanent sources of magnetic field. The structure is mounted on arc-shaped supports that evenly distribute the load on the supporting surface and protect it from snow drifts and water flows. A model of a two-channel system with the lower location of the controlled channel has been developed. Further improvement of the system involves the use of the "magnetic potential hole" effect and the principles of the planetary model construction.

1 Introduction

According to the type of trestle equipment, most of the implemented transport systems based on magnetic levitation (maglev) can be defined as single-channel - all equipment in one pipeline (pipe) or around one beam [1]. The overpass stator winding is divided into an accelerating winding and a winding which provides lifting or suspension of transport modules. Transport modules move when a moving magnetic field is generated in the accelerating winding.

The greatest progress in the development of maglev transport has been made in the China [2]. Currently, there is a tendency to consider maglev systems as a type of rail transport, namely high-speed trains [3]. In Italy, RFI and Nevomo are planning to combine maglev with the existing rail system [4].

The Hyperloop transportation system was tested, which is a pipeline inside which cargo and passenger capsules move at high speed in a highly rarefied atmosphere [5].

Maglev and Hyperloop train systems are large technical systems. The creation of large technical systems (industrial and transportation) poses threats to the Arctic environment. On the other hand, the operation of such systems in harsh Arctic conditions is also unsafe, especially in permafrost conditions. In addition, as industrial facilities or mineral deposits are completed, the transport routes to them may become redundant. For example, the operation of the Novy Urengoy – Yamburg railway line was discontinued.

In this regard, in the Arctic it may be advisable to use small-sized transport systems with the number of passengers 4-5 people or carrying capacity of 0.5 tons.

We propose to consider small-sized transportation systems, which can be connected to the main maglev and rail transport systems at hubs for subsequent transportation of passengers and cargo to hard-to-reach points [6]. At the same time, such systems will have some similarities with the Hyperloop system.

2 Small-sized maglev system

The proposed system is shown in Figure 1. The circular magnetic field sources are located in two conjugate channels. In the upper channel, the magnetic sources – movers are located. Movers can form an arbitrarily long sequence.

The upper channel will be called the control channel. the control channel has an oval shape in the cross section, which eliminates the overturning of the movers. The movers in the channel have a gap relative to the inner walls, which does not cause difficulties when moving.

* Corresponding author: jek-sun@mail.ru

Thus, the control channel is a pipe, inside of which ideal conditions for movement of movers are created, in particular, by analogy with Hyperloop, it can be a very rarefied atmosphere. But we are not trying, unlike Hyperloop, to put the entire transport module (TM) inside the pipe.

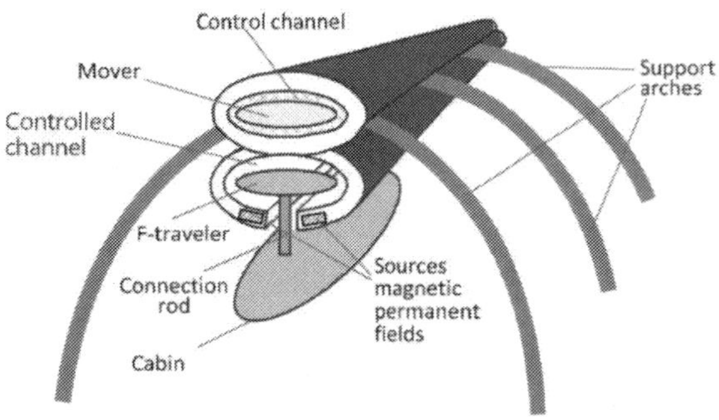

Fig. 1. General view of the small-sized maglev system.

The TM consists of a cabin and several magnetic field sources connected to the cabin, for example, by means of rods. The magnetic field sources of the TM are located in the lower channel (Fig. 1), which we call the controllable channel. In the case of connecting rods, the controlled channel has a notch, in addition, it is equipped with sources of rectangular-shaped magnetic constant fields (magnetic rails) for levitation of the TM.

3 Features of the small-sized maglev system

3.1 Advantages of the small-sized maglev system

Ideal conditions for the movement of movers (rarefied atmosphere, low temperature, etc.) can be created in the pipe, which allows not to use a large amount of electric current. The design of the overpass ensures a safe distance of the cabin from magnetic field sources, as well as overcoming hard-to-access terrain. Such transportation system with small passenger and freight flows will pay off faster than trains on magnetic suspension.

3.2 Features of the small-sized maglev system limiting its application

Tests of the transportation system model revealed several features that may limit its use. Specifically:

a) The box for f-travelers is made with a cutout, which weakens the strength of the structure.

b) The use of more than two connecting rods per cabin makes it difficult for the modules to move in roundabouts.

c) The connecting rods do not provide absolutely contactless movement of the transport modules, so at some points they will touch the edges the box cutout.

4 Possible solutions to problems

The next step in the development of such transportation systems could be the non-contact coupling of movers and f-travelers using the "magnetic potential hole" effect and superconducting magnetic field sources.

The "magnetic potential hole" effect in the interaction of ideal conducting ring structures was discovered by V.V. Kozoriz [7]. Dimensions of rings and distances between them are determined from relations (1) and (2):

$$\Psi_1 \Psi_2^{-1} \neq 1; \quad (1)$$

$$a_1 R^{-1}, a_2 R^{-1} \leq \frac{1}{2}, \quad (2)$$

where Ψ1 and Ψ2 are magnetic flows; a1 is the radius of a smaller ring; a2 is the radius of a bigger ring; R is the distance between the rings.

Let us also refer turn to the organization of movement in the universe [8]. Cosmic bodies with magnetic fields move in elliptical orbits relative to more massive bodies, so planets move relative to stars, and rotate around their axes.

This principle can be the basis for the functioning of multi-channel maglev systems (Figure 2).

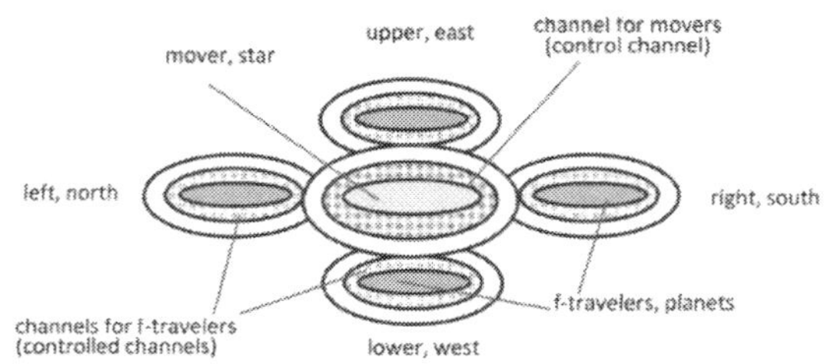

Fig. 2. Planetary model of maglev-system.

In the center of the system there is a control channel, inside of which movers (stars) move. The control channel is connected to the controlled channels, the number of which can reach four in accordance with the number of parts of the world: upper (east), right (south), lower (west), left (north). Together with the movers, satellites (planets) located in controlled channels move. In the planetary model with four controlled channels, the "magnetic potential hole" effect appears both when the magnetic sources are located in one plane (north – center – south) and when they are located in parallel planes (east; center; west).

Other researchers also consider magnet-interacting systems based on planetary models [9].

A possible implementation of a multi-channel maglev system based on a planetary model is shown in Figure 3, where denoted: 1 – cabins of transport modules; 2 – magnetic sources of transport modules; 3 – f-travelers; 4 – controlled channels; 5 – mover; 6 –

 https://doi.org/10.1051/e3sconf/202337802003

control channel; 7 – linear motor; 8 – trestle rack; 9 – supporting surface.

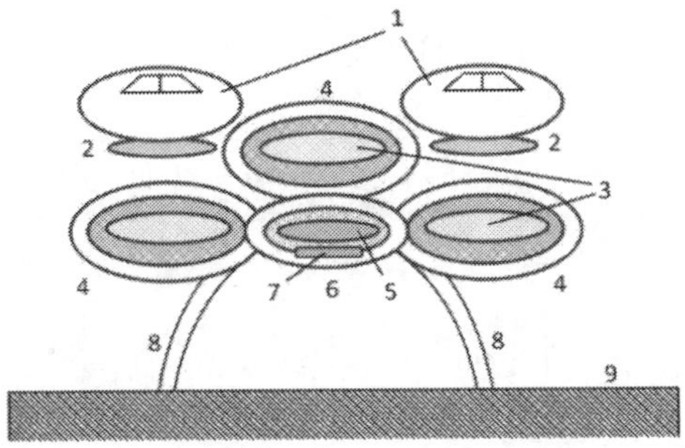

Fig. 3. One of the implementations of a multi-channel maglev system

The location of the magnetic sources in the longitudinal planes (top view) is shown in Figure 4.

To meet conditions (1) and (2), the rings of larger and smaller sizes must alternate with each other both in the sequence of movers and in sets of f-travelers and magnetic sources of transport modules.

controlled channels 4 in the form of a pipe may are physically absent. movers 5, when interacting with f-travelers 3, must form a stable structure (nest).

The cabins 1 of the transport modules with their magnetic sources 2 sort of to “sit” in this nest. When the sequence of movers 5 in the channel 6 moves, the nest together with cabins 1 also moves in a predetermined direction (in this case forward or backward).

We have not yet defined a specific method for setting the motion sequence of the movers. This can be a linear motor, pressure drop, liquid flow, and many other things.

To describe the processes, the development of Dr. G.A. Vitkov [10] may be useful in case of using liquid flow. In particular, he proposed models for the occurrence of resistance. These models can be used to calculate hydrodynamic, heat and mass exchange processes in any systems and conditions of force fields.

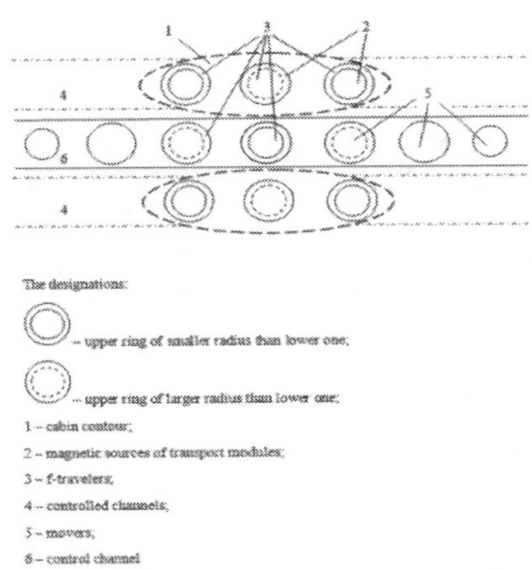

Fig. 4. The location of magnetic sources in longitudinal planes (top view).

5 Our mock-up studies

We discovered the resistance of the movers to each other when we tried to investigate the mock-up model. A mock-up made by 3D printing a three-channel system was tested (Figure 5).

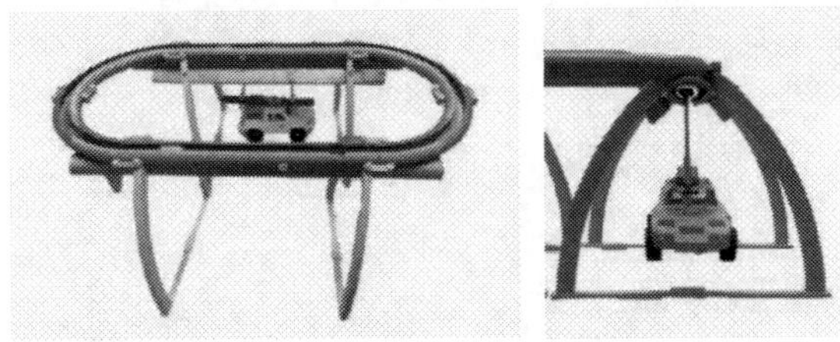

Fig. 5. The mock-up of the three-channel system

Niobium magnets with a diameter of 30 mm with countersink were used as movers. The control channel is a closed pipe of 160 cm length with an oval section and a slot in the upper part for observation of the movers. The controlled channel is a straight pipe with oval section and a cutout in the lower part for connecting rods of the transport module model. It was assumed that the sequence of movers in the control channel would move from one or more sources of rotational motion. When the movers interact with the f-travelers of the transport module model, it would have to move linearly. However, the action of the resistance forces manifested themselves.

To overcome this resistance, we have developed magnetic transmission (Figure 6).

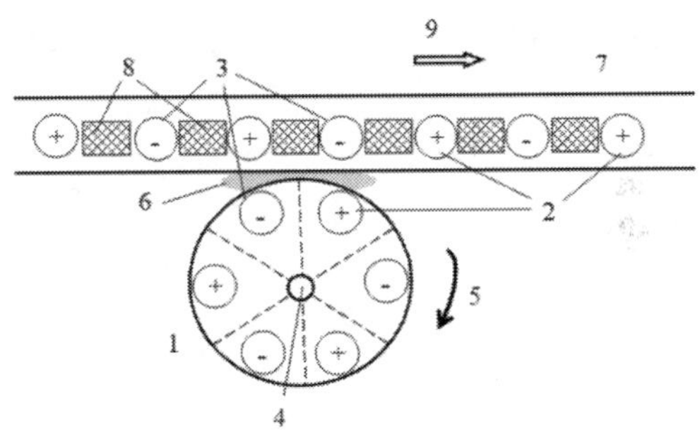

Fig. 6. Magnetic transmission

Fig. 6 shows: 1 – rotor; 2 – magnetic sources of positive polarity; 3 – magnetic sources of negative polarity; 4 – rotation axis; 5 – rotation direction; 6 – interaction sector; 7 - guiding beam; 8 – “soft” buffers; 9 – direction of TM movement. Magnetic sources 2 and 3, located in beam 7 and separated by buffers 8, form a sequence of movers.

Such magnetic transmission allows for recovery. Let us suppose that the unloaded movers move linearly, for example, with a linear stepper motor. By overcoming the interaction sector, they will force the rotor rotate. As a result, energy can be generated and accumulated. When the movers do useful work in the suspended transport modules, the rotors rotate due to the stored energy. This ensures the linear movement of the sequence of movers and the associated transport modules.

The long control channel can be multi-cascade. A system with a three-cascade control channel is shown in Figure 7. The controlled channels can be called working ones, since they provide motion of f-travelers and transport modules. The source of rotational motion can be called the accelerator of the movers.

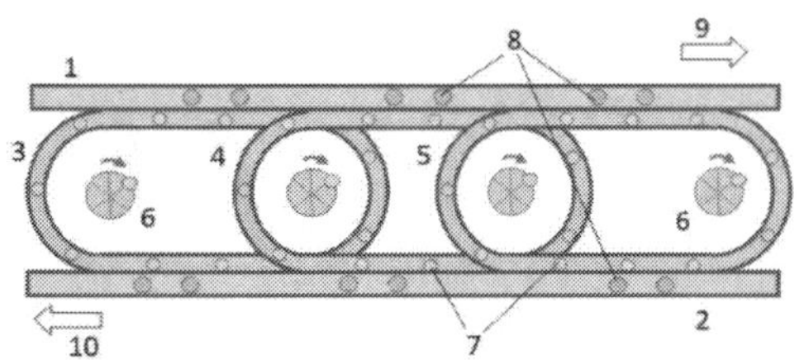

Fig. 7. The system with three-cascade control channel

In Fig. 7, 7 is marked as follows: 1 – working channel 1; 2 – working channel 2; 3 – first control channel cascade; 4 – second control channel cascade; 5 – third control channel cascade; 6 – rotors; 7 – movers; 8 – f-travelers; 9 – direction of f-travelers movement in working channel 1; 10 – direction of f-travelers in working channel 2.

Thus, taking into account the direction of the transport modules, each control channel cascade covers two accelerators: an initial and a final. The final accelerator of one cascade is initial for the next.

For example, accelerator 6 is the initial one for stage 3 of the control channel. It sets the acceleration of nearby movers 7. Under the action of this acceleration, movers 7 reach the interaction zone with the final accelerator of cascade 3, which is the initial accelerator of cascade 4. The process is then repeated for cascades 4 and 5. The movers 7 interact magnetically with the f-travelers 8 which are located in working channels 1 and 2. As a result, the transport modules move in directions 9 and 10.

6 Features of the multi-channel systems

1. The movers must be powerful magnetic field sources and have a circular shape. It is desirable that they are magnets on superconductors.

2. The fellow travelers (f-travelers) are constant magnetic field sources, which are part of the transport module, and also has a circular shape.

3. magnetic field sources are placed in channels. The movers are placed in the control channel. F-travelers are placed in controlled channels.

4. arc-shaped trestle allows for compact placement of the control channel and the controlled channels, to ensure their interaction.

5. ideal interaction of movers with f-travelers can be ensured through the "magnetic potential hole" effect.

6. In the planetary model of the maglev system, the electromagnet can be replaced by the motion of movers along an elliptical trajectory in the control channel. The movers will pull or push f-travelers in the working channels. The transport modules will also move along with the f-travelers.

7. Small-sized maglev systems of the trestle type can be used to create messages-bindings of remote industrial facilities with the reference nodes of the transport network of the Arctic zone of the Russian Federation.

In particular, in the Arctic, maglev systems can be used to connect remote facilities to backbone transportation systems, fill gaps in the supporting systems, connect remote marine terminals to the coastal infrastructure [11, 12], transport of hydrocarbons in mini tanks [13], etc.

A study of the weather parameters of the proposed maglev systems has not yet been carried out. However, there is reason to believe that they will be similar to the characteristics of the Yunitsky string transport systems [14].

It should be noted that when using superconducting magnets as movers, their operability is cheaper to ensure in climatic zones with prevailing low temperatures than in southern territories.

Installation of piles in permafrost conditions is a significant problem [15]. The trestle supports will essentially be piles whose bases rest deep into the permafrost, so that the seasonal thawing of its upper layer does not affect the stability of the structure. The gaps between the column supports do not prevent the free circulation of air under them, which minimizes the additional heat effect.

The method of pile installation is determined on the basis of physical and mechanical parameters of permafrost soils, the average annual soil temperature, the climatic area of construction, the time of the year and the requirements for the degree of accuracy of pile immersion into the permafrost. The average annual temperature of permafrost soils is determined at a depth of 10-15 m, where the change of seasons causes almost no change in the level of the average soil temperature. Given this indicator, frozen soils are divided into low-temperature (from -1.5°C) and high-temperature (0°C, not lower than -1.5°C) soil layers. Based on this, a certain method of installing piles is chosen.

In addition to the technical component, an important advantage of trestle maglev systems is the fact that they do not interfere with the free movement of local fauna under the highway. Thus, the negative effect of the inclusion of extraneous organisms in the ecosystem is minimized. The transport modules of the maglev systems themselves are capable of driving in almost any weather.

7 Conclusion

The events associated with the spread of coronavirus infection have led to a drop in production in many countries, accordingly affecting the development of new modes of transport around the world. There is growing skepticism towards high-speed transport projects due to significant investments in construction and long payback periods, their feasibility and economic efficiency are questioned. However, according to the authors [16], the main recommendation for creating the most sustainable multimodal transportation system (as a whole) is to change thinking, identify and evaluate approaches to the organization of transportation, especially for evaluating new concepts of transportation facilities with real potential for solving problems.

References

1. L. Blow et al., Energy consumption of track-based high-speed trains: maglev systems in comparison with wheel-rail systems Transportation Systems and Technology, **4(3.1)**, pp. 134-155, doi: 10.17816/transsyst201843s1134-155 (2018)
2. G. Lin and X. Sheng, Application and Further Development of Maglev Transportation in China, *Transportation Systems and Technology*, **4(3)**, pp. 36-43, doi: 10.17816/transsyst20184336-43 (2018)
3. A.A. Zaitsev and Yu.F. Antonov, *Magnetolevitation transport technology* (Moscow: FIZMATLIT, 2014), p. 476
4. DA FS NEWS - RFI e Nevomo studiano soluzioni per un futuro a levitazione magnetic. Available online: https://www.rfi.it/it/news-e-media/comunicati-stampa-e-news/2021/6/18/da-fs-news---rfi-e-nevomo-studiano-soluzioni-per-un-futuro-a-lev.html (2021)
5. Z. Kleinman, Virgin Hyperloop pod transport tests first passenger journey. Available online: https://www.bbc.com/news/technology-54838982 (2020)
6. Transport system with two-way levitation of modules moved relative to an arch-type overpass, by E.Yu. Sundukov, V.E. Sundukova, L.F. Selivonov, N.A. Tarabukina. (2018, December 14). Patent RU 2722256. [Online]. Available online: https://www.fips.ru/registers-web/action?acName=clickRegister&regName=RUPAT (by number of the invention) (2018)
7. V.V. Kozoriz, *Dynamic systems of the magnetic interacting free bodies* (Kiev: Sciences thought, 1981), p. 160
8. G. Bruno "On the Infinite Universe and Worlds (1584)" Available online: http://www.faculty.umb.edu/gary_zabel/Courses/Parallel%20Universes/Texts/On%20the%20Infinite%20Universe%20and%20Worlds.htm (1584)
9. O.N. Molokanov, *Development of calculation methods and analysis of promising designs of contactless gears. Dissertation for the degree of candidate of technical sciences* (Moscow, MPEI, 2012)
10. G.A. Vitkov, A new class of resistance forces in solid environments (Tver, TSTU, 1997), p. 352
11. S.B. Kryltcov, A.B. Makhovikov, M.A. Korobitcyna, Novel Approach to Collect and Process Power Quality Data in Medium-Voltage Distribution Grids, *Symmetry*, **13**, pp. 460-460 (2021)
12. A. Cherepovitsyn, A. Lebedev Drill Cuttings Disposal Efficiency in Offshore Oil Drilling, *Journal of Marine Science and Engineering*, 11(2), p. 317, doi: 10.3390/jmse11020317 (2023)
13. P.S. Tcvetkov, A.E. Cherepovitsyn, A.B. Makhovikov, Economic assessment of heat and power generation from small-scale liquefied natural gas in Russia, *Energy Reports*, **6**, pp. 391-402, (2020)
14. V. Litovsky, The concept of placing in the Arctic the productive forc es on the basis of A. E. Yunitsky infrastructure of the second level and the spatial model of the transport network "Polar lace" for "mobile settlements", *Bulletin of Murmansk State Technical University*, **19(2)**, pp. 431-442, doi: 10.21443/1560-9278-2016-2-431-442 (2016)
15. The construction and calculation of foundations on permafrost soils. Available online: https://kakfundament.ru/svajnyj/raschet-fundamentov-na-vechnomerzlyx-gruntax (2021)
16. V. Bourquin, et al., Transportation as an engineering system: towards sustainable transportation using existing technologies into innovative architectures to satisfy current and future needs, *The International Maglev Board, editor. Maglev Solutions for People, Cities, and Regions?* pp. 1195-204 (2016)

North Pole ice-resistant self-propelled platform as an innovative complex for research in the Arctic

Vladimir Alekseevich Likhomanov[1,*], *Yuri* Ugrumov[2], *Anna* Savitskaya[3], *Polina* Patseva[4], *Nina* Krupina[5], *Alexei* Chernov[6], *Ivan* Svistunov[7] and *Vladimir Antonovich* Likhomanov[8]

[1] Ph.D. in Engineering Science, Head of Department, Department of ship performance in ice, Arctic and Antarctic Research Institute, 199397, St. Petersburg, Russian Federation
[2]Deputy director for expeditionary work, Arctic and Antarctic Research Institute, 199397, St. Petersburg, Russian Federation
[3]Arctic and Antarctic Research Institute, 199397, St. Petersburg, Russian Federation
[4]Arctic and Antarctic Research Institute, 199397, St. Petersburg, Russian Federation
[5]Arctic and Antarctic Research Institute, 199397, St. Petersburg, Russian Federation
[6]Arctic and Antarctic Research Institute, 199397, St. Petersburg, Russian Federation
[7] Ph.D. in Engineering Science, Arctic and Antarctic Research Institute, 199397, St. Petersburg, Russian Federation
[8]Arctic and Antarctic Research Institute, 199397, St. Petersburg, Russian Federation

Abstract. The Russian presence in the Arctic Region and the development of the Arctic is one of the most important geopolitical interests of Russia. At the beginning of the 21st century, the problem of using innovative methods in Arctic research came to the fore. Based on the analysis of the unique work experience of drifting stations "Severniy Polus" ("North Pole") (1937-2015), Arctic and Antarctic Research Institute specialists have concluded that the stations should be replaced by a modern scientific complex, capable of solving a wider range of problems. As a result, it was proposed to create an ice-resistant self-propelled platform (IRSPP) – an engineering structure for permanent basing of scientific observatories. The IRSPP is designed to conduct year-round comprehensive scientific research in the high latitudes of the Arctic Ocean and should make a long drift of at least one year together with the surrounding ice massive. The scientific complex of the IRSPP includes 16 different laboratories, including an ice load monitoring laboratory. A unique ice load monitoring system (ILMS) has been developed for the IRSPP. The ILMS will be an essential part of the system of safe operation and persistence of the IRSPP and will provide the platform hull with a measurement tool for studying the mechanics of deformation and destruction of sea ice in its interaction with the engineering structures. In all respects, the IRSPP is unparalleled in the world. Its use can open a new chapter in the exploration of the Russian Arctic and in international collaboration aimed at studying the northern latitudes. The platform was put into operation in 2022.

1 Introduction

The Russian presence in the Arctic Region and exploration of the Arctic is one of the most important geopolitical interests of Russia. This is evidenced by the official documents of the session of the Russian Federation Government on "Protecting the interests of the Russian Federation in the high-latitude polar regions", the Security Council of Russian Federation and the Maritime Collegium.

It was decided that the Arctic is Russia's primary strategic resource base. At the same time, the Arctic has an exceptional military and strategic importance for defense purposes. Both the forces of the Northern Fleet and its operational zone, and the largest enterprises of the defense industry complex are based here.

This causes a growing interest in scientific research in the Arctic region. Over many years, the regular studies have been conducted on the Severniy Polus (English North Pole, hereinafter SP) drifting stations, a camp set up on an ice floe. The SP stations were Russian research and operational-production bases, where the nature and mechanisms of formation of hydrometeorological and geophysical processes in the atmosphere, the ice cover and hydrosphere of the central areas of the Arctic Ocean were studied year-round. They collected hydrometeorological information to transmitted it to the World Meteorological Organization, and conducted methodical research. In general, scientific research of drifting SP stations was determined by the programs of Russia's overall research system in the Arctic.

As a result of the research carried out at the stations, a large amount of data was obtained from observations of the state of the Arctic basin environment in the field of oceanography, study of sea ice, atmosphere, ozonosphere, biosphere and ecology.

* Corresponding author: likh@aari.ru

2 Life and works at drifting stations

The research data characterize the intensity of processes taking place in the central Arctic and reflect the transformation of its natural environment, which is influenced by the global climate change.

Traditionally, the research programs on the SP stations had to solve the following tasks:

- to carry out a program of standard meteorological, actinometric and aerological observations
- to carry out comprehensive work on background meteorological monitoring of the marine environment
- to study the dynamic-thermodynamic processes and evolution of the sea ice cover
- to identify the thermohaline and hydrochemical structure of water masses in the station drift area
- to evaluate seasonal variability of the components of dolomite system in surface or mixed ocean and the near-surface layer of the atmosphere
- to isolate morphometric parameters of undeformed ice and snow cover and study their changes over time
- to carry out a series of hydrographic studies, including profiling of the sedimentary section in the station's drifting area.

On average, the personnel of the SP stations consisted of 18-20 people, including a medical doctor and technical staff (4-6 people) ensuring the operation of the station. Sometimes, in case of favorable weather conditions and an additional work program, a seasonal team of 6-10 people was brought to the station to carry out construction works and to conduct scientific research under separate programs of the interested academic centers and ministries of the USSR (later the Russian Federation).

After some time, it became impossible to set up the SP stations due to the lack of ice fields, suitable for placing a camp.

Since the 1960s, the AARI (Arctic and Antarctic Research Institute) had been working on the option of using floating structures as long-term drifting bases for scientific observatories instead of the drifting SP stations, as a complex of mobile buildings on a drifting ice field. Finally, the idea of creating an ice-resistant self-propelled platform was presented.

Before starting work on the preliminary project design requirements, a thorough analysis of the organizational and scientific activities of the Soviet-Russian SP stations over time was conducted in order to establish the structural form of the platform. Three major indicators were considered: life safety of the station, cost efficiency and labor.

The financial costs of setting up drifting stations of the traditional type, which included searching for ice fields, relocation of personnel, preparation and transportation of necessary equipment, funding scientific and technical programs of the station, evacuation costs (including in emergency cases), etc. were analyzed. The results of the economic analysis were compared with the costs of construction and exploitation of different variants of an ice-resistant platform, to accommodate a new-type drifting observatory. This allowed to determine the preliminary dimensions of the platform, its displacement and the required usable area inside it.

Analysis of the working process at the drifting station showed that the most labor-intensive operations are: loading of the logistical equipment, building structures and materials, machinery, fuel, and lubricants, food and construction of service and accommodation quarters when establishing a new station on the drifting ice field. In addition, we can mention cases of emergency work, as they can occur at any moment of drifting.

For example, here is a calculation of the labor of the staff of the SP-36 station:

- 100 % of the personnel take part in the arrangement of the camp and its life support.
- 20 to 40 % of personal and working hours of the scientific personnel of the station is spent on station life support and emergency works.

Emergency station rescue operations are conducted due to the fact that the ice fields, on which the stations are based, are constantly under compression, so pressure ridges suddenly form in unexpected places, which can crush houses and equipment, cracks can appear, and the station's size can be reduced.

The labor intensity of work to ensure proper living and working conditions on the drifting ice floes increased due to the fact that it was necessary to organize reserve storage facilities, and first of all reserve stores of oils and lubricants and foodstuffs to ensure safety of material supplies.

It should be noted that due to selfless work of polar explorers and optimal organization of all work, distraction of scientific staff of the stations do the indicated work did not affect the quality and completeness of work under the research programs. However, it can be argued that the specified working conditions of the polar explorers exacerbated the physical and psychological strain on their bodies.

I.D. Papanin sent a telegram on February, 2 1938 during the last working period of the first Soviet SP station: "In the station area, fragments of ice fields not bigger than 70 m in size continue to collapse. Cracks of 1 to 5 m in size, droughts up to 50 m. Ice blocks are mutually shifting. Up to the horizon, the ice is 9 out of 10 tenth. It is impossible to land an aircraft within line of sight. We live in a silk tent on an ice block 50 m by 30 m. A second antenna mast is placed on another ice floe during communications. We have a three-month food supply, equipment and the result. Regards to all. Papanin."

A.N. Chilingarov, the head of the SP-19 drifting station, which unlike Papanin's station was set up on an ice island, wrote: "About twenty meters away from the cabin, parallel to our "main street" there is a wide crack. It passes very close to the food storage, cutting us off from the fuel supplies, gas and the tent with the household equipment. The food storage facility is under the biggest danger. It is a miracle that this whole pile of bags and boxes remained on the edge and did not fall down into the abyss."

Figures 1-4 illustrate the working and living conditions at drifting stations, with the example of the SP-35 station.

Fig. 1. An ice crack at the SP-35 station (Photo by V. Kustov)

Fig. 2. The SP-35 station: access to a residential building during ice and snow melting (Photo by V. Kustov)

Besides, it is important to mention the obvious danger of polar bears visits to the station. Such visits endanger life and health of polar explorers, as well as equipment and food supplies. In particular, dozens of cases of polar bears' presence next to the station and even in the camp were recorded during the drifting period of the SP-35 station (Fig. 3).

Fig. 3. A polar bear in the camp of the SP-35 drifting station (Photo by V. Kustov)

Fig. 4. Demobilization of the drifting station (Photo from the internet)

To summarize, it can be added that living conditions on the drifting ice field limit the possibilities of using modern scientific equipment, expand the duration of explication of research results and reduce their quality, and living conditions are unattractive for young specialists. The last two arguments reduce the effectiveness of scientific research in the balance of costs for creation and maintaining SP drifting stations.

3 IRSPP design

Having analyzed the work of SP stations, according to AARI specialists, the most optimal form is a floating engineering structure to base a research observatory, which has maximum efficiency at minimum construction and operating costs.

In 2015-2016, Vympel Design Bureau and Admiralty Shipyards started to work on the image of an advanced research platform vessel for a research station.

In April 2018, Roshydromet (Russian Federal Service for Hydrometeorology and Environmental Monitoring) and Admiralty Shipyards signed a contract for the design and construction of the ice-resistant self-propelled platform (IRSPP) "Severniy Polus". Construction work started in December 2018 and was put into operation in 2022. The IRSPP is a self-propelled displacement steel vessel (Fig. 5).

The IRSPP is designed to conduct year-round comprehensive scientific research in the high latitudes of the Arctic Ocean and must perform a long drift of at least one-year, together with the surrounding ice massif.

Generally, the platform has Arc5 ice class – autonomous navigation in the first-year ice up to one meter thick. At the same time, the hull strength along the boards and the bottom corresponds to the Arc8 class, which allows coping with the compression of multi-year ice.

The IRSPP has the following main dimensions: length overall/DWL – 83.1/76.7 m, width overall/DWL waterline – 22.5/21.8 m, side height at midship – 11.4 m, draft DWL/minimum – 8.6/7.5 m. The maximum displacement of the IRSPP is approximately 10,400 t. With main engine power of 4200 kW, the navigation speed in deep calm water is at least 10 knots.

Fig. 5. General view of the IRSPP: 3D-model (left) and the launched (right) (Photos by AARI)

The platform propulsion complex consists of azimuthal thrusters and an azimuthal water-jet propulsor in the bow that functions as a bow thruster. If needed, the IRSPP can autonomously move to a new initial drifting point in open water or in relatively light ice conditions.

Prior to construction of the IRSPP, the AARI conducted a wide range of ice and nautical trials with the platform model in the AARI ice tank [1]. Apart from traditional trials to assess the ice resistance of the moving of the platform, additional experiments were held to observe the effect of ice compression on the IRSPP hull (Fig. 6). Based on the results of the model tests, it was possible to select the optimal hull shape for its successful exploitation in ice conditions.

Fig. 6. Testing of the IRSPP model in the ice tank of the AARI during ice compression. (Photos by AARI)

The standard scenario for a self-propelled platform is as following: the vessel will have to move autonomously or with the help of an icebreaker to a certain point and go adrift. It would be able to freeze into the ice without damaging its own construction and float on with it. Such drifting can last up to two years. If necessary, the vessel can receive or send helicopters with cargo or passengers.

The IRSPP will start drifting in the eastern sector of the Arctic, as well as the SP stations. The best months for the IRSPP to move to the starting point of the drift are August-September, because at this time the extent of the concentrated ice area is minimal.

It is assumed that the personnel on the platform will work on a rotation basis. The platform will comfortably accommodate 54 people: 34 researchers, 14 crewmembers and 6 members of the helicopter team. The IRSPP is equipped with a helipad to receive Mi-38 helicopters. Safe and comfortable living and working conditions are ensured at temperatures down to –50 °C.

The IRSPP is a scientific complex designed to conduct various types of research and observation. It includes 16 different laboratories, including the laboratory of ice load monitoring, which is part of the ice-research complex [2]: meteorological laboratory, atmosphere laboratory, aerological laboratory, satellite hydro meteorological data reception laboratory, geophysical laboratory, ice load monitoring laboratory, sample preparation laboratory, geological laboratory, hydro chemical laboratory, environmental laboratory, oceanographic "dry" laboratory, oceanographic "wet" laboratory, ice research laboratory, ice-research (cold) laboratory, and a laboratory of special research.

Along with the traditional research complex, the creation of a self-propelled platform will broaden the spectrum of research and observation in this area [3]:

- ionospheric observations – based on the receiving station for inclined ionospheric sounding (modern digital ionosounde of new generation with linear frequency modulation) for operative diagnostics of HF radio wave propagation conditions
- geological research – to conduct a wide range of geological studies of bottom sediments in remote and hard-to-reach areas of the Arctic basin, based on a sample preparation complex for using piston ground pipelines and dredges, wet and analytical laboratories, core storage facilities
- chemical and environmental research based on the analytical complex
- acoustic tomography of the Arctic basin - monitoring of hydro physical fields of the Arctic Ocean
- studies of the boundary layer and free atmosphere in the central Arctic
- magnetic and gravitational research – monitoring, in order to determine magnetic and gravitational fields (maps) in high latitudes of the Arctic
- study of ice load and ice mechanics–previously never conducted research on the mechanics of deformation and failure of the sea ice, designed to improve methods for calculating local and global ice loads on vessels and other engineer structures.

4 Ice-load monitoring system (ILMS)

The ice-load monitoring system with which the IRSPP is equipped deserves particular attention. Currently, to ensure safe exploitation, almost all engineering structures and vessels, designed to work in ice conditions, are equipped with ice load monitoring systems (ILMSs). Generally, the main purpose of an ILMS is continuous monitoring of the stress-strain state of an object, when it is exposed to ice loads on it. If the value of measured parameters exceeds the set threshold values, the system generates a warning, based on which the system operator can take measures to reduce the load [4].

The IRSPP is an unconventional engineering structure with its own operating features, so a unique system of ice load monitoring was developed. The ILMS for the IRSPP is designed for scientific purposes, but it also ensures safe working of the platform in ice conditions [5].

The functions and configuration of the ILMS instrumentation were defined based on the operating condition of the IRSPP, analysis of ice conditions and scenarios of ice loading on the vessel's hull, and the results of model tests in the ice tank of the AARI. They were also determined during the Transarctic-2019 expeditions on the scientific diesel-electric research vessel "Akademik Tryoshnikov", where the technology for monitoring ice loads in drift conditions was tested [6].

The main measuring subsystem of the ILMS is the hull stress-strain monitoring subsystem, which is required to measure the response of the hull structure to the effects of the ice-loads in different regimes of platform operation.

Measurements are performed using an extensive system of fiber-optic deformation sensors, placed on the elements of the hull structure. Taking into consideration that the most interesting processes are interactions of the hull and the ice during ice compression, the majority of the sensors are placed in the central part of the hull [5].

The main feature of ILMSs for the IRSPP is that it includes a subsystem for estimation of the parameters of the stress-strain state of the ice adjacent to the IRSPP. This subsystem will be connected to the ILMS only during the drift of the IRSPP. At the first stage of IRSPP operation as a part of this subsystem, the measuring elements in this subsystem will be membrane type pressure sensors, which will be frozen into the ice cover in two mutually perpendicular destinations at different horizons along the ice thickness. The subsystem is necessary to study the processes of interaction between the ice field and the hull of the IRSPP, it will allow tracing the balance between the ice cover and the pressure which the ice exerts on the platform board. Also, measurements of the surrounding ice sheet (primarily thickness and physical and mechanical characteristics of the ice) will be made during the drift, which will help to supplement the data obtained by the ILMS.

Apart from that, the ILMS allows to collect data on kinematic parameters of the IRSPP motion (angle of heel, trim, yaw angle, linear-acceleration in three axes); meteorological data (air temperature, wind speed and direction); parameter of the propulsion complex (power, RPM, steering angle of the azimuth thruster); coordinates, speed and course of the IRSPP.

It should be noted that the currently existing implemented projects of ILMSs are not intended for measurements in ice compression conditions. Thus, the mentioned ILMS makes the IRSPP a unique scientific measuring complex for studying the processes of ice impact on structures obtaining new data on the parameters and ways of the ice-load action.

5 Conclusion

In all respects, the IRSPP is unparalleled in the world. Its construction may open a new chapter in the research of the Russian Arctic and in international collaboration aimed at studying the northern latitudes. A self-propelled platform as a base for a research station has its obvious advantages compared to an ice field. It is safe, predictable and manageable, it is possible to conveniently and effectively place there a big amount of research equipment, provide its power supply and operability Finally, it is easier to create comfortable working conditions for crew and scientists. Such a floating observatory will make it possible to monitor most of the Arctic basin and promptly obtain reliable scientific data on natural processes in the northern polar regions of the Earth.

References

1. I.A. Svistunov, et al. Experimental-analytical study of the North Pole platform stability under the conditions of intensive ice pressures, *Proceedings of the 25th International Conference on Port and Ocean Engineering under Arctic Conditions June 9–13, 2019, Delft, The Netherlands.* Available online http://poac.com/Papers/2019/pdf/POAC19-123.pdf (2019)
2. A.S. Makarov, et al. Conceptual principles for creation of the drifting polar research platform, *Arctic: ecology and economy*, 3(31), pp. 65–75 (2018)
3. A.S. Makarov, et al., Perspectives of using a drifting ice-resistant platform of the "North Pole" type, which is under construction, as a multifunctional research complex in the Arctic, *Proceedings of the 25th International Conference on Port and Ocean Engineering under Arctic Conditions June 9–13, 2019, Delft, The Netherlands*, Available online: http://poac.com/Papers/2019/pdf/POAC19-123.pdf (2019)
4. P.V. Maksimova, et al., Ice loads monitoring systems for ships and engineering structures. On the problem of creating an object state monitoring system, *Arctic and Antarctic Research*, **2**, pp. 101–112 (2016)
5. A.V. Chernov, et al., Methodology for developing the configuration of an ice load monitoring system for an ice-resistant self-propelled drifting platform, Arctic and Antarctic Research, 66(4), pp. 515-533 (2020)
6. P.V. Maksimova, et al. Research of R/V "Akademik Tryoshnikov" hull structures response to ice actions during the first stage of the "Transarktika-2019" expedition, *Arctic and Antarctic Research*, 66(1), pp. 82–101 (2020)

Reducing harmful air emissions from vehicles, ships and stationary facilities operating in the Arctic

Boris Antipov[1,*], *Konstantin* Shotidi[2] and *Denis* Novikov[3]

[1] Professor, Doctor of Technical Sciences, Department of Thermodynamics and Heat Engines, National University of Oil and Gas «Gubkin University», 119991, Moscow, Russian Federation
[2] Professor, Doctor of Technical Sciences, Department of Thermodynamics and Heat Engines, National University of Oil and Gas «Gubkin University», 119991, Moscow, Russian Federation
[3] Student, Department of Thermodynamics and Heat Engines, National University of Oil and Gas «Gubkin University», 119991, Moscow, Russian Federation

Abstract. The paper proposes a method for reducing harmful emissions into the Arctic atmosphere during the operation of diesel engines using electromagnetic waves of ultra-high frequency (UHF) and ultrasound. Experimental studies were conducted, which confirmed the reduction of NOx emissions by 50%. In order to implement the method, a multipurpose fuel unit is being developed.

1 Introduction

Severe climatic conditions, hard-to-access locations, and difficulties with power supply have led to widespread use of diesel engines in the Arctic. Vehicles, construction machinery, all-terrain vehicles, most ships, diesel locomotives, mobile and stationary power plants operating in the Arctic are equipped with diesel engines. Given the plans of the Russian Federation for the development of the Arctic region, it is necessary to take into account the following factors. First, there is no alternative to diesel-powered machinery operating today in the Arctic for the next decades. The Arctic ecosystem is very fragile and susceptible to anthropogenic impact [1]. The diesel engine is one of the main sources of harmful emissions into the atmosphere [2]. The combustion of fuel in the cylinders of diesel engines generates various substances that are relased into the atmosphere with exhaust gases. Their composition depends on the design of the diesel engine, its power and operating modes, as well as the quality of the fuel used [10]. Exhaust gases emit water (H2O), oxygen (O2), carbon dioxide (CO2), carbon monoxide (CO), unburned hydrocarbons (HC or CH) and nitrogen oxides (NOx). nitrogen oxides (NOx) are currently considered the most dangerous for humans, flora and fauna and the ecology of the area as a whole. They are dangerous when inhaled into the lungs of humans and animals; besides, when released into the atmosphere, they precipitate so-called acid rain - aqueous solutions of nitrogen oxides. Such precipitation negatively affects the entire fragile ecosystem of the Arctic, destroys a thin layer of soil in the tundra, and negatively affects the life of aquatic organisms. aqueous solutions of nitrogen oxides are also formed by contract of nitrogen oxides contained in the exhaust gases with snow and water.

2 Research

The employees of the Department of Thermodynamics and Thermal Engines of Gubkin National University of Oil and Gas are dealing with the issues of reducing harmful emissions into the atmosphere during the operation of diesel engines [9, 14]. Taking into account that engineers have been interested in studying the effects of electromagnetic waves on various substances, including fossil fuels, for more than a decade, the department has been conducting research and development work in recent years to assess the possibility of using electromagnetic waves of ultra-high frequency (UHF) and ultrasound to improve the efficiency of diesel engines. The expected scientific result of the study is an assessment of the possibility of practical application of the features of thermohydrodynamic and physico-chemical processes occurring in the cylinders of diesel engines using a water-fuel-mixture as fuel, when the fuel-water mixture is exposed to microwaves and ultrasound to reduce harmful emissions and fuel consumption. Analysis of works on the formation of nitrogen oxides allows to distinguish several mechanisms of NOx formation. Fast nitrogen oxides NOx are formed as a result of collisions of nitrogen molecules with hydrocarbon radicals in the combustion zone at pulsating temperature. If the fuel contains nitrogen-containing compounds, fuel nitrogen oxides are formed during their low-temperature oxidation. This process is accompanied by low temperatures of 1000-1200K. Thermal nitrogen oxides NOx are formed during the high-temperature oxidation of nitrogen in the air [16].

* Corresponding author: antipov-bn@mail.ru

Atmospheric nitrogen is oxidized at high temperatures (> 2000K) in the zone of combustion products behind the flame front. The main provisions of the thermal theory of nitrogen oxides formation are:

Nitrogen is oxidized by atmospheric oxygen in the zone of combustion products behind the flame front.

The amount of NOx is determined not only by the maximum possible combustion temperature, but by the number of nitrogen and oxygen atoms in the zone of combustion products, and does not depend on the type of fuel, provided that the fuel itself does not contain nitrogen.

The cooling rate of the exhaust gases significantly affects the content of nitrogen oxides in them.

The maximum combustion temperature affects the concentration of nitrogen oxides only in lean mixtures, while in rich mixtures this concentration is limited by the decomposition kinetics.

When the temperature exceed 1700 K, thermal nitrogen oxides become predominant over all others.

Analysis of the conditions of formation of nitrogen oxides leads to the conclusion that reducing the temperature in the combustion zone is the determining factor ensuring the the reduction in NOx formation. To solve the problem of reducing the temperature in the combustion zone, it is proposed to supply a certain amount of water to the engine cylinders together with the fuel, which, when evaporating, will ensure a decrease in temperature and, as a consequence, a decrease in NOx content in the diesel exhaust gases. Analysis of thermodynamic processes in the diesel engine cylinder insicates the possibility of using water in the combustion zone in order to reduce fuel consumption. For this purpose, it is proposed to treat the fuel water mixture before feeding it into the diesel engine cylinders with microwave waves and ultrasound [7, 9,].

It is commonly known that the fuel-air mixture is the smallest droplets of fuel in the air. Let us assume that, along with fuel droplets, water droplets are also entered in the fuel-air mixture, after which this three-component mixture was exposed to microwaves with a frequency of 2.45 GHz. When waves of this frequency affect water, water molecules in the form of an electric dipole begin to actively vibrate in the electromagnetic microwave field, which leads to the heating of water. The result of this effect will increase the temperature of water droplets and, subject to heat transfer, increase the average temperature of the mixture, which will ensure better evaporation of the fuel, and, as a consequence, increase the efficiency of the heat engine. Besides, an increase in water temperature will lead to an increase in internal pressure inside the water droplet surrounded by fuel on the outside. However, water droplets will retain their integrity up to a certain point under the action of surface tension forces. When placing such droplets in the cylinder of a diesel engine and disrupting the surface layer of the droplets (e.g., under the action of a flame front, mutual collision, additional treatment with microwave radiation), micro explosions of water droplets occur, which, after instantaneous explosion and evaporation, destroy and reduce fuel droplets, increasing the fuel burning surface. This will increase the pressure in the combustion chamber, and, as a consequence, increase the power and efficiency of the heat engine with the same amount of fuel consumed [5].

When running engines on the fuel-water mixture, it is especially important to ensure that water droplets suspended in the fuel are evenly distributed throughout the entire volume of the mixture. to achieve maximum homogeneity of the fuel-water mixture before feeding it into the cylinder, it is possible by pre-treating it with ultrasound. This will lead to the fragmentation of water droplets suspended in the fuel, their uniform distribution throughout the volume and the formation of a fuel-water emulsion.

The aim of the study was to assess the possibility of reducing nitrogen oxide emissions into the atmosphere and reducing fuel consumption when the fuel-water mixture treated with microwaves and ultrasound is fed into the diesel engine cylinders. For experimental verification of the proposed hypothesis, a laboratory bench equipped with a Kipor KM178 diesel engine was designed and manufactured. In addition to the standard fuel supply system, an experimental fuel supply system was designed and manufactured as a separate unit for preparation and feeding into the engine cylinder of the fuel-water emulsion treated with ultrasound and microwaves [4]. Schematic diagram of the experimental fuel supply system is shown in Figure 1.

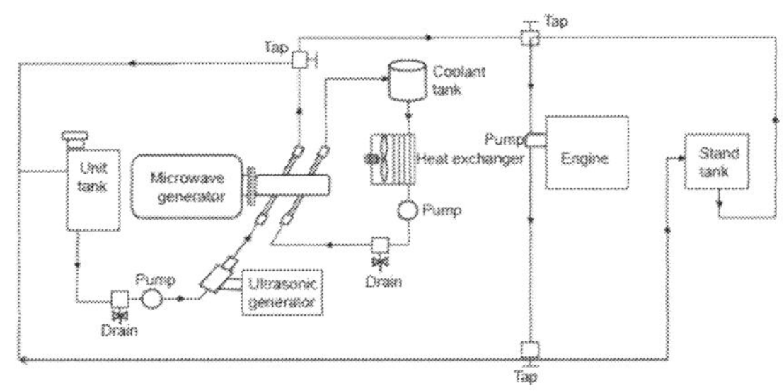

Fig. 1. Schematic diagram of the experimental fuel supply

General view of the bench with the engine, loading device and fuel supply unit is shown in Figure 2.

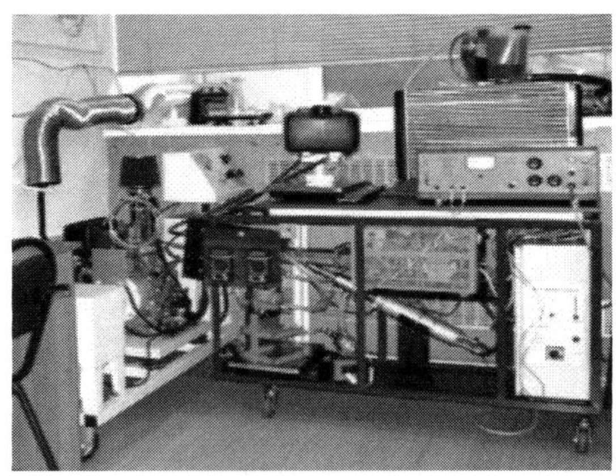

Fig. 2. General view of the bench

https://doi.org/10.1051/e3sconf/202337802005

In the course of the experiment, the main operating parameters of the diesel engine and the loading device were recorded by a system of sensors and displayed using a computer connected to the bench with the Research Engines 3.2 software as shown in Figure 3. The composition of exhaust gases was measured with a Testo t350XL gas analyzer.

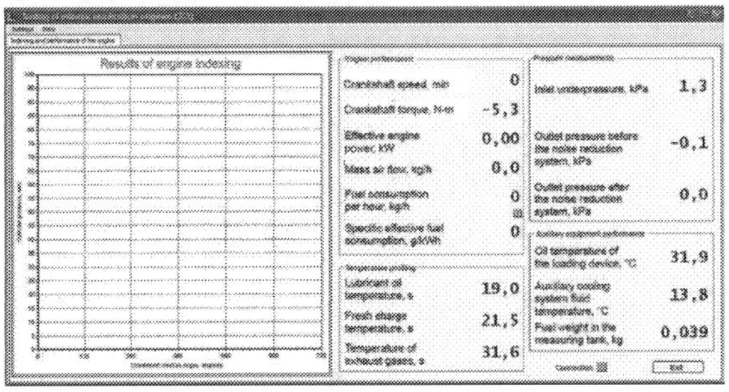

Fig. 3. Measurement results

Comparison of test results of diesel engine when idling with standard fuel and fuel-water mixture with different percentage of water treated with ultrasound and microwaves are presented in Figures 4 and 5.

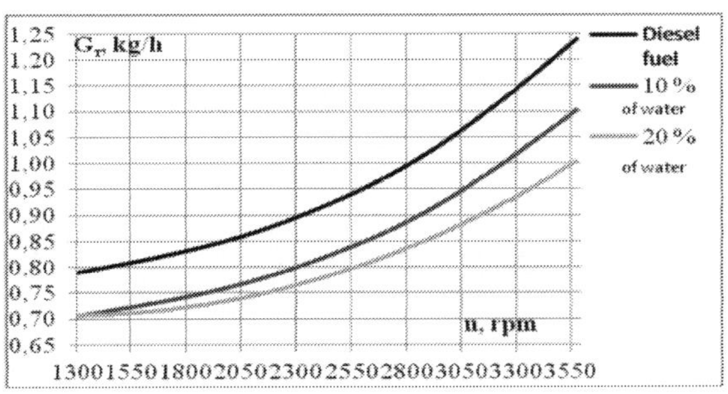

Fig. 4. Dependence of fuel consumption on the crankshaft speed

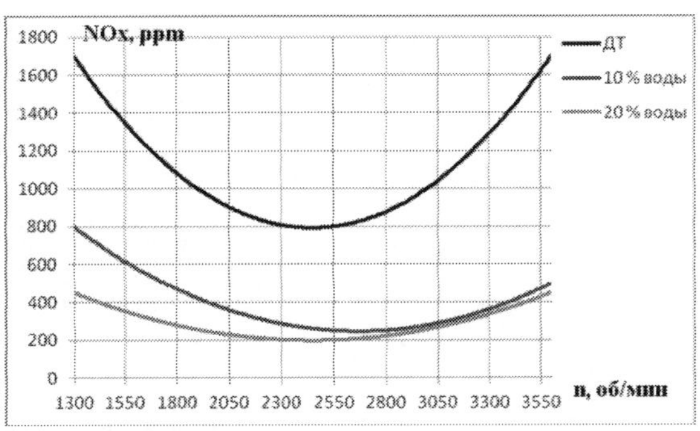

Fig. 5. Dependence of NOx concentration on the crankshaft speed

3 Conclusion

The main conclusions of the research work are as follows:

1. The use of fuel-water mixture treated with ultrasound and microwaves as fuel for the diesel engine can reduce the NOx content in exhaust gases by 40-50% and reduce fuel consumption by 3-10% when idling.
2. The use of the fuel-water mixture treated by ultrasound and microwaves significantly reduces the combustion temperature in the engine cylinder. When using pure diesel fuel, the temperature is 1900 - 2100 K, when using fuel-water emulsion containing 10 - 20% water, the temperature is 1500 - 1800 K. Calculations prove that lowering the combustion temperature to 2000 K and lower leads to a significant decrease in the concentration of nitrogen oxides. However, when the water content in the fuel-water mixture rises to 20%, the engine power decreases. Hence, it follows that it is preferable and optimal to run the engine on a fuel-water mixture containing 10% water.
3. The volume fraction of water equal to 10% ensures decrease in the calorimetric combustion temperature to 1700 - 1800 K, which is sufficient to normalize the content of nitrogen oxides in the exhaust gases.
4. In the course of experiments, it was noted that when using the fuel-water mixture, there is a decrease in the content of carbon monoxide, CO in the exhaust gases, as well as a decrease in soot formation.
5. All the studies were performed when the diesel engine was idling; therefore, further study of the load and speed modes of operation is of interest.

This paper presents a developed and experimentally tested method of reducing the content of nitrogen oxides in the exhaust gases of a diesel engine. The main advantages of the presented method include the possibility of its application at any facilities equipped with both new and long-operated diesel engines, without structural modifications in the engines themselves. The expected effect is achieved by installing a small-sized special fuel unit into the standard fuel system of the facility. An important factor in favor of this method is the possibility to modify facilities operating in the Arctic in the conditions of field workshops without sending them to manufacturers.

In conclusion, it should be noted that the development of technology will inevitably lead to the development of new engines that do not pollute the atmosphere. But achieving their widespread use will take a long time and during this period it is necessary to protect the Arctic atmosphere from harmful emissions of operating diesel engines.

References

1. N.B. Sedova and E.Y. Kochemasova, Environmental problems of the Arctic and their socio-economic consequences, *Journal «ECO»*, **5** (2017)
2. A.S. Atavov and M.M. Amanatov *The impact of motor transport on the ecology of the city in the Arctic region* Available online: https://nsportal.ru (2014)
3. A.V. Shevchuk and V.V. Kurteyev, On the development of the main directions of scientific

research Arctic zone of the Russian Federation, *Journal Arctic and North*, **22** (2016)

4. B.N. Antipov, et al., Patent for invention No.2581503 dated 10.03.2015, Stand for testing heat engines, *Bulletin*, **11** (2015)
5. B.N. Antipov, et al., Increasing the efficiency of heat engines through microwaves, *Bulletin by Central Development Commission Rosnedra,* p. 53 (2012)
6. B.N. Antipov, et al. Study of the characteristics of a diesel engine when fuel or fuel-water emulsion is exposed to microwaves and ultrasound, *Quarterly scientific and technical journal "Quality Management in the Oil and Gas Sector"*, 3, p. 36 (2015)
7. B.N. Antipov, et al., Patent for invention No 2527005 dated 29.03.2013 Method of preparation of liquid fuel for combustion in the combustion chamber, *Bulletin*, **24** (2013)
8. B.N. Antipov and S.M. Kuptsov, Increasing the efficiency of heat engines. Main and field pipelines. Design, construction, operation, repair, *Collection of Scientific and Technical Articles*, **1**, p. 24 (2018)
9. B.N. Antipov et al., Patent for invention No. 2266470 dated 14.04. 2004 Method for preparing the liquid fuel for spraying into the combustion chamber, *Bulletin*, **35** (2005)
10. R.Z. Kavtaradze, *Theory of piston engines. Special chapters* (Moscow, Publishing house of Bauman Moscow State Technical University 2008)
11. F. Bedford, et al., Effects of Direct Water Injection on DI Diesel Engine, *Combustion Society of Automotive Engineers* (2000)
12. Y.I. Dukhanin, et al., Patent for invention No. 2569779 dated 11.27.2015 Installation for producing a water-fuel emulsion, Bulletin, **33** (2015)
13. Z.R. Kavtaradze, Reducing the concentration of nitrogen oxides in combustion products high-speed diesel by improving the workflow, *Process Dissertation Candidate of Technical Sciences* (2006)
14. N.N. Kolenko and Y.I. Dukhanin, Water-fuel systems for diesel power plants, *Engine Scientific Technical Journal «Engine»*, **5** (2012)
15. R. Lanzafame, Water Injection Effects In A Single-Cylinder CFR Combustion in SI Engines SP-1436 (1999)
16. Ya.B. Zeldovich, *Oxidation of nitrogen in combustion* (Publishing house of the Academy of Sciences of the USSR, 1947)

https://doi.org/10.1051/e3sconf/202337802006

Implementation of an electrical network monitoring system based on a non-contact temperature sensor and a current sensor based on the Hall effect

Dmitriy Burylov[1,*], *Vladimir* Smirnov[2,*], *Vladimir* Dvoynikov[3,*] *Vadim* Shpenst[4]

[1] Candidate of Technical Sciences, Associate Professor, Department of Electronic Systems, Saint Petersburg Mining University, 199106, St. Petersburg, Russian Federation
[2]Department of Electronic Systems, Saint Petersburg Mining University, 199106, St. Petersburg, Russian Federation
[3]Department of Electronic Systems, Saint Petersburg Mining University, 199106, St. Petersburg, Russian Federation
[4]Doctor of Technical Sciences, Professor, Dean of the Energy Faculty, Saint Petersburg Mining University, 199106, St. Petersburg, Russian Federation

Abstract. The safety of people and the environment is an important issue at any enterprise. One of the main threats in the production environment is the occurrence of fire and the failure of electrical equipment. Fire alarm systems provide immediate evacuation of people and fight the source of fire. The article describes implementation of an electrical circuit system based on a non-contact temperature sensor and a Hall effect current sensor, which is necessary for fire prevention. The experiment showed that the data obtained is sufficient for timely change of current level in the conductors and prevention of emergency situation. The article consists of an introduction, theoretical information about the temperature sensor and current sensor based on the Hall effect, description of Bluetooth Mesh network topology, development of a wireless network diagram for the electrical network monitoring system, conclusion.

1 Introduction

One of the main ways to reduce damage from fires is the use of fire alarms; however, their disadvantage is that they are triggered immediately at the time of the fire, while there are systems that prevent its occurrence. there are sensors of temperature and short circuit in electric networks, their use in the domestic conditions and in the production could significantly reduce the number of fires, since according to statistics for 2020 in Russia, 34% of fires occur due to breakdowns in the electric networks and malfunction of electrical equipment [1]. No less damage is caused by fires in the Arctic, in particular in various fields and along pipelines, since this area is less developed by man, and, as a result, it is not always possible to prevent fires quickly. In addition, mining companies suffer significant losses due to equipment failures. Also, in the risk zone there are warehouses and tanks with various types of burning fuel [2]. A fire in such areas can lead to severe explosions or critical environmental pollution [3].

The purpose of the work is to analyze the operation of non-contact temperature sensor and Hall sensor and draw conclusions about the possibility of their joint use in a single electrical network monitoring system.

To achieve the goal, we will analyze the operation of a non-contact temperature sensor, consider options for using a Hall sensor to track the limit values of the current; develop a wireless communication scheme for a unified system for monitoring faults in an electrical network using a distributed Bluetooth Mesh network [4].

2 Non-constact temperature sensor

Non-constact temperature sensors are designed to monitor the temperature of remote or hard-to-reach objects. No need to touch allows non-contact sensors to measure very large temperature ranges.

Currently, non-constact temperature sensors are widely used in various security systems and alarm systems. The main advantage of such sensors is that they act directionally and do not require attachment directly to the monitored object. Also, there are devices that determine the temperature not only in a given direction, but also by sectors, which can be used in electronic intelligence and in everyday life, it is possible to implement the function of reading readings using modern laser systems [5, 6]. We suggest using sensors to detect faults in electrical circuit elements.

We use Omron D6T thermal sensor. Omron D6T MEMS thermal sensor is an ultra-sensitive infrared temperature sensor that makes full use of Omron's proprietary MEMS measurement technology. Since the D6T sensor can also track temperature in space, it can be also used to maintain optimal temperature levels at all times, instantly detecting unusual temperature changes, thereby identifying areas of overheating for early fire prevention [7].

* Corresponding author: burylov_DA@pers.spmi.ru, s171203@stud.spmi.ru, s170792@stud.spmi.ru

https://doi.org/10.1051/e3sconf/202337802006

While standard temperature sensors can only measure temperature at one specific contact point, the D6T can measure the temperature of an entire area without contact.

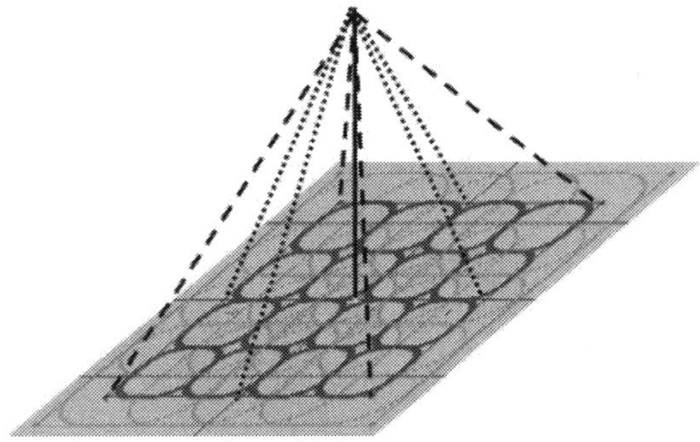

Fig. 1. Temperature measurement with a thermal sensor of the entire surface area.

We take the embedded microcircuit by its surface, and fix the sensor on the cover of the electrical network node. For the selected sensor the field of view is 90°. From this we get that the electrical node must correspond to the following ratio:

$$L \approx S * 0.81,$$

where L – the maximum length of the side of the microcircuit surface;

S – the distance to the sensor.

It can be concluded that the use of the selected sensor for monitoring the temperature on the surface of integrated circuits will be applicable in most nodes of the electrical network.

3 Hall Sensor

Hall-effect sensor is a measuring transducer for measuring the magnitude of a magnetic field.

This is a Hall-effect sensor, the essence of which is that when a conductor with a direct current is placed in a magnetic field, there is a transverse potential difference arises in this conductor, also referred to as the Hall voltage.

The direction of deflection of the electrons in the conductor is perpendicular to the direction of the magnetic field. On different sides of the plate the density of electrons will be different, which will cause a potential difference. This is detected by the Hall sensor.

It took 75 years for the Hall effect to be used in technology, when the production of semiconductor films with the desired properties was established. Hall sensors first appeared in automobiles to measure the angle of the camshaft or crankshaft position to determine the moment of ignition in internal combustion engines.

Later, with the development of microelectronics, it became possible to manufacture a miniature sensor containing everything necessary - a permanent magnet and a chip with a sensitive element. Such a device has a number of undeniable advantages.

Firstly, it is small in size.

Secondly, the electrical signal from the sensor has, according to the terminology of specialists, a rectangular form: when turned on, it immediately gains a definite and constant value and does not have a burst character. This is a big plus for electronic control.

The sensor has other advantages, but let's mention the disadvantages. The main one is inherent to any electronic circuit: the sensor is sensitive to electromagnetic interference arising in the power supply circuit. In addition, the Hall-effect sensor is more expensive than a magnetoelectric sensor and theoretically less reliable, since it contains an electronic circuit, but large-scale production and technological development minimize these factors.

We use the Allegro ACS37612 sensor in our work. It allows AC and DC measurements without the need for an external core or field concentrator shield. It is designed for electrical networks where hundreds of amperes are flowing through a busbar or printed circuit board.

The ACS37612 is available with 120 kHz and 240 kHz bandwidths to minimize response times. Fast response times allow timely detection of critical overcurrents on the electrical network. Ambient operating temperature range from –40°C to 150°C and a high level of ESD protection make it ready to work in harsh conditions [8].

Therefore, the use of the selected magnetic field measurement sensor for monitoring current drops is optimal.

4 Bluetooth Mesh

Bluetooth Mesh is a network topology used to link many-to-many (m: m) devices. The mesh topology available in Bluetooth LE enables creation of large-scale device networks and is ideal for control, monitoring and automation systems where tens, hundreds or thousands of devices need to communicate with each other reliably and securely [9].

5 Development of a monitoring device for an electrical unit

The monitoring device includes:

1. Contactless temperature sensor Omron D6T
2. Hall effect sensor ACS37612
3. Wireless module
4. Microcontroller.

The Pic16F15376 microcontroller by Microchip [10] was chosen for the implementation of our project. The choice was made based on its advantages:

1. Availability on the market
2. Relatively low price on the market
3. High reliability
4. Availability of the necessary periphery.

The PIC16F15376 microcontrollers contain analog, core-independent peripherals and communication peripherals combined with eXtreme Low-Power (XLP) technology for a wide range of general-purpose [11] and low-power applications.

The devices have multiple PWM, multiple communication, temperature and memory functions,

https://doi.org/10.1051/e3sconf/202337802006

such as a memory access section (MAP) to support clients with data protection and bootloader applications, and a device information area (DIA) that stores factory calibration values for improving the accuracy of the temperature sensor.

The contactless temperature sensor supports I2C communication. Let's connect it to the microcontroller using the appropriate inputs.

The selected Hall sensor is connected to one of the analog inputs on the microcontroller.

We will develop software for a microcontroller using the MPLABX IDE development environment. The algorithm of the program is as follows:

1. Collecting data from a contactless temperature sensor
2. Collecting data from the Hall sensor
3. Formatting the received data
4. Transferring data to an external device via the UART interface.

6 Development of a wireless network diagram for an electrical network monitoring system

It is necessary to implement the data transfer between the individual measuring devices, within one of the premises, the room in which the switchboard is located. The peculiarity of the location of the sensors is that some of them may be out of the range of wireless technology. Also, security is an important requirement for the system. Based on this, the use of a mesh topology is optimal.

The NINA-B312 module from u-blox was chosen to create the network. The device has support for the most recent version of Bluetooth v5.0 [13, 14], built-in software that allows to work with mesh topology, and allows you to configure the network using AT commands.

Here is a diagram of the wireless distributed network for the monitoring system (Figure 2).

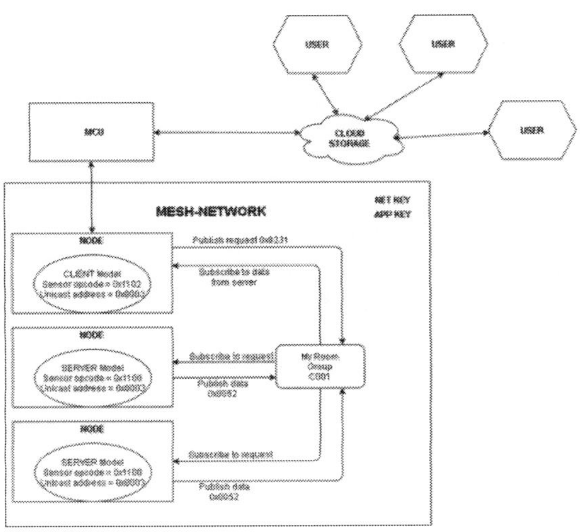

Fig. 2. Diagram of a mesh network of a fault monitoring system in an electrical network for one room

For the convenience of working with the results of measurement, the data from the client via a wireless Wi-Fi network are transmitted to a special cloud storage with its own authorization system.

7 Realization of the experiment

Experimental measurement of temperature and current level in conductors was carried out on an electricity meter. For the visualization of the experiment, a window application for Windows was developed in the Visual Studio development environment in the C# programming language using the ".NET" technology.

The temperature data and data from the Hall-effect sensor are displayed on the main panel in the form of a defined area divided into cells by temperature values and tabular data.

For the experiment, temperature indicators and current values were measured in a cabinet with an electricity meter and automatic switches.

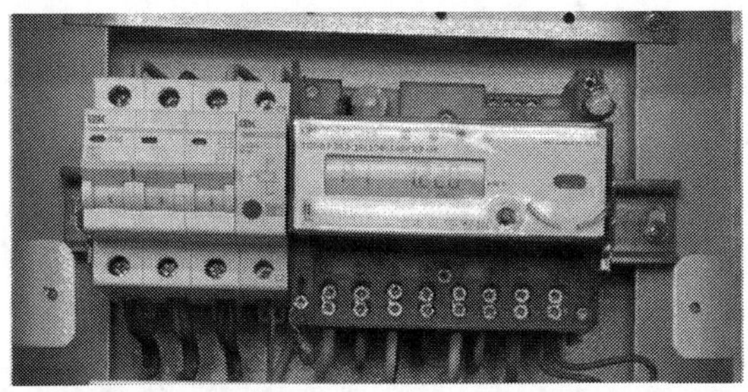

Fig. 3. Object to be measured

The temperature sensor was installed 10 centimeters from the measured area. measurements were carried out in the area where the conductors were connected to the machines and the meter (Figure 4). This area is the most fire-hazardous, as the conductors in it are exposed to air and, therefore, to corrosion.

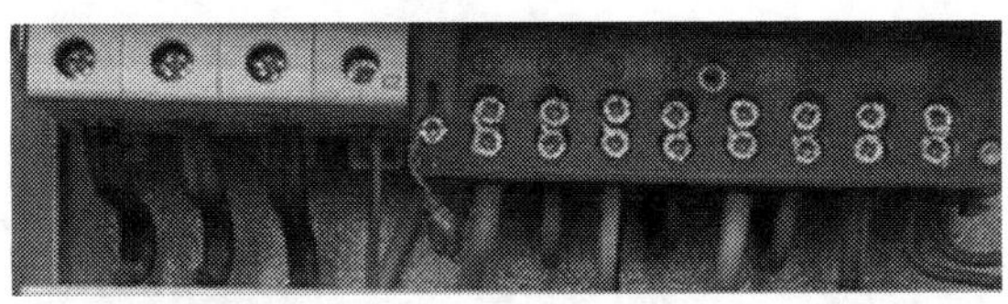

Fig. 4. Measurement area

Current readings were taken alternately in each conductor, which can considered a disadvantage of this system, since one Hall-effect sensor is not capable of measuring several objects simultaneously. In future developments the number of sensors will be equal to the number of measured conductors. The results of the experiment were visualized using the Windows.NET window application developed by us (Figure 5).

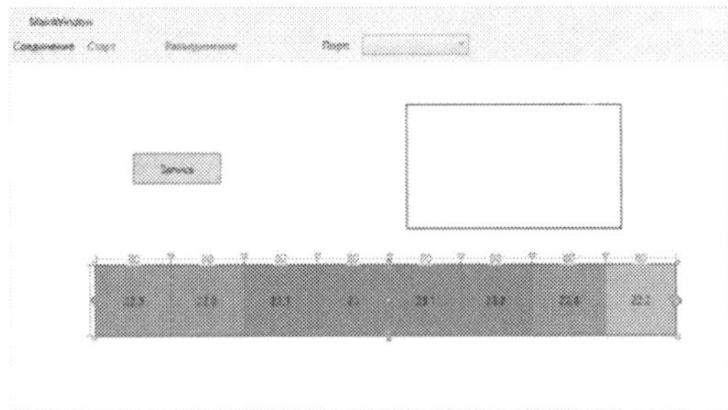

Fig. 5. The results of the experimental measurement of the electric meter

The temperature range was about twenty-three degrees. The current values in the conductors we took coincided with the electric meter readings.

Thus, we can conclude from the readings obtained that the system is operational.

8 Conclusion

The conclusions of this work are as follows:

- The operation of a non-contact temperature sensor was analyzed and it was found that its use will significantly reduce the risks of fire and damage to the electrical network.
- The operation of the Hall-effect sensor was considered andthe most suitable options for working with high currents, as well as with a short response time were chosen.
- A wireless network scheme for communication between monitoring devices using Bluetooth Mesh technology was developed.

When analyzing the operation of a non-contact thermal sensor and current sensor based on the Hall effect and testing them, it was concluded that the system is workable. Implementation of such a system will make it possible to avoid fires and their consequences not only in production or mining, but also in domestic conditions.

References

1. State report "On the state of protection of the population and territories of the Russian Federation from natural and man-made emergencies in 2020"
2. O. Y. Morozova, "Analysis of Typical Electric Power Facilities Requiring Remote Monitoring," 2020 IEEE Conference of Russian Young Researchers in Electrical and Electronic Engineering (EIConRus), 2020, pp. 1255-1257, doi: 10.1109/EIConRus49466.2020.9039389
3. V.S. Litvinenko, et al., Barriers to implementation of hydrogen initiatives in the context of global energy sustainable development, *Journal of Mining Institute*, **244**, pp. 428-438 (2020)
4. S.D. Lozada Montero, A.A. Belitsky, F.R. Salao Rey, D. Stalin Moya Roman, Helmetooth - Device of Increasing Vision and Communication for Motorcyclists, *2022 IEEE Conference of Russian Young Researchers in Electrical and Electronic Engineering (EIConRus)*, pp. 170-175 (2022)
5. Y.A. Mordvintseva, A.A. Belitsky, Training the Haar Cascade Classifier in Sorting Radio-Electronic Components. Automating the Collection of Images for Teaching, *2021 IEEE Conference of Russian Young Researchers in Electrical and Electronic Engineering (EIConRus)*, pp. 1019–1022, 9396135 (2021)
6. K.G. Karapetyan, O.V.Denisova, Crowding of neodymium ions as a structural feature of fluorophosphate glasses,*Tsvetnye Metally*, **5**, pp. 48–53 (2022)
7. MEMS Thermal Sensors D6T User's Manual. Cat. No. A284-E1-03
8. Coreless, High Precision, Hall-Effect Current Sensor IC with Common-Mode Field Rejection and High Bandwidth (240 kHz). Allegro Micro Systems
9. Bluetooth Mesh networking FAQs. Available online: https://www.bluetooth.com/learn-about-bluetooth/bluetooth-technology/topology-options/le-mesh/mesh-faq/ (2021)
10. PIC16(L)F15356/75/76/85/86 Full-Featured 28/40/44/48-Pin Microcontrollers. Datasheet
11. Y. Morozova and V.A. Shpenst, Research of the Energy Radiation Signatures and Display of General Facilities of Electrical Power Industry, *2020 IEEE Conference of Russian Young Researchers in Electrical and Electronic Engineering (EIConRus)*, pp. 1258-1260, doi: 10.1109/EIConRus49466.2020.9039378 (2020)
12. O.A. Gogina, A.A. Belitsky, A.A. Churyumov, Automatic customisation of temperature sensor based on piecewise-line interpolation, *2019 IEEE Conference of Russian Young Researchers in Electrical and Electronic Engineering (EIConRus)*, pp. 511–513, 8657199 (2019)
13. Martin Woolley and Sarah Schmidt, "Bluetooth 5. Go Faster. Go Further"
14. M. Woolley, S. Schmidt, Bluetooth mesh networking. Available online: https://www.bluetooth.com/wpontent/uploads/2019/03/Bluetooth-Mesh-Paving-the-Way-for-Smart-Lighting_FINAL.pdf (2017)

Study of water purification reactor operating modes

Alexander Chuikov[1,*], *Andrey* Volkov[2,*], *Igor* Barskii[3,*] and *Dmitrii* Shabanov[4,*]

[1]Technician, Peter the Great St. Petersburg Polytechnic University, 195251, St. Petersburg, Russian Federation
[2]Professor, Graduate School of Automation and Robotics, Peter the Great St. Petersburg Polytechnic University, 195251, St. Petersburg, Russian Federation
[3]Master's degree student, Graduate School of Automation and Robotics, Peter the Great St. Petersburg Polytechnic University, 195251, St. Petersburg, Russian Federation
[4]Teaching assistant, Graduate School of Automation and Robotics, Peter the Great St. Petersburg Polytechnic University, 195251, St. Petersburg, Russian Federation

Abstract. Wastewater treatment is one of the main problems of development and urbanization of the Arctic regions. This article is devoted to the development of a method for assessing the degree of water purification in order to determine the optimum operating parameters of treatment reactors for water treatment by ferrates from industrial and household waste resulting from the development of Arctic regions. Experiments reflecting the influence of the amount of air supplied to the purification plant on the processes of mixing and distribution of water were carried out and analyzed. Based on the experimental data, a method for determining the optimum operating parameters of a new mobile water purification plant has been developed and applied. The operation of the plant is based on the use of the ferrate method. equations and dependencies describing the relationship between the reactor volume, liquid flow rate and the degree of water purification are given. The degree of water pollution is estimated as its fraction which has left the reactor before the time required for complete treatment of the liquid from the moment it enters i the reactor. Optimal parameters of the ferrate reactor, such as volume and air supply, for a mobile water treatment plant were determined.

1 Introduction

The relevance of this study is due to the need to find, select and apply an efficient, safe and cost-effective water treatment method in a harsh Arctic climate [1].

One of the most suitable methods of water treatment for this task is the ferrate method, which uses sodium ferrate. Sodium ferrate has an oxidizing and disinfecting effect [2]. It is able to decompose most of the harmful chemicals into low-toxic products (dead organic matter, salts, etc.). In addition, the ferrate decomposition product itself is low-toxic (iron hydroxide).

This method solves the technical problem of the impossibility of highly efficient treatment of wastewater of different nature, representing a mixture of water with inclusions of different types and concentrations [3]. The method also makes it possible to reduce the influence of factors causing the need for reclamation of disturbed land with industrial wastewater sludge [4].

The development of new water treatment plants is often a rather complex process, involving many areas of engineering and chemical activities. One of the stages of development is the design stage of reactors, tanks and other containers, in which the main physical and chemical processes for water treatment and purification take place. When designing new reactors, it is convenient to use pre-developed mathematical models of processes to find optimal parameters (reactor volume, treated liquid flow rate, air flow rate fed into the reactor, reaction rate, reagent feed rate, etc.) of reactor operation. The construction of mathematical models is used in conditions when the mechanism of the phenomenon being described is poorly understood Observing the reaction of the system under study, a hypothetical model of the phenomenon is drawn up, which is then empirically tested under different conditions in order to clarify its individual parameters [5].

The purpose of this article is to develop a methodology for assessing the degree of water purification during its time in the reactor, in order to find the optimal parameters of water purification plants reactors, including in relation to the ferrate method.

2 Experiments to determine the degree of water mixing in a flow reactor

For high-quality purification water must be in contact with the reagent in the reactor for a certain time Td (desired time). In a flow reactor there is mixing of the layers of the fed water, so the volume of liquid fed at a certain time t0 will leave the reactor in the time interval tA-tB. If the value of Td is within this range, the quality of water purification can be estimated by the amount of water released in the time interval tA-Td. This

* Corresponding author: a Chuikov.as16@gmail.com, Volkov-and-1@yandex.ru, barskiiigor66@gmail.com, Dsb956@yandex.ru

distribution is influenced by the speed of mixing of the liquid and the amount of air supplied to the reactor.

Experiments were carried out on a previously developed experimental unit for water purification. The experiments aim to determine the distribution of the amount of water leaving the reactor at the time interval tA-tB.

In the experiments, colored water was used to simulate polluted water. The colored liquid was fed with a membrane pump with a capacity of 7.5 L/h into the reactor pre-filled with clean water. The air was supplied using a compressor.

The experiments were carried out with the following parameters:

– liquid supply to the unit in the absence of air supply
– liquid supply to the unit with an air flow rate of 3.25 L/min
– liquid supply to the unit with an air flow rate of 6.5 L/min
– liquid supply to the unit with an air flow rate of 13 L/min.

The process of mixing and distribution of the colored liquid was recorded using optical sensors (Fig. 1).

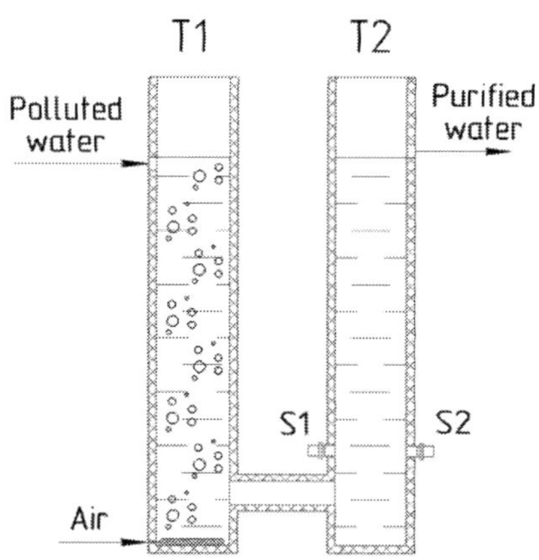

T1 - first reservoir, C2 - second reservoir, S1 - optical emitter, S2 - optical receiver.

Fig. 1. Diagram of water purification technical equipment

After the experiments, the graphs of the readings from the S2 sensor were plotted (Fig. 2). To eliminate the noise on the graphs, the moving average filtering method was applied.

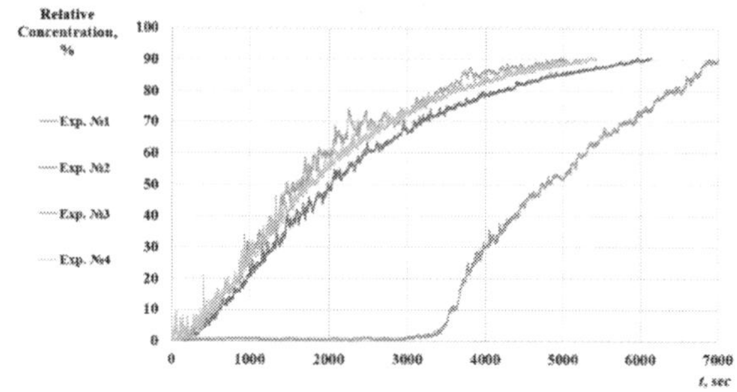

Fig. 2. Changes in dirty water concentration

The mixing process of dirty water entering the reactor can be described using a normal distribution (Fig. 3).

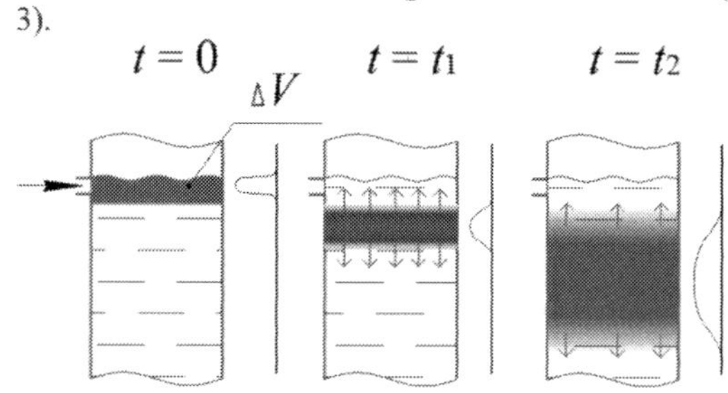

Fig. 3. Mixing liquid distribution

Let a quantity of liquid equal to ΔV be supplied at each moment of time, then the normal distribution will show what part of the liquid from the volume ΔV and at what time moment t will leave the reactor.

The normal distribution is characterized by the following function:

$$F = \frac{1}{2}\left[1 + erf\left(\frac{x-\mu}{\left(2\sigma^2\right)^{1/2}}\right)\right], \quad (1)$$

where μ is the mean of the distribution,
σ is the standard deviation.

The more time a water particle spends in the reactor, the more likely the particle will escape into the distant layers and the more "smeared out" the distribution will be. In this regard, it is necessary to use the variable σ, which will depend on the time that the liquid entering the reactor will spend in it.

$$\sigma(t) = k \cdot t, \quad (2)$$

where k is the empirical coefficient.

After selecting the necessary parameters, the graph corresponding to experiment No. 1 is plotted (Fig. 4).

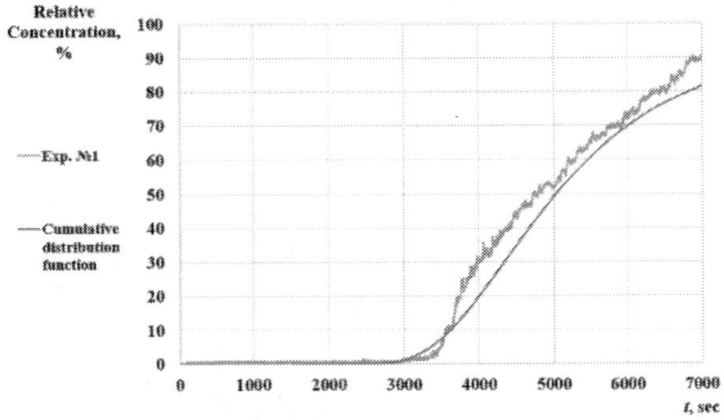

Fig. 4. Graph of the relative concentration of dirty water of experiment No. 1 and the distribution function

The graph above shows that the distribution function practically coincides with the obtained data.

The distribution graph f (t) was obtained by differentiating the plotted distribution function (Fig. 5).

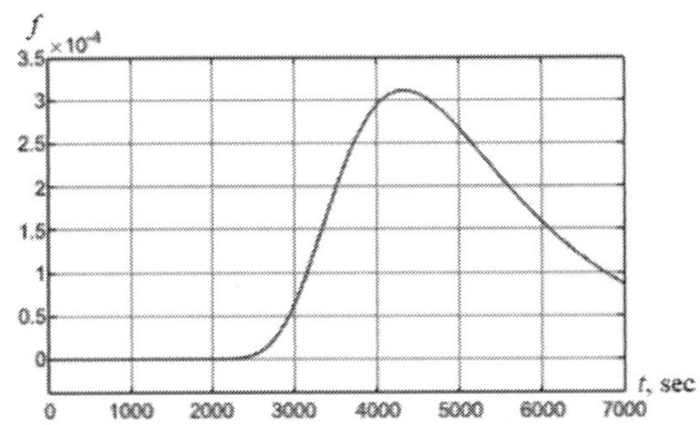

Fig. 5. Dirty liquid distribution at the reactor outlet, without air supply

The data of experiments 2, 3 and 4, in which air was supplied to the reactor, are almost identical. For one of them, we plot the corresponding distribution function. the necessary parameters are selected in advance (Fig. 6).

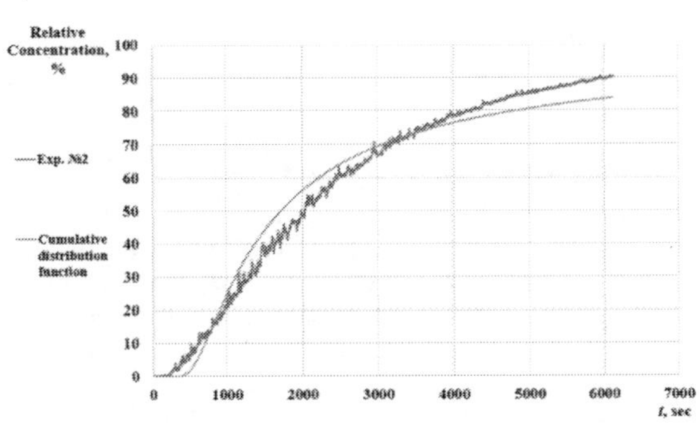

Fig. 6. Diagram of the relative concentration of dirty water of experiment No. 2 and the distribution function

A similar mixing description problem has also been described for reactive transport problems [6].

3 Mathematical estimation of the time of water passage through the reactor with perfect mixing

Based on the fact that for a given unit the supply air flow rate of more than 3.25 l/min has almost no effect on the results, we can assume that with further increase in air flow rate the mixing of liquid also occurs instantly and "perfectly".

The mathematical description of instant water mixing is shown below.

The amount of dirty water in the reactor can be found as described below.

When dirty water of volume ΔV enters the reactor, the amount of dirty water in the reactor will be equal to the sum of the amount of dirty water remaining in the reactor Vrem and the volume ΔV (Fig.7).

Dirty liquid concentration in the reactor is:

$$C_{dir}(t) = V_{dir}(t)/V \quad .(3)$$

Thus, the amount of dirty water remaining in the reactor will be equal to the product of the dirty liquid concentration and the volume of all the water in the reactor.

$$V_{rem}(t) = C_{dir}(t-dt)\cdot(V_{full}-\Delta V) = V_{dir}(t-dt)\cdot\frac{V_{full}-\Delta V}{V_{full}} \quad .(4)$$

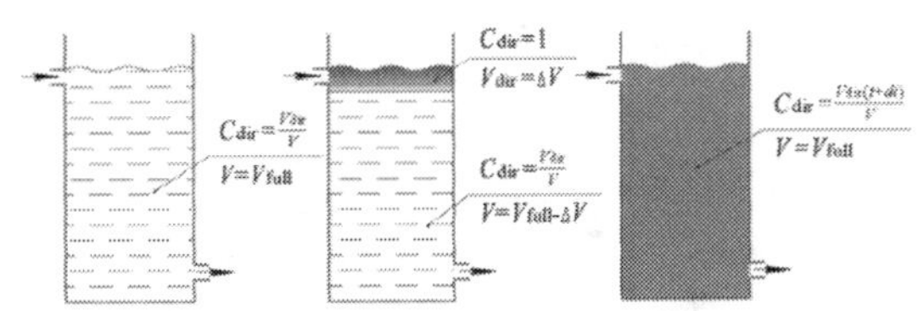

Fig. 7. Mixing the liquid in the reactor

The volume of dirty liquid in the reactor after adding ΔV can be calculated as:

$$V_{dir}(t+dt) = V_{dir}(t)\cdot\frac{V-\Delta V}{V}+\Delta V = V_{dir}(t)-V_{dir}(t)\cdot\frac{\Delta V}{V}+\Delta V \quad (5)$$

The resulting differential equation is:

$$dV_{dir}(t) = V_{dir}(t+dt)-V_{dir}(t) = -V_{dir}(t)\cdot\frac{\Delta V}{V}+\Delta V = \Delta V\cdot(1-\frac{V_{dir}(t)}{V}) \quad (6)$$

$$V_{dir}(t) = \int dV_{dir}(t)dt = \int \Delta V\cdot(1-\frac{V_{dir}(t)}{V})dt \quad . (7)$$

The necessary calculations were performed in the MATLAB program (Fig. 8).

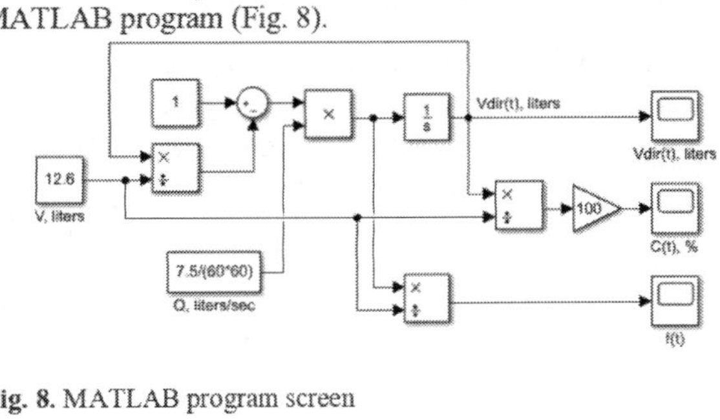

Fig. 8. MATLAB program screen

https://doi.org/10.1051/e3sconf/202337803001

The obtained dependence of the amount of dirty water in the reactor on the time t is presented (Fig. 9).

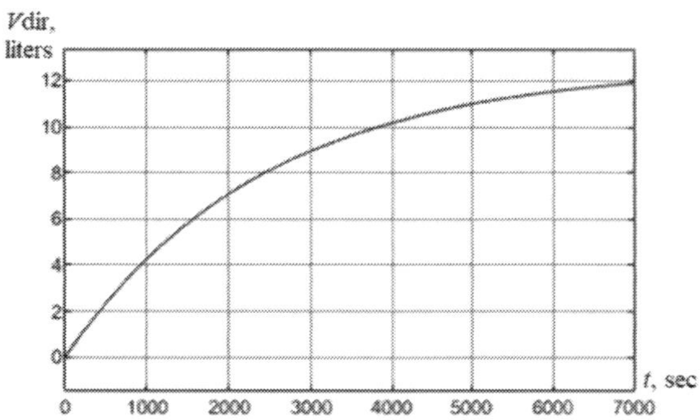

Fig. 9. The volume of dirty water in the reactor

The obtained dependence of the amount of dirty water at the reactor outlet from the moment of its supply (t = 0) on the time t in the presence of air feeding into the reactor is presented (Fig. 10).

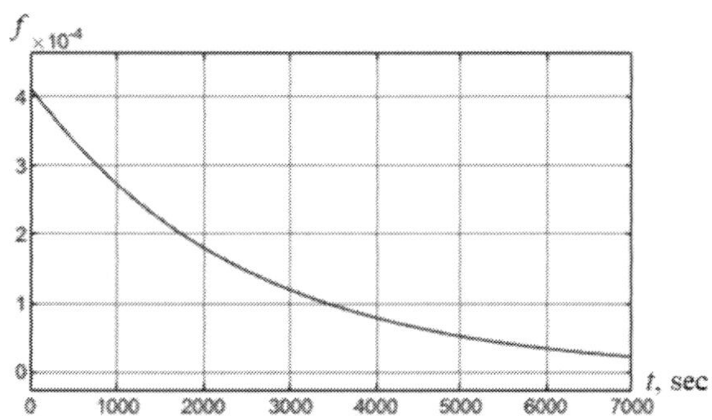

Fig.10. Distribution of dirty liquid at the reactor outlet (with air supply)

From the obtained distribution function graph it is possible to determine the fraction of particles from the fed volume of dirty water that will leave the reactor at time t. With the help of this it is possible to estimate the degree of water purification, knowing the desired time during which the water should be treated in the reactor with the known type of contamination.

4 Estimation of optimal parameters of the reactor for water purification using the ferrate method

The information obtained from the experiments is useful in the design of water treatment plants. The above equations allow us to describe the relationship between reactor volume, flow rate and quality of treated water, taking into account the air supply to the reactor.

Currently, a mobile wastewater treatment plant is being developed as part of the OneDrop project. Using the described technique, it is possible to estimate the optimal volume of the reactor in which the water is purified with ferrates [7]. One of the components of the mobile plant is a ferrate module, which allows processing of pre-purified liquid. This module consists of a ferrate reactor, an oxidation reactor and a coagulation tank (Fig. 11).

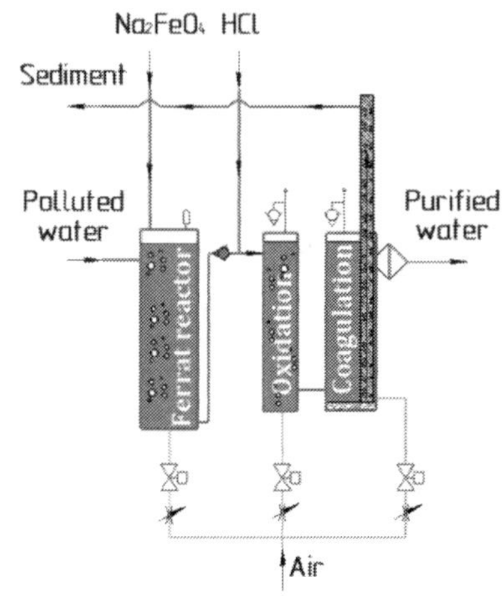

Fig. 11. The ferrate module diagram

Dirty water from the previous modules is fed from above into the ferrate reactor, and sodium ferrate (Na2FeO4) is dosed from above. During the downward circulation of the liquid, there is an interaction between the dirty substances contained in the water and the ferrate An air supply is provided in the lower part of the reactor. Air bubbles move in the direction opposite to the main flow, providing mixing of the liquid to avoid sedimentation of the reagent.

At the next stage, an oxidizing agent, hydrochloric acid (HCl), is added to the liquid before it enters the oxidation reactor. This agent separates the particles formed during the reaction between the ferrate and the water pollutants. there is also an air supply for mixing, which prevents the deposition of the resulting particles.

In the next tank, coagulated particles with a high specific gravity move down, settle and are removed from the reactor by an airlift. The purified water is filtered and discharged to the following purification stages.

The previously developed technique was applied to calculate the required reactor volume and the effect of the presence of air supply.

The average design flow rate of water is taken Q = 1 L/min; experimentally it was determined that the optimal time that water should spend in the reactor for purification is Td = 180 sec. To estimate the optimal reactor volume, two graphs are plotted as a function of the degree of water pollution at the reactor outlet on the reactor volume: with the use of air for mixing and without mixing (Fig. 12, 13). The degree of pollution C is the fraction of water (from 0 to 1) that left the reactor before the time Td after entering it.

https://doi.org/10.1051/e3sconf/202337803001

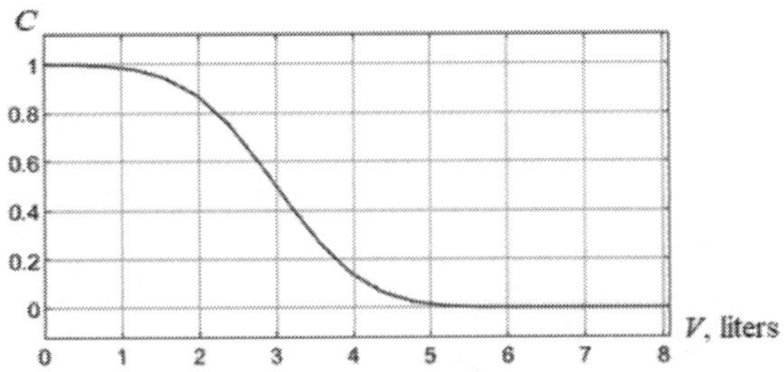

Fig. 12. Relationship between the degree of water pollution and the volume of the reactor (without air supply)

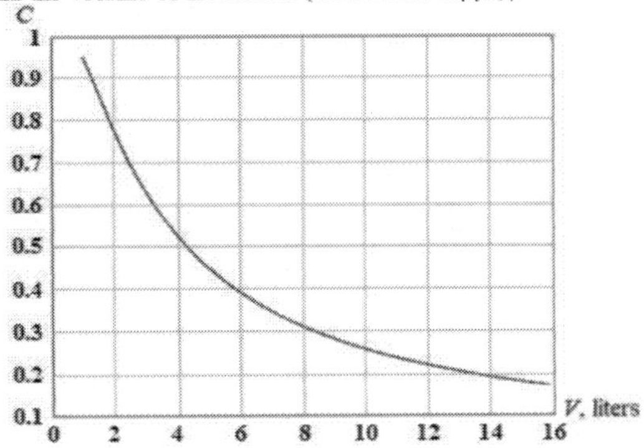

Fig. 13. Relationship between the degree of water pollution and the volume of the reactor (with air supply)

Due to the design features, it is difficult to use a reactor with a volume of more than 7 liters. The graph shows the degree of pollution of treated water when using a 7-liter reactor without air is practically 0, when using air it is about 35%.

Fig 14 also shows the graph of the dependence of the degree of pollution of the water treated in this reactor C on its flow rate and when air is used for mixing.

The graph of the dependence of the degree of water pollution C on the time Td required for water purification, when using a reactor with of 7 liters and without using air (Fig. 15). The planned flow rate is 0.6–1.3 L/min.

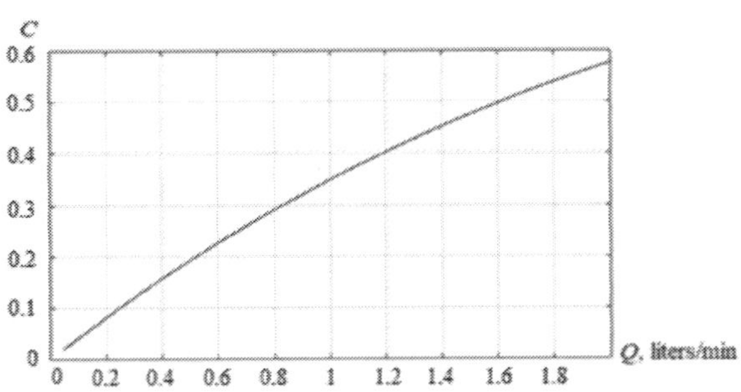

Fig. 14. Dependence of the degree of water pollution on its consumption (with air supply; reactor volume is 7 liters)

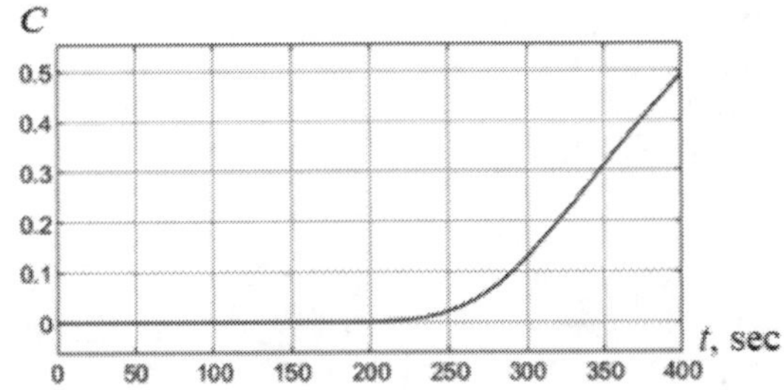

Fig. 15. Dependence of the degree of water pollution on time Td (without air supply; reactor volume is 7 liters).

5 Conclusion

Ferrate technology of water treatment is acceptable to operate in the Arctic.

The proposed methodology makes it possible to estimate the expected purity level of water with and without air supply. If air supply is not used, the mixing process is described with a distribution similar to a normal distribution. The mixing process with air supply is described by differential equation (5). Also, the relationship between water purity level, reactor volume and water flow rate allows us to choose the optimal parameters of the plant.

Graphs plotted for water purification in a ferrate reactor show that the use of air reduces the degree of water purification. In this regard, it is proposed to test the plant without air supply - perhaps the entire reagent will have time to react without settling to the bottom. mixing is then not required. Alternatively, an experiment with intermittent, periodic air supply can be performed.

Note that the k factor used in the air-free mixing calculations must be chosen individually for each reactor design. This factor is influenced by the geometry of the reactor and its size ratio.

The work was performed within the thematic project "South-East Finland - Russia CBC Program2014-2020 - OneDrop Project - KS1648".

References

1. P.E. Jensen, et al., Water, sanitation, pollution, and health in the Arctic, *Environ. Sci. Pollut. Res,* **25**, pp. 32827–32830 (2018)
2. E.N. Arakcheyev, et al., Experimental substantiation of the practicability of disinfection and purification of water and wastewaters with potassium ferrate, *Hygiene and Sanitation*, **96(3)**, pp. 216-222 (2017)
3. A.V. Karim, et al., *Advanced Oxidation Processes - Applications, Trends, and Prospects ed Bustillo-Lecompte C F* (London, IntechOpen, 2020), **chapter 5**
4. T.A. Petrova, et al,. Rehabilitation of Disturbed Lands with Industrial Wastewater Sludge, *Minerals*, **12(3)**, p. 376 (2022)

5. B. Savun-Hekimoğlu, On the Use of Mathematical Models for Wastewater Treatment: A Review and Analysis of Activated Sludge Models ASM1 and ASM3, *Int. J. Environ. Geoinformatics*, **8(1)**, pp. 1-18 (2021)
6. J. Soler Sagarra, et al., Water Mixing Approach (WMA) for reactive transport modelling, doi: 10.1002/essoar.10504136.1 (2020)
7. R.E. Nekrasov, et al., Apparatus for water purification of natural water, Patent Russian Federation no. RU 175092 U1 (1997)

Investigation of the effect of algorithms of positioning robots on their power and energy consumption

Andrey Volkov[1], *Anna* Kolesnikova[2], *Anastasia* Kornilova[3] and *Olga* Matsko[4,*]

[1]Professor, Higher School of Automation and Robotics, Peter the Great St. Petersburg Polytechnic University, 195251, St. Petersburg, Russian Federation
[2]Senior lecturer, Higher School of Automation and Robotics, Peter the Great St. Petersburg Polytechnic University, 195251, St. Petersburg, Russian Federation
[3]Assistant lecturer, Higher School of Automation and Robotics, Peter the Great St. Petersburg Polytechnic University, 195251, St. Petersburg, Russian Federation
[4]Associate professor, Higher School of Automation and Robotics, Peter the Great St. Petersburg Polytechnic University, 195251, St. Petersburg, Russian Federation

Abstract. The paper is devoted to the study of motion algorithms that determine the power and energy consumption of process equipment and positioning robots, which are important factors affecting the environment and environmental safety of the technosphere. Based on the study of mathematical models of the degrees of mobility of the gantry robot drives, it is found that under the traditional laws of motion, a fairly significant peak in power consumption occurs at the end of acceleration. The paper shows that by separating in time the moments of the end of acceleration of different drives, it is possible to significantly reduce the maximum power of the robot. A comprehensive approach to the choice of an algorithm and laws of motion that provide an acceptable combination of energy consumption and maximum power consumption has been developed. The authors have shown that there are such laws of motion that ensure the constancy of power for the duration of the acceleration stage and its minimization. The relationship between the minimum instantaneous power and power consumption for such robots has been established. The results obtained can be used to optimize the operation algorithms of both of existing and newly +-robots.

1 Current state of the problem

In the modern world, energy consumption in the technosphere is a key factor influence on the biosphere that determines the consumption of natural resources and the degree of environmental pollution. There are works in which the authors show that the law of motion determines the engine heating losses [4]. However, in the absence of recuperation heat losses on braking resistors are many times greater than heat losses in windings. There are known approaches to energy saving for transport robots, when the positive effect is determined by the choice of the motion path [1, 3]. Some authors [2] use a modified PSO method to optimize the trajectory of robots in the absence of resistance forces operating in zero gravity. However, all these works the problems only from one side: from the side of reducing energy consumption. energy consumption also determines the maximum power value. Reduction of energy consumption is often accompanied by an increase in the maximum power consumption, and vice versa [5]. In its turn, the maximum power consumption determines the size of the drive, its overall dimensions and weight [6], which is often a key factor in the design of robots and their power supply systems.

2 Problem statement

Considering the above, it is necessary to investigate the influence of motion laws and operating algorithms of the degrees of mobility of positioning robots at a given time of motion and movement on the energy costs and maximum power consumption, as well as to evaluate the effectiveness of optimization of motion algorithms for different motion laws and inertial load.

Let us construct a mathematical model of a gantry robot with two degrees of mobility having rectangular laws of acceleration variation. For example, we consider a gantry robot with mass m (Figure 1), when the motion vector of point A is equal to s and is implemented in time T.

The motion of the gantry along the X-axis can be considered as translational motion, and along the Y-axis as relative motion. To simplify the calculations and without violating the generality of the approach, a rectangular law of motion is considered for each degree of mobility. In this case, the dimensionless laws of change in acceleration, change in velocity and, and change in displacement variation have the following form:

* Corresponding author: onmatsko@gmail.com

$$\rho''(\alpha,\zeta) := \begin{cases} \dfrac{1}{\zeta\cdot(1-\zeta)} & \text{if } 0 \le \alpha \le \zeta \\ 0 & \text{if } \zeta < \alpha < 1-\zeta \\ \dfrac{1}{\zeta\cdot(\zeta-1)} & \text{if } 1-\zeta \le \alpha \le 1 \end{cases} \quad (1)$$

$$\rho'(\alpha,\zeta) := \begin{cases} \dfrac{\alpha}{\zeta\cdot(1-\zeta)} & \text{if } 0 \le \alpha \le \zeta \\ \dfrac{1}{1-\zeta} & \text{if } \zeta < \alpha \le 1-\zeta \\ \dfrac{1-\alpha}{\zeta\cdot(1-\zeta)} & \text{if } 1-\zeta < \alpha \le 1 \end{cases} \quad (2)$$

$$\rho(\alpha,\zeta) := \begin{cases} \dfrac{\alpha^2}{2\zeta\cdot(1-\zeta)} & \text{if } 0 \le \alpha \le \zeta \\ \dfrac{(\zeta-2\alpha)}{2\cdot(\zeta-1)} & \text{if } \zeta < \alpha \le 1-\zeta \\ \dfrac{(1-\alpha)^2}{2\zeta\cdot(\zeta-1)}+1 & \text{if } 1-\zeta < \alpha \le 1 \end{cases} \quad (3)$$

where $\alpha = t/T$ is the dimensionless time; $\zeta = \tau/T$ is the dimensionless acceleration time equal to the deceleration time. Figure 2 shows the corresponding dependence diagrams.

The absolute values of accelerations and velocities of motion are related to the relative values by the following equations:

where s is the stroke value for the corresponding degree of mobility.

$$\begin{aligned} a(t,\tau) &= k''\cdot\rho''(\alpha,\zeta) \\ v(t,\tau) &= k'\cdot\rho'(\alpha,\zeta) \\ x(t,\tau) &= k\cdot\rho(\alpha,\zeta) \end{aligned} \quad (4)$$

$$k'' = \frac{s}{T^2};\ k' = \frac{s}{T};\ k = s, \quad (5)$$

where s is the stroke value for the corresponding degree of mobility.

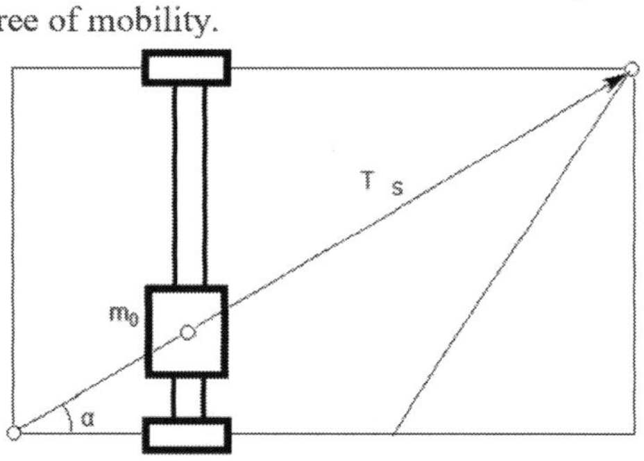

Fig. 1. Design pattern

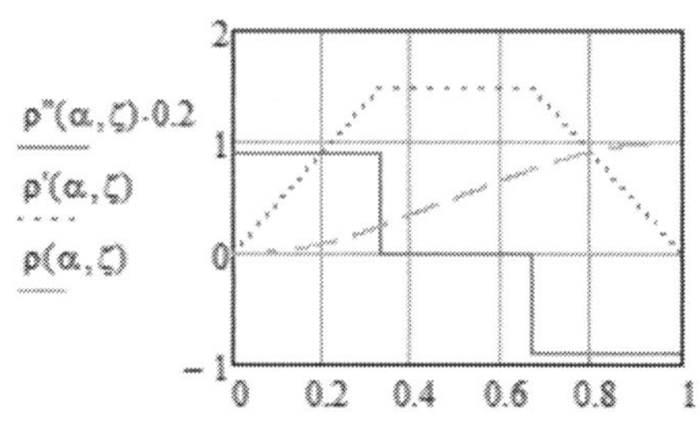

Fig. 2. Dependence diagrams for dimensionless acceleration, velocity and displacement on dimensionless time

The total instantaneous power is determined by the expression:

$$\eta_s(\alpha,\zeta_1,\zeta_2) = \rho'_1(\alpha,\zeta_1)\cdot\rho''_1(\alpha,\zeta_1) + \rho'_2(\alpha,\zeta_2)\cdot\rho''_2(\alpha,\zeta_2) \quad (6)$$

Hereafter, the digital index "1" means the ratio of the parameter to the translation degree of mobility of the manipulator, and "2" - to the relative degree of mobility.

The absolute and dimensionless values of the instantaneous power of the robot are related to each other by the following equation:

$$N_s(t,T,\tau_1,\tau_2) = N_1(t,T,\tau_1) + N_1(t,T,\tau_1) = k_{\eta 1}\cdot\eta_1(\alpha,\zeta_1) + k_{\eta 2}\cdot\eta_2(\alpha,\zeta_2) \quad (7)$$

The following designations for the coefficients of power conversion are introduced here:

$$k_{\eta 1} = \frac{m(S\cdot\cos(\varphi))^2}{T^3} \qquad k_{\eta 2} = \frac{m\cdot\mu\cdot(S\cdot\sin(\varphi))^2}{T^3} \quad (8)$$

where m is the mass of the moving parts of the robot, and μ is the coefficient that determines the fraction of the moving parts of the robot attributable to the relative degree of mobility.

In the absence of energy recovery, when the kinetic energy is dissipated on the braking resistor during deceleration, the energy cost for one cycle of motion (acceleration, uniform motion and deceleration) is determined by the time integral of the instantaneous power in the interval when the latter takes positive values:

$$p(\zeta_1,\zeta_2) = \int_0^1 \eta'_s(\alpha,\zeta_1,\zeta_2)\, d \quad , \quad (9)$$

where the following designation is introduced:

$$\eta'_s(\alpha,\zeta_1,\zeta_2) = \begin{cases} \eta_s(\alpha,\zeta_1,\zeta_2) & \text{if } \eta_s(\alpha,\zeta_1,\zeta_2) \ge 0 \\ 0 & \text{if } \eta_s(\alpha,\zeta_1,\zeta_2) < 0 \end{cases} \quad (10)$$

3 Study of a mathematical model of a robot with rectangular motion laws of degrees of mobility

It is obvious that duri synchronous operation of the drives according to the same types of motion laws selected above, the maximum instantaneous power will occur at the end of acceleration. The graph of the power changes is shown in Figure 3 (solid line). In this case, the minimum possible power will occur at each degree of mobility, but the total maximum power will be determined by the sum of the two minimum possible powers. The dotted line shows the law of variation of the robot power for the non-minimum possible powers of the degrees of mobility, the peaks of which are separated in time. Thus, in the considered example, the choice of rational laws of motion for each degree of mobility leads to an increase in the maximum power consumption by 41% as compared to the optimal value.

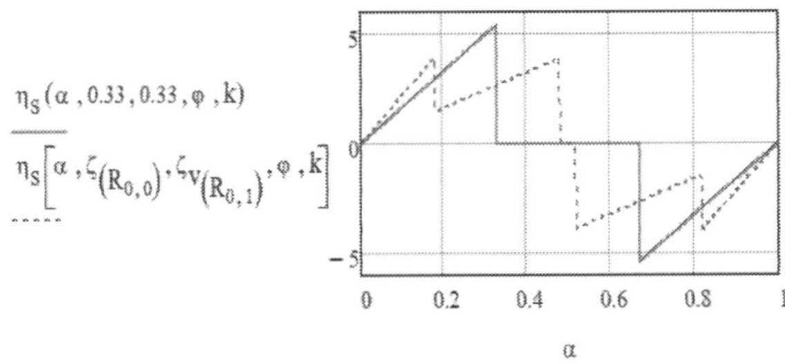

Fig. 3. Time dependences of the instantaneous power of the robot

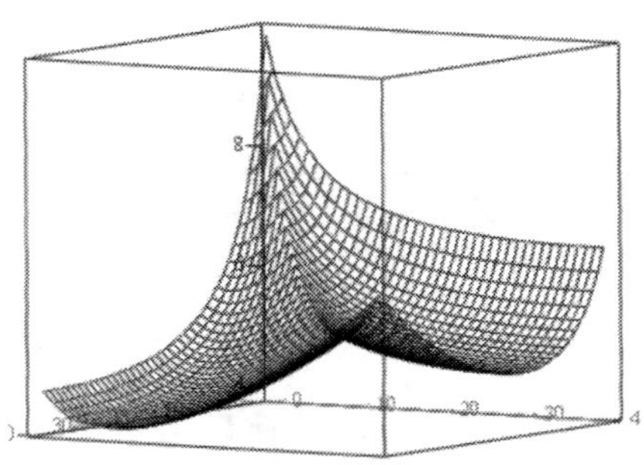

Fig. 4. Dependences of the maximum instantaneous power on the acceleration time of the degrees of mobility

To evaluate the efficiency of optimizing the acceleration time of translatory and relative degrees of mobility according to the criteria of minimum power the following formulas are obtained:

$$\delta_\eta(\varsigma_2) = \frac{\eta_m(\varsigma_1, \varsigma_2) - \eta_{min}}{\eta_{min}} \cdot 100\% \qquad (11)$$

$$\delta_p(\varsigma_2) = \frac{p(\varsigma_1, \varsigma_2) - p_{min}}{p_{min}} 100\% \qquad (12)$$

where $\eta_m(\zeta_1, \zeta_2)$ is the maximum value of the instantaneous power at the optimum value ζ_{10} and arbitrary value ζ_2; η_{min} is the minimum possible value of the maximum power per cycle; $p(\zeta_1, \zeta_2)$ is the energy consumption at optimum power value ζ_{10} and arbitrary value ζ_2; p_{min} is the minimum possible value of energy consumption.

Figure 5 and Figure 6 show the dependences of the efficiency of optimizing the acceleration time of the relative mobility degree at the optimal acceleration time of the translatory mobility degree according to the criteria of the minima of instantaneous power and energy consumption, respectively. The optimization efficiency shows by how many percent the instantaneous power and energy consumption will increase when the acceleration time of the relative degree of mobility changes compared to the optimal value. Similar graphs can be plotted for the translatory degree of mobility. It follows from the graphs that for the example under consideration, with an increase in acceleration time, power increases by 57% and power consumption increases by 40%.

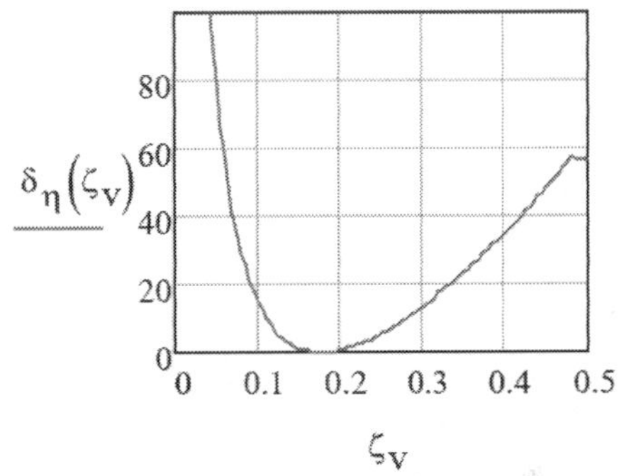

Fig. 5. Dependence of optimization efficiency ζ_{20} on the criterion of minimizing instantaneous power consumption

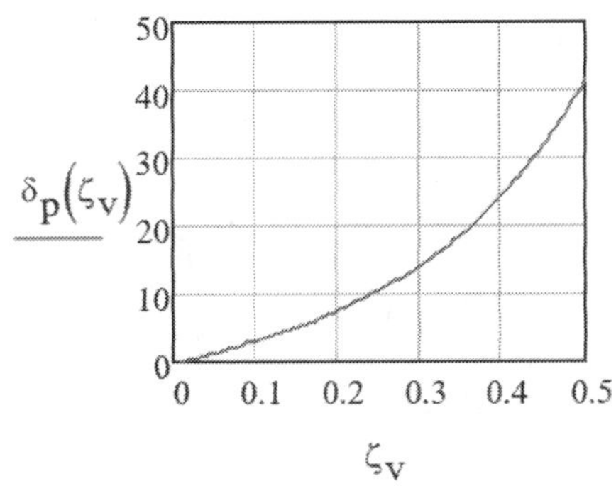

Fig. 6. Dependence of optimization efficiency ζ_{20} on the criterion of minimizing energy consumption

Figure 7 shows the parametric dependence of the optimization efficiency by the criterion of minimizing power on the optimization efficiency by the criterion of minimizing energy consumption. The following conclusions can be drawn from the graph. Firstly, if the acceleration time ζ_v lies in the interval [0.17; 0.5] a decrease in power leads to the decrease in energy consumption and, conversely, an increase in power leads to the increase in energy consumption. Secondly, in the interval [0; 0.17] a decrease in power is accompanied by an increase in energy consumption, and vice versa. Consequently, depending on the formulation of the problem it is possible to set weight coefficients for each criterion and determine the optimal acceleration time.

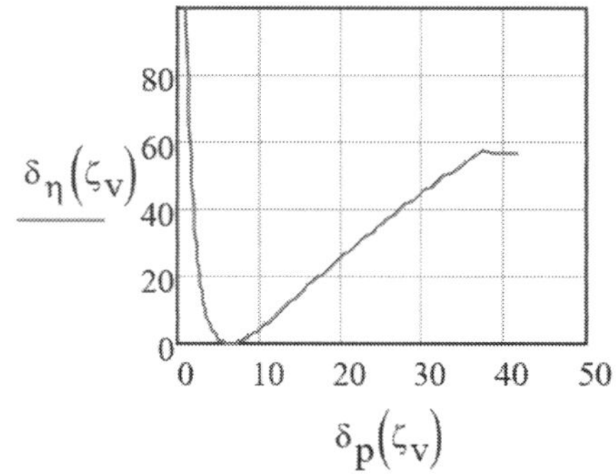

Fig. 7. Dependence of $\delta_n(\zeta_2)$ on $\delta_p(\zeta_2)$

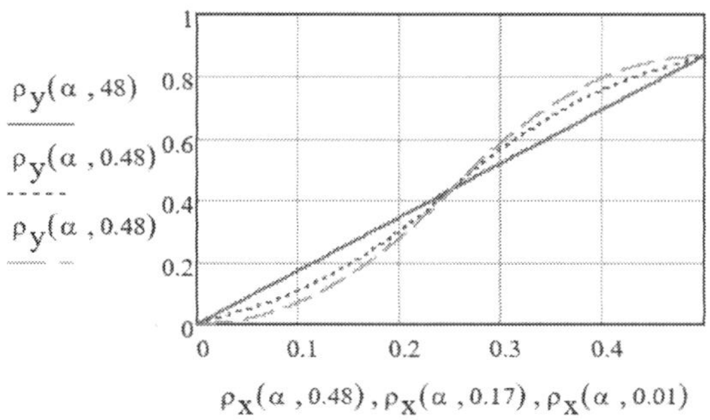

Fig. 8. Trajectories of the robot point A for different algorithms of movement of mobility degrees

Parametric equations of the trajectory of the robot output link in the plane have the following forms:

$$\rho_x(\alpha, \zeta_1) = \cos(\varphi) \cdot \rho(\alpha, \zeta_1) \quad (13)$$

$$\rho_y(\alpha, \zeta_2) = \sin(\varphi) \cdot \rho(\alpha, \zeta_2) \quad (14)$$

With the same laws of motion for each degree of mobility the trajectory of movement is a segment (Figure 8). Changing the dimensionless acceleration times for each degree of mobility leads to a curvature of the trajectory.

4 Mathematical model of a robot having mobility degrees moving according to the optimally synthesized law of motion

An optimally synthesized law is understood as a law of motion that ensures the constancy of power consumption in the acceleration stage under inertial load and, therefore, satisfies the equation:

$$m_0 \cdot u' \cdot u'' = N \quad (15)$$

where u' and u'' are the velocity and acceleration of the moving mass m_0.

5 Investigation of the mathematical model of a robot having mobility degrees moving according to the optimally synthesized law of motion

The solution of equation (15) has the following form:

$$u = \frac{2}{3} \cdot \sqrt{\frac{2N \cdot t^3}{m_0}} \quad (16)$$

Differentiating equation (16) by time it is possible to obtain the following expressions for velocity and acceleration:

$$u' = \sqrt{\frac{2N}{m_0} t} \quad (17)$$

$$u'' = \sqrt{\frac{N}{2m_0 \cdot t}} \quad (18)$$

Substituting the displacement value $s/2$, the acceleration time τ and half the motion cycle time $T/2$ into the solution of equation (16) we determine the power N required for such a displacement:

$$N = \frac{m_0 \cdot s^2}{8 \cdot \left[\frac{2}{3} \cdot \sqrt{\tau^3} + \sqrt{\tau} \cdot \left(\frac{T}{2} - \tau\right)\right]^2} \quad (19)$$

The time-constant powers that are necessary for the translatory and relative degrees of mobility are determined by the equations:

$$N_1 = \frac{m \cdot s^2 \cdot \cos(\varphi)^2}{8 \cdot \left[\frac{2}{3} \cdot \sqrt{\tau^3} + \sqrt{\tau} \cdot \left(\frac{T}{2} - \tau\right)\right]^2} \quad (20)$$

$$N_2 = \frac{m \cdot \mu \cdot s^2 \cdot \sin(\varphi)^2}{8 \cdot \left[\frac{2}{3} \cdot \sqrt{\tau^3} + \sqrt{\tau} \cdot \left(\frac{T}{2} - \tau\right)\right]^2} \quad (21)$$

The total, constant in magnitude, power consumed by the robot when moving on distance s during the time T is:

$$N_S = \frac{m \cdot s^2}{8 \cdot \left[\frac{2}{3} \cdot \sqrt{\tau^3} + \sqrt{\tau} \cdot \left(\frac{T}{2} - \tau\right)\right]^2} \cdot \left(\cos(\varphi)^2 + \mu \cdot \sin(\varphi)^2\right) \qquad (22)$$

In dimensionless form the total power equals:

$$\eta_{s0}(\zeta) := \frac{1}{8 \cdot \left[\frac{2}{3} \sqrt{\zeta^3} + \sqrt{\zeta} \cdot \left(\frac{1}{2} - \zeta\right)\right]^2} \cdot \left(\cos(\varphi)^2 + \mu \cdot \sin(\varphi)^2\right) \qquad (23)$$

Figures 9 and 10 show the dependences of power and energy consumption of the acceleration time. For the above example $\mu = \pi/4$ and $k = 0.6$ with the acceleration time for each degree of mobility equal to 0.5; the power and energy consumption are 1.8 and 0.9 in dimensionless units, respectively. In this case the minimum power consumption and maximum energy consumption are ensured. This corresponds to a reduction in power by 54% and in energy consumption by 18%.

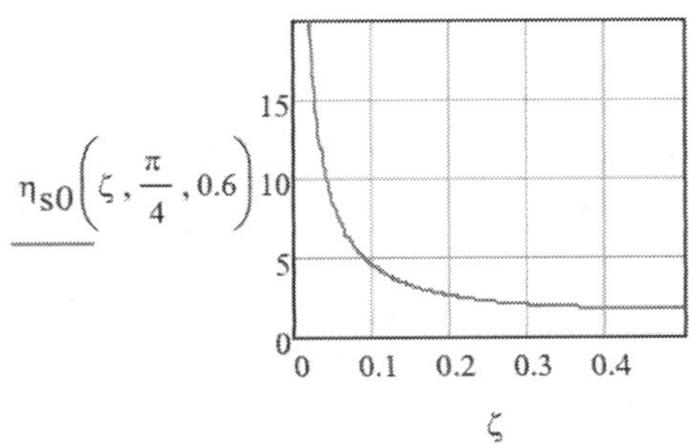

Fig. 9. Dependence of power on acceleration time for the synthesized law of motion

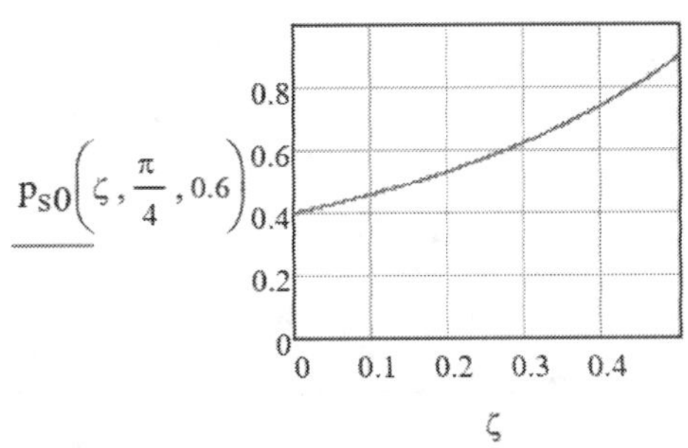

Fig. 10. Dependence of energy consumption on acceleration time for the synthesized law of motion

5 Conclusion

As a result of the analysis of the developed mathematical models of the two-axis gantry robot, it was found that the algorithm and the law of motion of its drives are able to change the power consumption and the energy consumption by tens of percent at the same travel distances and motion cycle duration. Under typical laws of motion, there are intervals of acceleration time, in which a decrease in power leads to a decrease in energy consumption, and intervals in which a decrease in power is accompanied by an increase in energy consumption.

Intentionally synthesized laws of motion make it possible to significantly reduce the power and energy consumption, however, a decrease in power is always accompanied by an increase in energy consumption and, conversely, an increase in power leads to a decrease in energy consumption throughout the whole interval of the acceleration time. In this case, the range of power variations is infinite, and the maximum possible change in energy consumption does not exceed 100% ... 150% and depends on the moving masses and degrees of mobility.

The results obtained can be used in the development of control algorithms for process equipment and gantry robots for various purposes. In addition, based on the results obtained, it can be argued that they are suitable for all types of robots with a large number of degrees of mobility. Ultimately, reducing energy consumption in the technosphere at a relatively low cost leads to the saving of natural resources and the reduction of environmental pollution. In the future, the results obtained analytically can be used in theoretical and experimental studies of energy consumption of specific types of motors and machines in general.

References

1. A.G. Egorov and O.S. Zakharova, J. *Comp. and Sys. Sci. Int.*, **54**, 495 (2015)
2. D. Qiao, *Int. J. Metrol. Qual. Eng.*, **10**, 2019011 (2019)
3. A. Seilla, et al., *MATEC Web Conf.*, **197**, 11016 (2018)
4. H. Ying, et al., *Hindawi Math. Probl. in Eng.*, Article ID 7247093 (2018)
5. A.N. Volkov, et al., Static balancing and power consumption of industrial robots and process machines for various application, *J. Phys.: Conf. Series*, 1753 (2021)
6. A. N. Volkov, et al., Optimization of the robot motion law by the criterion of minimizing maximum instantaneous power and electric motor size, *IOP Conf. Series: Earth and Env. Sci.*, **539**, 0121228 (2020)

Thermodynamic modeling of the reaction of simultaneous interaction of nickel (II), cobalt (II) and iron (III) oxides with carbon

Olga Kolesnikova[1,*], *Farida* Nurbaeva[2], *Guizhan* Sabyrbayeva[3] and *Lazzat* Nurshakhanova[4]

[1] Master of Engineering and Technology, Junior Researcher, Department of Science, Production and Innovation, M. Auezov South Kazakhstan University, 160012 Shymkent, Kazakhstan
[2] Candidate of Technical Sciences, Associate Professor, Department of Ecology and Geology, Sh. Yessenov Caspian University of Technology and Engineering, 130002, Aktau, Kazakhstan
[3] Candidate of Technical Sciences, Associate Professor, Department of Ecology and Geology, Sh. Yessenov Caspian University of Technology and Engineering, 130002, Aktau, Kazakhstan
[4] Candidate of Technical Sciences, Associate Professor, Department of Petrochemical Engineering, Sh. Yessenov Caspian University of Technology and Engineering, 130002, Aktau, Kazakhstan

Abstract. This paper considers thermodynamic modeling of electric smelting of Allarechenskoye deposit dumps for nickel, cobalt, and iron extraction. Thermochemical transformations in the reaction of interaction of carbon with oxides of nickel (II), cobalt (II) and iron (III) are considered. The investigations of thermochemical transformations of the chemical-phase composition of the system were carried out using the software complex "Astra-4" developed by a team of scientists at the Bauman Moscow State Technical University. Thermochemical transformations and chemical-phase composition of the system as a function of T (temperature) and P (pressure) were studied. In particular, the temperature was in the range from 973K to 1773K and the pressure range was taken from 0.001 MPa to 0.1 MPa. According to the results, it was found that iron in the system transitions to condensed iron, with the degree of transition (αFe) from 29.0% at T = 1173K to 99.1% at 1773K, the degree of transitions of nickel to condensed nickel is 100% before T = 1273K, and with an increase in temperature from 1273K to 1773K decreases to 99,98%; the transition of cobalt to the gas phase is distributed similarly to nickel.

1 Introduction

Nickel production involves complex processing of oxide-silicate nickel/cobalt ore as well as sulfide copper-nickel ores. Key impurities, such as gold, copper, and platinum group metals, and the demand for them influence the result of nickel processing [1-6].

Nickel is used (as an additive) in the production of corrosion-resistant and special heat-resistant alloys and in the production of stainless steels. These metallurgical products are used in mechanical engineering, instrument engineering, household appliances and electronics.

The U.S. Geological Survey claimed in early 2014 that the world's nickel reserves were about 130 million tons. However, these figures out to be overstated. Over the last decade, proven reserves are estimated at 74 million tons of ore. Australia, Brazil, New Caledonia and Russia are known to have huge explored and proven nickel reserves. Cuba, Indonesia, Canada, China and South Africa also add to that list [1–12].

Oxide-silicate nickel/cobalt deposits form the basis of the global nickel industry. The rest of the deposits, the processing of which extracts nickel, account for 0.1-0.2% of the total metal reserves [1-12].

More than half of the world's proven nickel reserves are in the form of oxide deposits. It is known that only 40% of the metal was extracted from lateritic ores at the beginning of 2014. This is explained by the significant content of nickel and copper in the sulfide ores. Additionally, impurities of rare and noble metals are of special value in ore processing, as this is an additional and substantial profit compared to the target product (nickel).

The sulfide ore deposits are mainly located in the Russian Federation, South Africa, Canada, and China. 50% of Australia's nickel ore deposits produce as well. However, the mineral resource base of Cuba, Indonesia and New Caledonia, and Kazakhstan is mainly represented by oxide and silicate nickel ore deposits [1-5, 7-12].

The world's leading Ni producers use ores that contain at least 1.4% of the metal. However, the use of various methods and technologies related to the processing and enrichment of poor nickel ores is growing in intensity and popularity [1- 23].

Today, the Arctic zone of Russia appears to be one of the priority areas of development in economic and environmental terms. Over the years of industrial

* Corresponding author: ogkolesnikova@yandex.kz

development of the Arctic zone of Russia, a significant number of industrial enterprises of petrochemical, gas, mining and processing, mining and metallurgical and a number of other industries have been put into operation [13 - 43], which during their operation have been able to accumulate hundreds of millions of tons of various industrial waste, which today contribute to the negative impact on the environment of the Arctic region [55].

At the present stage, when there is a decrease in the content of metals in mined ores, an urgent problem is the involvement of man-made formations in the form of waste, dumps, etc. as secondary mineral raw materials, depending on their chemical composition and content.

One of these dumps is the dumps of the Allarechenskoe deposit (Pechenga District, Murmansk Region), which contains over 12 million tons of waste copper-nickel ore. The deposit was mined using the open-cut method, but it was completed in 1971. The main minerals mined at the deposit in the Murmansk region were nickel, copper, and cobalt. The dumps consist of overburden and mostly ore-free rocks, amphibolites, and granite-gneiss. The granulometric composition is extremely heterogeneous. Moreover, under the influence of precipitation and other natural phenomena, the primary ores in the dump for a long stay were subjected to hypergenic processes, which led to the formation of oxidized ores. As a result, a considerable part of the ore has lost its original quality [13].

Studies show that the most hazardous are sludge and tailings, as well as metallurgical wastes due to their fine-grained structure. However, a number of Russian researchers argue that coarse-grained sulfide-containing wastes from mining enterprises can be no less dangerous. Thus, based on the chemical composition of the Allarechenskoe deposit dumps (the presence of nickel and cobalt oxides) and the theoretical possibility of their use as secondary raw materials for the extraction of useful metals, studies have been conducted. They included the method of thermodynamic modeling of chemical and phase transformations under conditions of simultaneous interaction of nickel (II), cobalt (II) and iron (III) oxides with carbon. The studies were conducted with the help of the complex program "Astra-4"[44]. The purpose of this work is to evaluate the effect of temperature and pressure in the thermodynamic simulation of reductive electric melting [29, 42, 45-50] under conditions of simultaneous interaction of nickel Ni(II), cobalt Co(II) and iron oxides Fe2O3(III) with carbon and subsequent extraction of Ni, Co and Fe.

2 Materials and Methods

The HSC Chemistry software package designed by Outokumpu (Finland) was used to calculate the Gibbs energy. Its essence is the possibility of reduction of metals in the reaction by simultaneous interaction of Ni (II), Co (II), and Fe (III) oxides with C [52]. The software allows the calculation of thermodynamic functions such as entropy (ΔS), enthalpy molar heat capacity (ΔH), Gibbs energy (ΔG) for each substance involved in the reaction, as well as changes of these thermodynamic functions during the chemical reaction [52-64].

According to the chemical transformation: NiO+CoO+Fe2O3+5C=2Fe+Ni+Co+5CO, the change in Gibbs energy (ΔG) of the reaction was calculated. The results of the mathematical calculation are shown in Table 1. The obtained Gibbs energy values showed the possibility of the reaction and reverse reduction of metals. Further work on the simulation of the chemical process was continued in the program "Astra-4". Thermodynamic modeling of metal reduction under the conditions of the above main chemical reaction was performed at a pressure of 0.001-0.1 MPa and temperature 973-1773K.

The "Astra-4" algorithm is based on the universal thermodynamic method for determining the equilibrium characteristics of heterogeneous systems. application of this method is based on the fundamental principle of maximum entropy. The "Astra-4" program allows to formulate quite simply the modeling problem statement for the thermodynamic method of research [44, 65 -74].

3 Results and Discussion

At a temperature of 873K and a Gibbs energy ΔG of -22314 kJ/mol the beginning of the reaction was recorded. A further increase in temperature up to 1773K gives more negative Gibbs energy values, which reach -775345 kJ/mol. These calculations are shown in Table 1.

Table 1. Results of the calculation of the Gibbs energy (ΔG) of the reaction NiO+CoO+Fe2O3+5C=Ni+Co+2Fe+5CO.

T, K	Enthalpy (ΔH), kJ	Entropy (ΔS), kJ/K	Gibbs energy (ΔG), kJ/mol	Constant equilibrium (K)	Logarithm of a constant (LogK)
573	738.685	875.730	236.892	2.53E-022	-21.597
673	733.720	867.750	149.725	2.38E-012	-11.622
773	723.210	860.14S	63.315	5.26E-005	-4.279
873	721.967	852.556	-22.314	2.16E+001	1.335
973	715.842	845.897	-107.216	5.70E+005	5.756
1073	713.827	843.923	-191.702	2.15E+009	9.333
1173	710.464	840.932	-275.949	1.94E+012	12.289
1273	707.504	838.566	-359.990	5.92E+014	14.773

Subsequent thermodynamic modeling in the software package "Astra-4" shows that the degree of iron (αFe) conversion in the interaction of NiO, CoO and Fe2O3 with carbon varies with pressure (at P = 0.001 and 0.01 MPa). However, it is independent of temperature and ranges from 99.9% to 100% condensed magnetite k • Fe3O4, where k - condensed phase (Tables 2, 3).

 https://doi.org/10.1051/e3sconf/202337804001

Table 2. Transition degree results for Ni, Co, Fe, C, O2 at 973-1773 K and 0.001MPa pressure

Elem.	Temperature, K				
	1773	**1573**	**1373**	**1173**	**973**
Degree of nickel distribution (αNi), %					
k•Ni	8.780929	8.834213	8.88642	8.9247534	8.948606
k•NiO	91.2142	91.16565	91.1135	91.075246	91.05139
NiO	9.67E-05	1.09E-06	0	0	0
Ni	0.004775	0.000135	1.28E-06	0	0
Total	*100*	*100*	*100*	*100*	*100*
Degree of cobalt distribution (αCo), %					
k•CoO	99.41675	99.99761	100	100	100
Co	0.082338	0.002395	0	0	0
CoO	0.500909	0	0	0	0
Total	*100*	*100*	*100*	*100*	*100*
Degree of iron distribution (αFe), %					
Fe	0.003696	8.08E-05	0	0	0
FeO	0.000581	0	0	0	0
k•Fe_3O_4	99.99572	99.99992	100	100	100
Total	*100*	*100*	*100*	*100*	*100*
Degree of carbon distribution (αC), %					
CO	4.336	2.990	1.767	0.867	0.317
CO_2	95.664	97.010	98.233	99.133	99.633
Total	*100*	*100*	*100*	*100*	*100*
Degree of oxygen distribution (αO_2), %					
O	0.000144	2.11E-06	0	0	0
O_2	0.000182	0.000182	0	0	0
CO	0.194118	0.13379	0.079038	0.0387774	0.014161
CO_2	8.566385	8.682608	8.787827	8.8647597	8.912373
CoO	0.004508	0	0	0	0
FeO	1.04E-05	0	0	0	0
k•Fe_3O_4	2.387805	2.386699	2.385583	2.3846229	2.384195
k•CoO	0.894704	0.894992	0.894564	0.8942036	0.894043
k•NiO	87.95205	87.90173	87.85299	87.817636	87.79523
Total	*100*	*100*	*100*	*100*	*100*

Table 3. Transition degree results for Ni, Co, Fe, C, O2 at 973-1773 K and a pressure of 0.01 MPa

Elem.	Temperature. K				
	1773	**1573**	**1373**	**1173**	**973**
Degree of nickel distribution (αNi), %					
k•Ni	*8.776904*	*8.834228*	*8.88642*	*8.9247534*	*8.948606*
k•NiO	*91.22261*	*91.16576*	*91.11358*	*91.07524*	*91.05139*
NiO	*9.65E-06*	*0*	*0*	*0*	*0*
Ni	*0.000476*	*1.35E-05*	*0*	*0*	*0*
Total	*100*	*100*	*100*	*100*	*100*
Degree of cobalt distribution (αCo), %					
k•CoO	*99.99124*	*99.99976*	*100*	*100*	*100*
Co	*0.008257*	*0.000239*	*0*	*0*	*0*
CoO	*0.000502*	*0*	*0*	*0*	*0*
Total	*100*	*100*	*100*	*100*	*100*
Degree of iron distribution (αFe), %					
Fe	*0.000369*	*0*	*0*	*0*	*0*
FeO	*5.8E-05*	*0*	*0*	*0*	*0*
k•Fe_3O_4	*99.99957*	*100*	*100*	*100*	*100*
Total	*100*	*100*	*100*	*100*	*100*
Degree of carbon distribution (αC), %					
CO	*4.335591*	*2.989643*	*1.767016*	*0. 8672793*	*0.316769*
CO_2	*95.66441*	*97.01035*	*98.23293*	*99.13272*	*99.63323*
Total	*100*	*100*	*100*	*100*	*100*
Degree of oxygen distribution (αO_2), %					
O	*1.44E-05*	*0*	*0*	*0*	*0*
O_2	*0.000997*	*1.82E-05*	*0*	*0*	*0*
CO	*0.194124*	*0.13379*	*0.079038*	*0.0337774*	*0.014161*
CO_2	*8.566657*	*8.682621*	*8.78785 1*	*8.8647597*	*8.912373*
CoO	*4.5E-06*	*0*	*0*	*0*	*0*
FeO	*1.04E-06*	*0*	*0*	*0*	*0*
k•Fe_3O_4	*2.38796*	*2.386729*	*2.385582*	*2.3846229*	*2.384195*
k•CoO	*0. 895399*	*0.894993*	*0.894563*	*0. 8942036*	*0.894043*
k•NiO	*87.95434*	*87.90185*	*87.85297*	*37.817636*	*87.79523*
Total	*100*	*100*	*100*	*100*	*100*

When the pressure rises to P=0.1 MPa, iron goes into the k • Fe phase. The chemical compound k • Fe3C at temperature 973K contains 100% Fe. When the temperature is further increased to 1773K the iron content decreases and reaches 0.94% (Table 4).

Table 4. Transition degree results for Ni, Co, Fe, C, O2 at 973-1773 K and a pressure of 0.1 MPa.

Elem.	Temperature, K				
	1773	**1573**	**1373**	**1173**	**973**
Degree of nickel distribution (αNi), %					
k•Ni	99.99488	99.99985	100	100	100
Ni	0.005123	0.000146	1.36E-06	0	0
Total	100	100	100	100	100
Degree of cobalt distribution (αCo), %					
k•Co	99.99555	99.99988	100	100	100
Co	0.004454	0.000123	1.25E-06	0	0
Total	100	100	100	100	100
Degree of iron distribution (αFe), %					
Fe	0.005118	0.00017	2E-06	0	0
k•Fe_3C	0.942797	2.21578	9.50843	71.02594	100
k•Fe	99.05208	97.78405	90.49152	28.97406	0
Total	100	100	100	100	100
Degree of carbon distribution (αC), %					
CO	99.74867	99.40921	97.4645	81.05911	44.70088
CO;	0.12562	0.295346	1.267711	9.470441	27.64956
k•Fe_3C	0.125709	0.295442	1.267785	9.470441	13.33314
k•C	0	0	0	0	14.31643
Total	100	100	100	100	100
Degree of oxygen distribution (αO2), %					
CO	99.74376	99.40931	97.46453	31.05911	44.70033
CO_2	02251241	0.590695	2.535424	13.94039	55.29912
Total	100	100	100	100	100

The degree of carbon transformation (αC) in the temperature interval from 973K to 1373K and pressure equal to 0.1 MPa is presented in the system by the condensed phase with the content of condensed carbon (k • C). a change from 14.31% to 0% at temperatures of 973K and 1173K is observed, respectively, and k • Fe3C with a maximum transition rate of up to 13.33% at 973K. The carbon (αC) gas phase is represented as CO compounds ranging from 44.7% to 99.74% at 973K and 1773K, respectively, and CO2 with a carbon transition rate up to 27.64% at 973K with a decrease to 0.125% at 1773K. (Table 4).

Reducing the pressure to 0.001-0.01 MPa allows us to see how the transition process of carbon (αC) occurs. It should be noted that carbon is represented by mono- and dioxide, which are distributed in the following ratio: CO2 from 95.66 to 99.683% and CO from 4.336 to 0.317% at P = 0.001 MPa. Increasing the pressure to P = 0.01 MPa gives a CO2 distribution in the range of 95.664-99.68% and CO: 4.335 -0.316%.

The distribution of the degree of cobalt transition (αCo) under the conditions of interaction of nickel (II), cobalt (II) and iron (III) oxides with carbon is shown in Figures 2-4.

According to figure 2.3, αCo presented by k • CoO in the pressure range of 0.001-0.01 MPa is 99.9%. Increasing the pressure to P=0.1 MPa shows the degree of conversion of cobalt to k • Co and is 100% in the temperature range 973-1373K. A further increase in temperature to 1773K shows that αCo decreases to

99.99% as the transition to the gas phase begins, similar to the behavior of Ni (Table 4).

Oxygen in the system at a pressure equal to 0.1 MPa goes to CO^2 up to 55.29% at T=973K, and further decreases to 0.25% at T=1773 K. In addition, oxygen is distributed to CO, reaching 44.7 - 99.74% at T=973-773 K (Table 4).

Thermodynamic studies performed in the NiO-CoO-Fe_2O_3-C system allowed us to make chemical equations at temperatures of 1173K:

$$NiO+CoO+Fe_2O_3+5C=k\cdot Ni+k\cdot Co+0.58k\cdot Fe+0.47k\cdot Fe_3C+0.47CO_2+4.05CO \quad (1)$$

$$NiO+CoO+Fe_2O_3+5C=k\cdot Ni+k\cdot Co+1.9k\cdot Fe+0.006k\cdot Fe_3C+0.006CO_2+4.9CO \quad (2)$$

4 Conclusions

Thus, studies conducted by thermodynamic modeling of the reaction of simultaneous interaction of nickel (II), cobalt (II), and iron (III) oxides with carbon in the temperature range 973-1773 K and pressure 0.001-0.1 MPa give the following results and conclusions:

- technogenic dumps of overburden and substandard oxidized ores of the Allarechenskoye deposit contain such valuable components as nickel and cobalt at the level of mined ores, and can be used as secondary raw materials for non-ferrous metal extraction and reduction of the anthropogenic load on the environment of the Arctic region.
- joint reduction of iron, nickel and cobalt in the temperature range 1373-1773K is possible. At these temperatures and pressures equal to 0.1 MPa the following chemical transitions are demonstrated: Fe from 90.49% to 99.05%, Ni from 99.994% to 100%, Co from 99.995% to 100.
- the thermodynamic calculations performed for the joint reduction of metals allow obtaining a ferroalloy in the form of ferronickel.
- with the organization of production for the processing of man-made waste from the Allarechensk dump with the production of ferronickel using the above technology, the socio-environmental and economic climate in the Arctic region will be improved, which will further contribute to the development of the Arctic.

References

1. I.D. Reznik, G.P. Ermakov, Ya.M. *Shneyerson Nikel* (Moscow, OOO "Nauka i tekhnologiya", 2001)
2. V.F. Borbat, I.Yu Leshch, *New processes in nickel and cobalt metallurgy Novyye protsessy v metallurgii nikelya i kobalta* (Moscow, Metallurgiya, 1976)
3. S.K. Dutta, D.R. Lodhari, *Nickel, In: Extraction of Nuclear and Non-ferrous Metals. Topics in Mining. Metallurgy and Materials Engineering* (Singapore, Springer, 2018)
4. K.A. Alibhai, A.W. Dudeney, D.J. Leak, FEMS – Microbiol, *Rev.*, **11**, (1993)
5. L.B. Sukla, V.V, Panchanadikar. R.N. Kar. World, *J. of Microbiol. and Biotech.*, **2** (1993)
6. A.B. Zhivayeva, T.V. Bashlykova, et al., *Tsvet. met.*, **3** (2007)
7. M. Seggiania, S. Vitoloa, S. D'Antoneb, *Hydrometallurgy*, **81.1** (2006)
8. S. Mohapatra, Ch. Sengupta, Nayak B. D., et al., *Korean J. of Chem. Eng.*, **25** (2008)
9. Sh. Zhou, et al., Wang. *Metallurg. and Mat. Trans. B.*, **47** (2016)
10. K.D. Kim, W.W. Huh, D.J. Min, *Metallurg. and Mat. Trans. B.*, **45** (2014)
11. Q. Zhao, J. Xue, W. Chen, *Trans. of the Indian Inst. of Met.*, **72** (2019)
12. O. Kadnikova, M. Toretayev, G. Altynbayeva, et. al., *J. Chem. Technol. Metall.*, **55(3)** (2020)
13. V.A. Masloboev, et al., *J. of Min. Sci.*, **50** (2014)
14. O. Kolesnikova, et al., *Materials*, **15(6980)** (2022)
15. B. Ye Zhakipbaev, et al., *Rasayan J. of Chem*, **14(2)** (2021)
16. T.M. Khudyakova et al., *Refract. Ind. Ceram.*, **60(1)** (2019)
17. V. P. Bondarenko, K. S. Nadirov, V. G. Golubev, *Neftyanoe Khozyaystvo - Oil Industry*, **1** (2017).
18. B.K. Kenzhaliyev et al., *Int. J. of Mechan. and Product. Eng. Res. and Develop*, **9(6)** (2019)
19. O.S. Baigenzhenov, et al., *Min. Proces. and Extract. Metal. Rev.*, **4** (2015)
20. E.V. Siziakova, P.V. Ivanov, A.V. Boikov, Application of calcium hydrocarboaluminate for the production of coarse-graded alumina, *Journal of Chemical Technology and Metallurgy*, **issue 1. vol. 54**, pp. 200-203 (2019)
21. C. Yang, et al, *Hydrometallurgy*, **106** (2011)
22. N. Koteleva, et al., Digitalization in open-pit mining: A new approach in monitoring and control of rock fragmentation, *Applied Sciences (Switzerland)*, **21100829268(11)** doi: 10.3390/app112210848 (2021)
23. I.E. Volokitina, et al., *Materials*, **15(7)** (2022)
24. A. Nedosekin, et al., Strategic Approach to Assessing Economic Sustainability Objects of Mineral Resources Sector of Russia, *Journal of Mining Institute*, **237**, pp. 354–360. doi:10.31897/pmi.2019.3.354. (2019)
25. N.V. Vasilyeva, et al., *Materials*, **15(3975)** (2022)
26. I.I. Beloglazov, et al., The Concept of Digital Twins for Tech Operator Training Simulator Design for Mining and Processing Industry, *Eurasian Mining*, pp. 50–54, doi:10.17580/em.2020.02.12. (2020)
27. A.A. Kul'chitskii, D.A. Kashin, The choice of a method for non-contact assessment of the composition of briquetted charge materials, *Journal of Physics: Conference Series*, **17426588(1399)**, doi: 10.1088/1742-6596/1399/4/044108 (2019)
28. M.K. Zhantasov, et al., *Chem. tod.*, **34(1)** (2016)

29. I.V. Sergeeva, et al., *St. in Transl.*, **47(3)** (2017)
30. E.K. Rodnichenko, et al., Augmented Reality Techniques in Industrial Warehouse Logistics in Mining Industry, *IOP Conference Series: Earth and Environmental Science*, **688**, pp. 1 - 7 (2021)
31. E. R. Fedorova, Mathematical model of red sludge sedimentation in single-level circular thickener, *Journal of Physics: Conference Series*, **issue 3, vol. 1333**, doi: 10.1088/1742-6596/1333/3/032018 (2019)
32. A.A. Kulchitskii, et al., Evaluating the Error of a System for Monitoring the Geometry of Anode Posts in Electrolytic Cells with Self-Baking Anode / Russian, *Journal of Nondestructive Testing*, **56(3)**, pp. 268-274 (2020)
33. N. Vasilyeva, E. Fedorova. et al., Big Data as a Tool for Building a Predictive Model of Mill Roll Wear, *Symmetry*, **13(5)** (2021)
34. E. Siziakova, et al., *Materials,* **15(4930)** (2022)
35. D.D. Guerra, et al., Alternative measures to reduce carbon dioxide emissions in the republic of Cuba, *Journal of Ecological Engineering*, **21(4)**, pp. 55–60 (2020)
36. N.V. Vasilyeva, P.V. Ivanov, Development of a control subsystem to stabilize burden materials charging into a furnace, *Journal of Physics: Conference Series*, **1210(1)**, 012158 (2019)
37. R. Klyuev, et al., Energy Indicators of Drilling Machines and Excavators in Mountain Territories. In: Murgul V.. Pukhkal V. (eds), *International Scientific Conference Energy Management of Municipal Facilities and Sustainable Energy Technologies EMMFT 2019. Advances in Intelligent Systems and Computing, Springer,* 1258 (2021)
38. Yuri L. Zhukovsky, et al., (2020) Potential for electric consumption management in the conditions of an isolated energy system in a remote population, *Sustainable Development of Mountain Territories*, **issue 4. vol. 12**, pp. 583-591.
39. B.T. Taimasov, B.K. Sarsenbayev, *Euras. Chem.-Technolog. J*, 19(4) (2017)
40. N. Zhangabay, et al., *Materials*, **15(4996)** (2022)
41. A.V. Nitsenko, et al., *Inorganic Materials*, **54(7)** (2018)
42. G.S. Kenzhibaeva, et al., *Refract. Ind. Ceram.,* **61(3)** (2020)
43. I. Romanova, *IOP Conference Series: Materials Science and Engineering*, **032016** (2018)
44. G.V. Sinyarev, et al., *Primeneniye EVM dlya termodinamicheskikh raschetov metallurgicheskikh protsessov. [The use of computers for thermodynamic calculations of metallurgical processes]* (Moskow, Nauka, 1962).
45. B.E. Serikbaev, et al., *Refr. and Ind. Ceram.*, **62(4)** (2021)
46. B.O. Sapargaliyeva, *Ras. J. of Chem.*, **15(3)** (2022)
47. I.K. Andrianov, A.V. Stankevich, *International Science and Technology Conference "East Conf 2019"*, 8725322I (2019)
48. I.V.Sergeeva., et al., *Izv. Ferr. Metall.*, **60(9)** (2017)
49. Aigul Mamyrbekova, *Rus. J. of Phys. Chem. A.*, **92(3)** (2018)
50. A.E. Filin, et al., *Ugol*, **9** (2022)
51. A. Roine, *Outokumpu HSC Chemistry for Windows. Chemical Reaction and Equilibrium loftware with Extensive Thermochemical Database* (Pori, Outokumpu Research OY, 2002)
52. I.K. Andrianov, S.V. Belykh, *International Science and Technology Conference "East Conf 2019"*, 8725410 (2019)
53. A.B. Agabekova, et al., *Ras. J. of Chem*, **15(3)**, (2022)
54. A. P. Vlasyuk, et al., *Math. Model. and Comput.*, **1** (2020)
55. A. Donayev, et al., *News of the Nation. Acad. of Sci. of the Repub. of Kazakhstan. Series of Geol. and Tech. Sci.*, **4** (2022)
56. V.G. Golubev, et al., *Ras. J. of Chem*, **15(3)** (2022)
57. S. Syrlybekkyzy, et al., *Ecol. and Indus. of Russ.*, **27(1)** (2023)
58. A. Naizabekov, et al., *J. of Chem. Tech. and Metallur.*, **57(4)** (2022)
59. S. Auyesbek, et al., *J. Compos. Sci.*, 7, 124 (2023)
60. Y. Amanbek, et al., *Energies*, **15 (7)**, 2404 (2022)
61. Z.N. Moldamuratov, et al., *Nanotechnol. in Construc.*, **14(3)** (2022)
62. Z. Aubakirova, et al., *Technobius*, **2(3)**, 0024 (2022)
63. V. Zhukovskyy, et al., *CEUR Worksh. Proceed.*, **3309** (2022)
64. I. Volokitina, *J. Chem. Technol. Metall.* **55** (2020)
65. S.V. Balovtsev, et al., *Min. Inform. and Analyt. Bull.*, **2-1** (2021)
66. I.Volokitina. *Met. Sci. Heat Treat.* **61,** 2 (2019)
67. S. Lezhnev, A. Naizabekov, E. Panin, et. al., *Proced. Eng.*, **81** (2014)
68. Ye. Utepov, *Computers and Concrete.* **25**, (2020)
69. L.V. Sokolovskaya, S.A. Kvyatkovskiy, S.M. Kozhakhmetov, A.S. Semenova, R.S. Seisembayev, *Metallurgist*, **65(5-6)** (2021)
70. R. Fediuk, M. Amran, A. Klyuev, et. al., *Materials.* **15(7)**, 2542 (2022)
71. E.A. Sit'ko, B.M. Sukurov, B.A. Omirzakov, et. al., *Metallurgist*, **63(9-10)** (2020)
72. A. Volokitin, A. Naizabekov, et. al., *J. of Chem. Technol. and Metall.*, **57(4)** (2022)
73. A.D. Akbassova, G.A. Sainova, I.O. Aimbetova, D.K. Sunakbaeva, M.M. Akeshova. *Fresen. Environ. Bull.*, **25(9)** (2016)
74. T. I. Ovchinnikova, O. M. Zinovieva, A. M. Merkulova, et. al., *Gornyi Zhurnal*, **3** (2020)

https://doi.org/10.1051/e3sconf/202337804002

Quality analysis of voxel models obtained with remote sensing

Mikhail Vystrchil[1,*], *Arseniy* Sukhov[1], *Anton* Rybakov[2], *Mikhail* Chura[3], *Galina* Artemova[4]

[1]The department of Mine Surveying, Saint Petersburg Mining University, 199106, Saint Petersburg, Russian Federation
[2]Ports department, NovomorNIIproekt Engineering Co, 353905, Novorossiysk, Russian Federation
[3]Mechanics Department, Ushakov Maritime State University, 353918, Novorossiysk, Russian Federation
[4]Transport college, Ushakov Maritime State University, 353918, Novorossiysk, Russian Federation

Abstract. Use of the resource potential of the Arctic is impossible without reliable information about the spatial and temporal characteristics of the developed territories. The topographic and bathymetric information are needed to solve engineering and logistic tasks. Furthermore, surveying, geodetic surveying and geological surveys in these territories are limited by climatic conditions and their infrastructural remoteness. That means, that it's needed to replace classical methods of performing geodetic works and surveys with the presence of workers on the object with modern automated methods of remote sensing of territories. That methods include the technologies of multipath echo-location, aerial or terrestrial laser scanning, photogrammetry and space sensing of the Earth. Despite the differences between these technologies, they are united by the result, the geospatial data, often called a "point cloud". Such point clouds are not regular in their structure and form 3D-model of the object due to the redundant data, which makes processing, systematizing and storing this kind of the information more difficult. The paper considers the method of generalization of geospatial data, that allows to reduce these disadvantages by approximating local sections of the model with planes that fit into a fragment of a point cloud using the least squares method. The model described in the paper allow to statistically assess the quality of the initial data, to simplify next mathematical processing for solving specific engineering problems, including geomechanical, environmental and glaciological monitoring.

1 Introduction

The development of mankind, which has determined the content and form of social relations, requires stable sources of natural resources for its preservation. The limit of the economic markets and occupied territories motivates society to use lands and resources, which accepted as technologically and economically unprofitable for development in the past. The development of the Arctic and the adjacent territories of the far North will preserve the economic and raw material balance, develop logistics and trade routes with strengthening stability in the world [1]. The existing experience of developing the territories of the far North connected with the specific nature of the climate and the fragility of the ecological systems.

Responsible and consistent development of the Arctic is impossible without knowledges about the nature of the relief of the land and the seabed of the Arctic Ocean. Despite the apparent simplicity and naturalness of the formulated task, its solution is not complete. For example, the volume of the Arctic territories of Canada that meet modern requirements is in the region of 6% of the total water area [2]. The land areas northern than 60 altitude are completely covered by the digital relief model ArcticDEM, but a number of published studies prove the limit of applicability in solving local problems. The actual errors of this model are about several meters in height [3], which implies the necessary additional surveying and geodetic surveys.

The solution of the problems of mine surveying, geodetic and hydrographic study of the Arctic is associated with difficulties caused by the poor development of the infrastructure of the region, systems of points of state geodetic networks [4] and the general severity of the climate. This implies the limited economic and production efficiency of using traditional methods of performing topographic and bathymetric works. Modern technologies of remote sensing, including laser scanning [5, 6], photogrammetric [6, 7] and space surveys [8] are increasingly used in solving various problems associated with the development of the Arctic. Monitoring of the state of glaciers [9-11], the thickness of the snow cover [12] and landslides caused by the melting of permafrost [13, 14] are increasingly performed using the above technologies.

2 Problem statement

The result of remote survey methods (in particular, laser scanning and photogrammetric surveys), produce huge irregular arrays of spatial data that form a model of the surveyed object as a "point cloud". Such clouds can

* Corresponding author: vystrchil_mg@pers.spmi.ru

E3S Web of Conferences **378**, 04002 (2023) https://doi.org/10.1051/e3sconf/202337804002
IIRPCMIA 2021

contain hundreds of millions of individual points, while remaining discrete. That means that the solution of practical problems requires the construction of surfaces that interpolate the space between adjacent points.

The solutions in applied practice consist in the construction of three-dimensional TIN surfaces that fill the space between the points with triangular polygons or DEM models - raster images, where the average elevation of the relief determines the color of the pixel for the corresponding area of the terrain. A TIN surface uses an excessive number of polygons, being not robust to errors in the initial data, which does not allow assessing the quality of the resulting model. A DEM model, on the other hand, does not differentiate relief changes within each pixel, allowing for the possibility of serious errors in case of a sharp relief change. The proposed method for processing point clouds is a kind of compromise between the approaches, reducing the disadvantages of each of the methods, while retaining most of their advantages.

3 The idea of work

The idea of work is to use a voxel model to structure data. This approach is often used in tasks requiring generalization of initial 3D data and allows to reduce the volume of points processed at a time. For example, in [15], it is used to segment a scan of a tree and highlight points related to its individual leaves, and in [16] to automate the monitoring of cultural heritage objects. The efficiency of voxel models directly depends on their scale - the size of the voxels used in the construction [17, 18]. The smaller the voxel size, the higher the detail and detail of the model, but the more difficult it is, in turn, to ensure its filling with a sufficient amount of initial data.

Dividing the original point cloud into relatively small areas will allow us to make an assumption of the linear nature of the change in the height of the points relative to their planned position. This assumption will allow replacing the original set of points within a single voxel with a plane that approximates it using the least squares method. The high density of the initial data makes it possible to guarantee the saturation of each voxel with points in an amount sufficient for calculating the plane and for assessing the accuracy of the parameters defining it. In that way, it is possible to assess the quality of the resulting model and the initial data.

4 Description of the experiment

The previously described calculation algorithm was implemented as a separate package of classes in the Python. An artificial relief model was used as an object of research. A three-dimensional point cloud was obtained photogrammetrically using the AgiSoft Metashape program.

The stability of the model to the influence of random factors (such as errors in determining points, their distribution within the model, errors in the external orientation of the model, etc.) was controlled statistically, for which the model was calculated using ten independent sets of images.

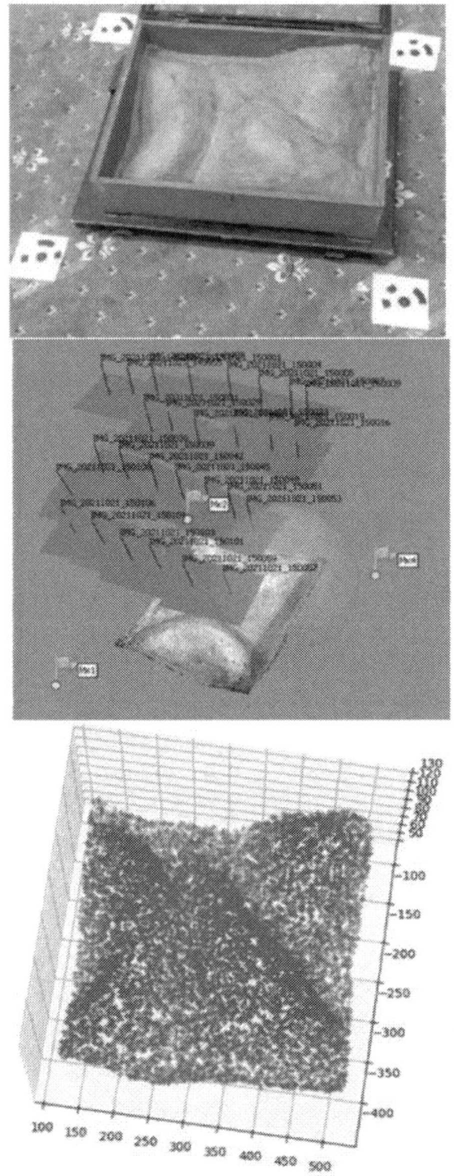

Fig. 1. Photo of the used model, an example of calculating a point cloud in the AgiSoft Metashape program, a cloud sparse to 10,000 points

The number of images in each series of measurements ranged from 15 to 20. The models were brought to a single coordinate system using four static exterior orientation marks. To increase the clarity of the experiment, the point clouds were brought to a general scale of 1000: 1. The number of points in each model was about 600-800 thousand points.

For each model, the calculation was performed for different sizes of the voxel grid. The grid spacing was equal to: 1 m, 5 m, 25 m, 50 m, 100 m and 250 m.

 https://doi.org/10.1051/e3sconf/202337804002

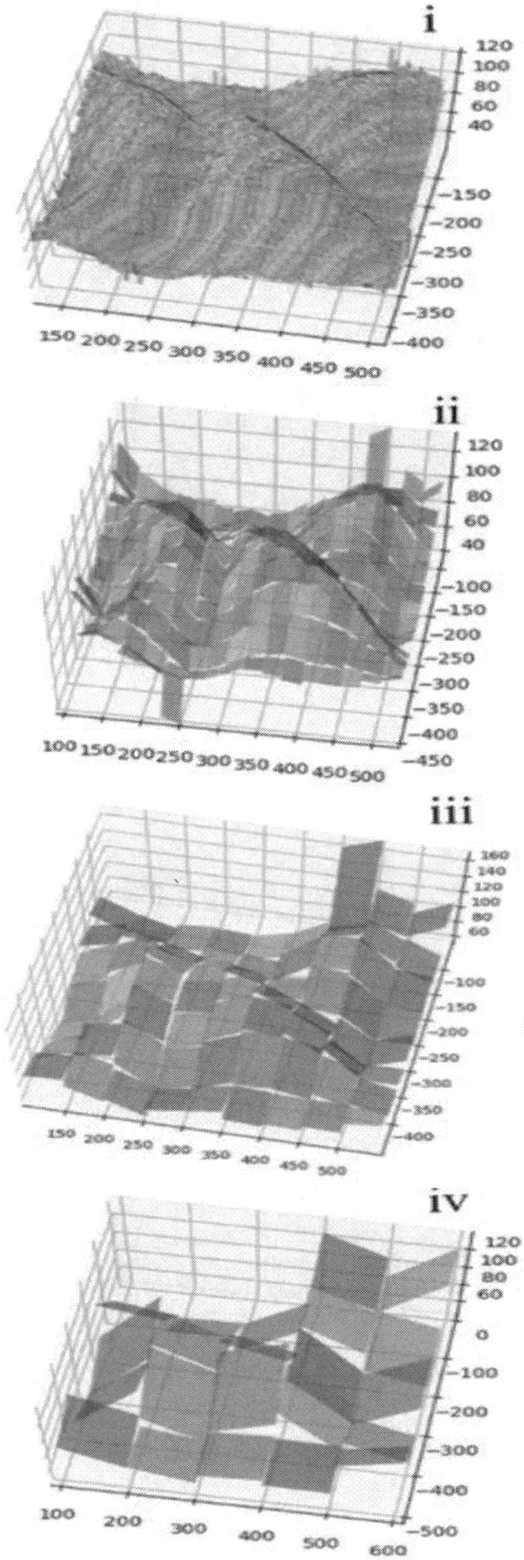

Fig. 2. An example of the calculated models at various scales of the voxel model: i) 5m; ii) 25m; iii) 50m; iv) 100m

For each voxel, the standard deviation (RMS) of points from the approximating plane was calculated, with an error map creation.

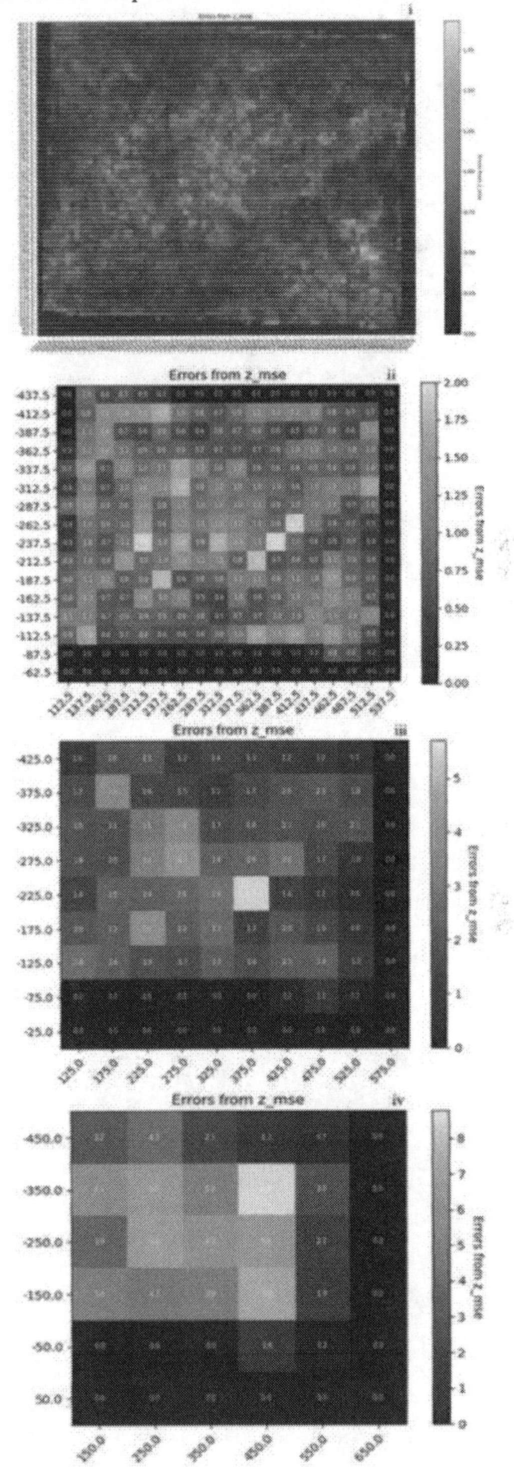

Fig. 3. The RMS of the approximating plane in height at various scales of the voxel model: i) 5 m; ii) 25 m; iii) 50 m; iv) 100 m

5 Experimental results

From the Figure 3, it can be seen that the root-mean-square error of the approximation of points in the voxels is not evenly. Large error values fall on voxels where the nature of the relief change is not linear, with an increase in the density of the voxel grid and a decrease in the size of an individual voxel, the absolute value of the RMS approximation decreases.

The simulation results for mesh sizes 1 and 250 m are not presented due to the low clarity of the result. In particular, the model, built on a grid of squares of 250 meters, contains only 6 voxels for the considered area of the site, which does not allow to represent the nature of the relief of the original cloud. The model built on a 1 m grid, on the contrary, almost accurately approximates the original point cloud, but due to the low saturation of each of the voxels with the initial data, it does not allow to reliably evaluate its accuracy.

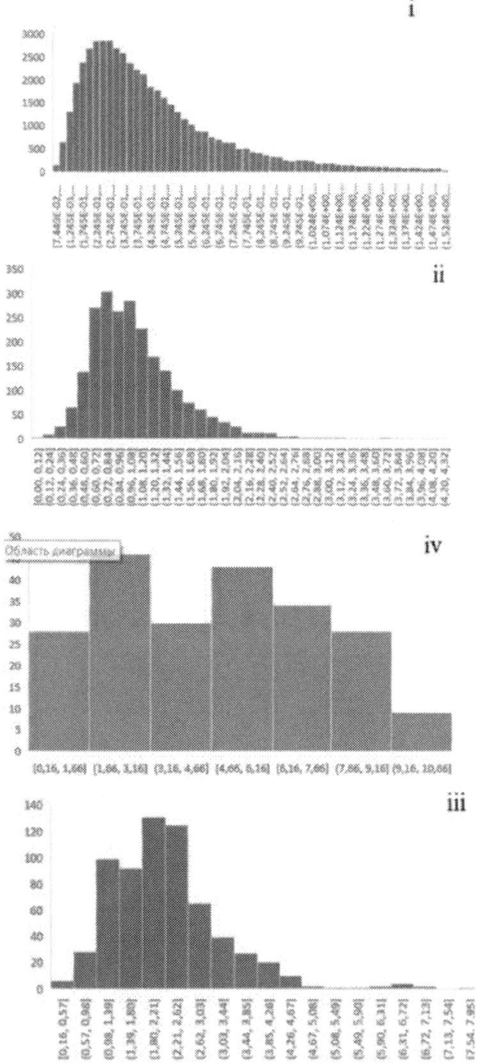

Fig. 4. Histograms of distribution of SKP approximation for different scales of the voxel model: i) 5 m; ii) 25 m; iii) 50 m; iv) 100 m

To check the robustness of the resulting models with respect to individual series of measurements, a one-way ANOVA was performed. The analysis results showed the equality of the mean value of the model error between the series for each of the voxel grid scales. The confidence level of the result ranged from 0.16 to 0.97. The hypothesis adopted in this way about the equality of the quality of models of the same scale with each other made it possible to combine them into one statistical sample for further analysis.

The nature of the RMS distribution of the model is shown on the frequency distribution histograms shown in Figure 4.

The histograms presented in Figure 4 show that the distribution of the approximation error within every individual voxel is a random variable, the distribution of which is consistent with the Pearson $\chi 2$ distribution. The asymmetry of the obtained distributions can be explained by the dominant influence of the change in the relief on the final quality of the approximation.

Reducing the model errors to a uniformly distributed approximation error is achievable by recursively refining the voxel grid in those areas where the error exceeds the specified tolerance. The depth of recursion can be roughly predicted from the dependence of the average value of the RMS approximation on the size of an individual voxel, shown in Figure 5.

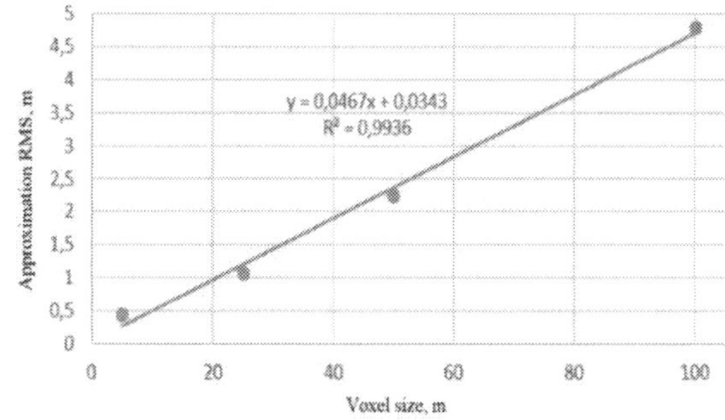

Fig. 5. Dependence of the average value of the RMS approximation on the size of an individual voxel

6 Conclusions and conclusion

The experiment proved the potential efficiency of using the proposed method for constructing a digital elevation model. The obtained regularities demonstrate a high correlation between the quality of the resulting model and the scale of the voxel grid used.

Assuming a constant voxel size and a fixed number of parameters describing the plane equation in three-dimensional space, the final size of the calculated model ceases to depend on the initial density of the processed point cloud. Each point of the model obtained during the survey will contribute to the improvement of the quality and reliability of the result, replacing the direct discharge of point clouds that is currently used. The predictability of the amount of data occupied will simplify the subsequent storage, processing and use of geospatial data.

The robustness of the obtained models with respect to input data errors allows the proposed method to be used to generalize and compare surveys at different times and to use them to solve the problems of geomechanical, ecological and glaciological monitoring.

References

1. V. Urak, A. Dushin, L.Mochalova *Vs sustainable development: scenarios for the future*. Journal of Mining Institute, **242** (2020)
2. R.Chénier, M. A. Faucher, R. Ahola, Y. Shelat, M. Sagram, *Bathymetric photogrammetry to update CHS charts: Comparing conventional 3D manual and automatic approaches*. ISPRS International Journal of Geo-Information, **7(10)** (2018)
3. A. V. Novikova, , A. P. Vergun, E. A. Zelenin, A. v. Baranskaya, S. A. Ogorodov, *Determining dynamics of the Kara Sea coasts using remote sensing and UAV data: A case study*. Russian Journal of Earth Sciences, **21(3)** (2021)
4. K. Anders, S. Marx, J. Boike, B. Herfort, E. J. Wilcox, M. Langer, P. Marsh, B. Höfle *Multitemporal terrestrial laser scanning point clouds for thaw subsidence observation at Arctic permafrost monitoring sites*. Earth Surface Processes and Landforms, **45(7),** 1589 (2020)
5. C. Flener, M. Vaaja, A. Jaakkola, A.Krooks, H. Kaartinen, A. Kukko, E. Kasvi, H. Hyyppä, J. Hyyppä, P. Alho, *Seamless mapping of river channels at high resolution using mobile liDAR and UAV-photography*. Remote Sensing, **5(12)** 6382 (2013)
6. A. A. Mogstad, Ø. Ødegård, S. M. Nornes, M. Ludvigsen, G. Johnsen, A. J. Sørensen, J. Berge, *Mapping the historical shipwreck Figaro in the high arctic using underwater sensor-carrying robots*. Remote Sensing, **12(6)** (2020)
7. M. Hodúl, R. Chénier, M. A. Faucher, R. Ahola, A. Knudby, S. Bird, *Photogrammetric Bathymetry for the Canadian Arctic*. Marine Geodesy, **43(1),** pp. 23–43 (2020)
8. V. Akovetsky, A. Afanasyev, *Space observations in the tasks of geoecological researches of coastal arctic shelves*. IOP Conference Series: Earth and Environmental Science, **539(1)** (2020)
9. J. Huber, R. McNabb, M. Zemp, *Elevation Changes of West-Central Greenland Glaciers From 1985 to 2012 From Remote Sensing*. Frontiers in Earth Science, **8** (2020)
10. E. A. Bash, B. J. Moorman, *Surface melt and the importance of water flow-an analysis based on high-resolution unmanned aerial vehicle (UAV) data for an Arctic glacier*. Cryosphere, **14(2)** 549–563 (2020)
11. J. C. Ryan, A. L. Hubbard, J. E. Box, J. Todd, P. Christoffersen, J. R. Carr, T. O. Holt, N. Snooke, *UAV photogrammetry and structure from motion to assess calving dynamics at Store Glacier, a large outlet draining the Greenland ice sheet*. Cryosphere, **9(1),** pp. 1–11 (2015)
12. É. Bernard, J. M. Friedt, M. Griselin, *Snowcover survey over an arctic glacier forefield: Contribution of photogrammetry to identify "icing" variability and processes*. Remote Sensing, **13(10)** (2021)
13. K. W. Turner, M. D. Pearce, D. D. Hughes, *Detailed characterization and monitoring of a retrogressive thaw slump from remotely piloted aircraft systems and identifying associated influence on carbon and nitrogen export*. Remote Sensing, **13(2)** 1–26 (2021)
14. T. B. Barnhart, B. T. Crosby, *Comparing two methods of surface change detection on an evolving thermokarst using high-temporal-frequency terrestrial laser scanning, Selawik River, Alaska*. Remote Sensing, **5(6),** pp. 2813–2837. (2013)
15. S. Li, L. Dai, H. Wang, Y. Wang, Z. He, S. Lin, *Estimating Leaf Area Density of Individual Trees Using the Point Cloud Segmentation of Terrestrial LiDAR Data and a Voxel-Based Model*. Remote Sensing, **9(11),** p. 1202. (2017)
16. G. Bitelli, G. Castellazzi, A. M. D'altri, S. de Miranda, A. Lambertini, I. Selvaggi, *Automated voxel model from point clouds for structural analysis of cultural heritage*. *International Archives of the Photogrammetry*, ISPRS Archives, **41,** pp. 191–197 (2016).
17. B. Guan, S. Lin, R. Wang, F. Zhou, X. Luo, Y. Zheng, *Voxel-based quadrilateral mesh generation from point cloud*. Multimedia Tools and Applications, **79(29–30)** 20561–20578 (2020)
18. X. Zhang, C. Fu, Y. Zhao, *An improved volumetric grid deep network model for point cloud segmentation*. Systems Science and Control Engineering, **9(S1),** pp. 161–167 (2021)

Peculiarities of climate change in the Arctic and implications for environmental management

Lyudmila Drachkova[1,*], *Nikolai* Kondratov[1,*] and *Anna* Trofimova[2,*]

[1]Candidate of Geographical Sciences, Associate Professor Higher School of natural sciences and technologies, Northern (Arctic) Federal University named after M.V. Lomonosov, 163002, Arkhangelsk, Russian Federation
[2] Department of Geography and Hydrometeorology, Higher School of natural sciences and technologies, Northern (Arctic) Federal University named after M.V. Lomonosov, 163002, Arkhangelsk, Russian Federation

Abstract. The Arctic is the high-latitude part of the Far North, including the land area and the Arctic Ocean marine environment within the exclusive economic zone of the Russian Federation. The importance of the Arctic issues is due to the fact that in the Russian Federation the formation and scientific rationale of priorities for the development of circumpolar territory are ones of the key objectives of Russia's economic development. The study of various areas and problems of Arctic development shows the particularly crucial role of geography in the formulation and solution of national economic objectives. The role of geography is becoming increasingly important not only because of the huge impact of the geographical factor in the Far North and the Arctic, but also due to remarkable differentiation of natural and social conditions of economic activity. In recent decades, the natural and socio-economic environment in the Arctic has been undergoing rapid transformation. One example is climate change, which may have an impact on environmental management on a global scale. The purpose of the article is to demonstrate the current trends of climate change based on sources and to identify the impact of these processes on the processes of environmental management.

1 Introduction

In the 21st century, the Arctic remains one of the most discussed topics in the scientific, economic, socio-humanitarian, and political spheres. The crucial tasks of economic the development of the Russian Federation include setting and scientifically substantiating priorities in the study and exploration of the Arctic territories. A well-founded choice of development priorities and the respective ways of their implementation requires clarification of Russia's position in the Arctic, specific knowledge about the current state of the Arctic environment and the perspectives of developing the region's resources in the context of climate change and globalization.

The Far North regions occupy more than a half of Russia's territory and are relatively well studied in geographical, geopolitical, historical and cultural, socio-demographic aspects. Such concepts as 'Arctic', 'North', 'Far North', while not exactly defined, are often used as synonyms in literature and state documents (development strategies). Among the Arctic states, Russia has the largest Arctic Sector. Taking the Arctic Circle as its southern border, the area of the Russian Arctic covers more than 9 million square kilometers, including 7 million square kilometers of the marine area, which corresponds to nearly a half of the area of the Arctic Ocean. The length of the coastline of the Russian Arctic is more than 22 thousand kilometers. The territory of the Russian Arctic covers significant developed areas near and above the Arctic Circle, including the areas of traditional natural resource use by indigenous peoples and the largest mineral deposits. Tourist recreational zones and wildlife reserves are also located here [1].

The interest to the Arctic stems from several reasons. One reason is that the Arctic region is currently undergoing irreversible transformations, the causes and consequences of which have not been fully understood. For example, air temperatures in the Arctic are rising faster than the global average. The Climate Doctrine of the Russian Federation defines climate change to be one of the global challenges of the 21st century, which must be addressed through an interdisciplinary approach comprising environmental, economic and social aspects of sustainable development [2]. The Strategy for the development of the Arctic zone of Russia until 2035 defines climate change as an important challenge to the socio-economic development of the region. The document says that climate change contributes to the emergence of both new economic opportunities and risks for economic activities and the environment [3]. Adaptation to climate change requires significant expenditures and international efforts. The media contribute substantially to the formation of public awareness, creating images of the consequences of shrinking sea ice coverage and ice depth in the Arctic

* Corresponding authors: drachkova@narfu.ru, n.kondratov@narfu.ru, a.trofimova@narfu.ru

Ocean, threats to biodiversity, resource wars of the 21st century. Secondly, countries both within (Denmark, Sweden, Norway, Finland, Iceland, Canada, the USA) and outside (China, Japan, the Republic of Korea, Singapore, India, Germany, Great Britain) the Arctic region, their trade unions and industrial corporations take an increased interest in the Arctic. the increasing role of the Arctic region in international cooperation stems from the fact the region is becoming key at the global level. The enhanced role of the area results from globalization, the reduction in deposits of mineral resources in traditional places of extraction, the development of transportation routes, the necessity to preserve the use of natural resource of indigenous peoples of the North, the development of tourism, and scientific research on climate change [4].

2 Main part

In natural and geographical terms, the Arctic has been identified by the Intergovernmental Panel on Climate Change (IPCC) as one of the most vulnerable world areas (along with island states, Africa and the African and Asian river deltas). The Arctic climate system exhibits forward and backward linkages that include ice, the cryosphere, permafrost, cloud cover, stratification of the Arctic Ocean, and geological substrate. The Arctic is a region of dynamic and unstable climatic processes, the imbalance of which leads to climatic shifts in the whole northern hemisphere. The urgent response to the climate challenge is conditioned by ongoing transformation of environmental problems into economic and political problems in the Arctic region. International problems may become a destabilizing factor for maritime activities (including naval activities) in the region. Analysis of the climatic features of the Arctic region is a problem of global importance. The need for analysis s due i to the fact that climate change has a double-edged impact on natural resources and the livelihood of the population around the world, including in the Arctic.

The Sixth Special Assessment Report of the IPCC (2018, Incheon, Republic of Korea) concludes that "... limiting global warming to 1.5°C will require rapid and far-reaching transition processes related to land, energy, industrial systems, and buildings, transportation and cities. Global carbon dioxide emissions caused by human activities will need to be reduced by almost 45% by 2030 compared to 2010 levels, reaching "net zero" by about 2050. This means that all remaining emissions must be balanced by removing carbon dioxide from the air" [5].

The first ice maps of the Greenland Sea and the Barents Sea were made at the Danish Meteorological Institute according to the data of based on ship observations in the spring-summer period. V. Yu. Wiese concluded (1937) that the warming process at the beginning of the 20th century was probably caused by a change in atmospheric circulation which was accompanied by an increase in the west and southwest winds over the Atlantic and the Norwegian Sea and an by increase in the Atlantic water inflow to the Arctic along with a simultaneous strengthening of the outflow to the Greenland Sea. The first third of the 20th century was noted by Wiese as a period of maximum warming in the Atlantic Arctic from Greenland to the Kara Sea in winter and a decrease in the ice cover of the Barents Sea and the Kara Sea in summer [6]. Modern concepts of temperature rise in the Arctic refers to a wider range of influencing factors.

Temperature trends in the Arctic have been changing over the 20th - 21st centuries, and the lack of regular instrumental monitoring does not allow to draw unequivocal conclusions about the direction of climate change." There is a trend of increasing temperature, which in certain areas of the world (Alaska, Northern Canada, Siberia) increased by 3°C over a period of 30 years (1971-2000). Though the magnitude and direction of the trend has varied, in recent decades the trend for the Arctic has been showing warming which is two times higher than global temperature increase" [7].

The Fifth IPCC Report states the fact that changes in the climate system during the industrial era are undeniable. Since the 1950s climate changes have been unprecedented from decades to millennia. Since 1850, the Earth's surface temperature has been higher than the previous decade, every decade the last three ones. The period of 30 years from 1983 to 2012 was the warmest in the last 1400 years in the northern hemisphere [8].

The rise in ocean temperature is the essential element of the increase of energy of the climate system. The oceans accounted for 90 percent of energy accumulated in the Earth's climate system from 1971 to 2010. The increase in ocean temperature was most pronounced globally near the surface, with a temperature increase of 0.11 [0.09 – 0.13] °C over the decade from 1971 to 2010. The distribution of observed and expected changes in global ocean surface temperature did not change qualitatively throughout the history of pattern implication for different scenarios of anthropogenic intervention. Data from the Hadley Centre SST2 set (since 1870) and CMIP3 models of ocean and atmosphere circulation point to elevated sea surface temperatures in the Russian Arctic, near the coast of Alaska and Greenland [8].

Atmospheric concentrations of major mixed greenhouse gases have increased and become unprecedented over the past 800 thousand years. Carbon dioxide concentrations have increased from their pre-industrial value (conventionally before 1750) by 40%, mainly due to CO_2 emissions from burning of fossil organic fuel and net emissions resulting from changes if the land use.

IPCC experts conclude that the state of the Earth's glacier systems is deteriorating everywhere, and the ice mass balance is negative almost everywhere. The area of sea ice is diminishing, which can be particularly seen in the Arctic during the warm season. In spring, the snow cover area decreases in the northern hemisphere. Sea level continues to rise at a rate that has exceeded the average of the previous two millennia since the middle of the 19th century [8].

The Arctic Council reports note that in the 21st century the average global temperature may increase by 2.8^{0}C (at present it is about 0.4-0.6^{0}C), while at the most part of the land surface it may increase by 3.5^{0}C and in the Arctic – by 7^{0}C. Warming processes in the Arctic are twice as fast as in the rest of the world, which was especially noticeable in 2012. The area of Arctic land covered by snow in the beginning of summer has shrunk by 18 percent since 1966. There is a decline in the number of days with snow cover in coastal Alaska and the northern Scandinavian Peninsula. Snow cover thickness has decreased in the North American Arctic and increased in the Russian North. The authors of the report assume that in the 21st century, the Arctic Ocean will become completely ice-free in summer. However, this has not happened yet and navigation in the eastern seas of the Arctic Ocean is possible only for nuclear icebreakers [9].

According to the Snow and Ice Data Center (USA), the ice-covered area in the Arctic diminished from 5.92 million sq. km to 4.76 million sq. km in 2006-2012. NASA and Roshydromet have registered the absolute minimum of ice over the past 30 years. The area covered by ice decreased from 7.3 million sq. km in 1979 to 5 million sq. km in 2007 (5.1 million sq. km in 2013).

The amount of precipitation in the Arctic region increased by 15 percent between 1970 and 2012, while the area of sea ice decreased by 10-15 percent, the area covered by snow decreased by 10 percent, and the temperature of the top permafrost layer rose by 2^0. The southern boundary of Russia's permafrost territories moved about 30-80 km between 1970 and 2007. The freezing period of northern rivers has decreased, the river flow to the Arctic Ocean and the inflow of warm water from the Pacific Ocean increased. Thickness of sea ice in the Arctic Ocean over the past 30 years decreased by 45%, while in the seas of Siberia, the Far East and Alaska (the Laptev Sea, the East Siberian Sea, the Chukchi Sea, the Beaufort Sea) it has diminished by 87.5 percent. The glaciation area of Novaya Zemlya and Franz Josef Land decreased by more than 720 sq. km, while glaciation extent has reduced by 250 cu m (1.5 percent). There is no trend of slow warming in the territory of Russia. "Shrinking snow and ice surfaces, which reflect the significant part of the light give way to darker land and ocean surfaces that absorb more solar energy. It intensifies the heating of soil and air. ...This can lead to a considerable increase in emissions of such greenhouse gases as carbon dioxide and methane in the Arctic... Melting glaciers and ice sheets contribute the most to global sea level rise. Arctic glaciers, ice caps and the ice sheet of Greenland produced 1.3 mm of the total annual global sea level rise by 3.1 mm between 2003 and 2008, which is 40 percent. Consequently, the Arctic's contribution to global sea level rise was much greater than previously expected" [10,12].

Climate change is a stressor for Arctic biological diversity, which includes 21 thousand species of plants and animals. About 10 percent of the world's catch of fish and seafood is origin from the Arctic. There is an invasion of trees in tundra ecosystems has been observed. The number of reindeer *(Rangifers)* and caribou *(Rangifer tarandus)* has dropped by a third in just one decade.

Assessing the impact of climate change on human economic activities in the Arctic is methodologically difficult. First of all, predictions of climate change based on physical and climate models are made for a period of 50-100 years, which does not correspond to long-term economic projections, which are usually made for a period not exceeding 10-20 years. Secondly, climate research trends focus on catastrophic global change scenarios. Economic calculations consider the temperature increase in the Arctic in terms of direct losses in the economy and social sector of individual states and the entire world. According to the calculations of foreign researchers (Shiklomanov, 2017), the consequences of permafrost degradation may result in the total renovation wal of the housing stock of Norilsk, Yakutsk, Vorkuta and Salekhard by 2050. The total volume of the housing stock of these cities are about 10 million sq. m and will require more than RUB 400 billion of investments. Across the Arctic, such investments could reach USD 160 trillion by 2100, or USD 2 trillion per year. It could account for 1.2 percent of the global GDP. Using the cost approach for Russia, the cost of cumulative (direct and indirect) damage from climate change by 2030could reach 3 percent of GDP per year, and in specific areas this index could reach be 5-6 percent of GDP. In total it may reach RUB 20 trillion (in 2011 prices). This estimate does not determine only the worst-case scenarios for the development of the Russian economy development under conditions of global climate change (global warming). If the economic situation worsens for some industries, on the contrary, it may improve for other industries. In the Russian Arctic, the temperature rise may result in an increase in the length of the growing season, expansion of farming and livestock areas; facilitating access to natural resources; lengthening of the navigation period in the seas of the Arctic Ocean, which may have a multiplicative effect on related sectors of the economy. It has been estimated that the ice-free navigation period may last for 90-120 days by 2025, more than 150 days by 2040 and over 200 days by 2090. As a result, taking into account mega-projects, the probable growth of the cargo turnover may exceed 50 million tons per year in the medium term. Direct income from ice escorting of ships along the Northern Sea Route could amount to RUB 30 billion annually. On the contrary, decrease in ice cover and the ice depth reduces the need for icebreakers. As the temperature rises, sea disturbance intensifies and the probability of iceberg occurrence increases. These conditions can have a negative impact on navigation and activities of oil and gas production facilities [11,14].

The water levels in reservoirs may rise as a result of the increased river flow into the Arctic Ocean basin. The risk of flooding and waterlogging of inhabited localities will increase, air humidity, frequency of fog and visibility will deteriorate.

As shown [11] with regard to the energy sector in the Russian Arctic zone, based on thermal power stations, due to the average annual air temperature increase the heating period is reduced. Consequently, expenses for fuel and heating systems are cut.

The cumulative effect of conditionally favorable impacts of climate change on the national economy by 2030 is projected to be characterized by an access of costs over expected benefits [11,15]. It should be kept in mind that these benefits will require new workforce competencies [16], innovative approaches in research activities, and technological innovations. This implies the need to attract investments [17], increase of expenses for education and scientific, research and experimental development, improvement of public administration. The uncertainty of climate change projections should be taken into account, as it has a negative impact on the accuracy of decisions made at the government level in the field of climate policy and subsequent investments into megaprojects, as well as in the determination of scenarios of economic development for the long term. In this regard, the estimates predicting that the Russian economy will get external impacts of climate change appear to be unreasonably optimistic. Assessment and projections of climate change impact on the dynamics and sectoral and territorial structure of the Arctic macro-region economy should take into consideration its territorial peculiarity which is strongly pronounced (due to the size of the macro-region) and the specifics of functioning of individual economic sectors of the Russian Arctic in the conditions of climate change [13]. Maintaining the reliability of energy supply will be provided by the existing types of energy systems in the short and long term. The challenge for the development of the Arctic could be the development of new industries characterized by an increased power demand. This problem will be solved by specialized energy companies within their competences. Reduced consumption of heat may be observed in the sphere of energy consumption. However, this process can also illustrate, firstly, a reduction in the number of energy consumers. Secondly, it is a consequence of the routine work on upgrading of heat and power supply systems, improving the quality of residential and public buildings, replacement of power supply schemes and equipment [11].

Negative impacts of climate change in the Arctic include growing public health risks (e.g., an increased infectious diseasies); increased frequency, intensity and duration of extreme precipitations, as well as floods, storms and other natural disasters; soil overwatering; increased fire hazards in subtundra forests; environmental imbalances (e.g., species displacement); increased energy consumption for air conditioning in summer.

3 Conclusion

The synergistic effect of climate change should be noted. Such effect is illustrated by the aggravation of anthropogenic risks and threats to Arctic ecosystems, which is the result of improved access to natural resources. This will lead to an increase in emissions polluting the environment. The specifics of the Russian Arctic lies in the predominant importance of social risks compared to environmental risks. This situation is explained by the fact that over 2.4 million people live in the Arctic, which exceeds the population in other Arctic regions.

References

1. D.A. Dodin, *Sustainable development of the Arctic (problems and prospects)* (St. Petersburg, Nauka, 2005)
2. Presidental Resolution, No. 861-rp validating Climate Doctrine of the Russian Federation. Available online: http://en.kremlin.ru/supplement/4822 (2009)
3. *The Strategy for the Development of the Arctic Zone of the Russian Federation and National Security up to 2035*, 26 October (2020)
4. A.I. Chistobaev and N.A. Kondratov, Arctic: geopolitical and geoeconomic interests of foreign countries, *Geopolitics and Security*, pp. 1-25 (2014)
5. An IPCC special report on the impacts of global warming of 1.5 °C above preindustrial levels and related global greenhouse gas emission pathways, in the context of strengthening the global response to the threat of climate change, sustainable development, and efforts to eradicate poverty. Available online: https://www.ipcc.ch/report/sixth-assessment-report-working-group-i/ (2021)
6. G.V. Alekseev et al., Evolution of the area of the Arctic sea ice cover under the conditions of modern climate changes, *Research of the Earth from Space* **2** (2015)
7. Yu.S. Tsaturov and A.V. Klepikov, Current climate change in the Arctic: results of a new assessment report of the Arctic Council Arctic: ecology and economics, 4-8 (2012)
8. Climate Change 2014: Synthesis Report. Contribution of Working Groups I, II and III to the Fifth Assessment Report of the Intergovernmental Panel on Climate Change (IPCC, Geneva, Switzerland, 2014), p. 151.
9. AMAP Climate Change Update 2019: An Update to Key Findings of Snow, Water, Ice and Permafrost in the Arctic (SWIPA) 2017. Arctic Monitoring and Assessment Programme (AMAP) (Oslo, Norway, 2019), p. 12.
10. Second Assessment Report on Climate Change and its Consequences in Russian Federation (Moscow: Roshydromet, 2014)
11. B.N Porfirev et al., The consequences of climate change for economic growth and development of individual sectors of the economy of the Russian Arctic Arctic: ecology and economics, 4-28 (2017)
12. G. Benoit, Agriculture, land, water and climate: Solutions for a world in transition Futuribles: Analyse et Prospective October, **438**, pp. 5-28 (2020)
13. B. Swynghedauw et al., 'Climat et Sante' de la Commission XIV Consequences of climate change on human and animal health [Article@Rapport 20-

07. Conséquences du changement climatique sur la santé humaine et animale] *Bulletin de l'Academie Nationale de Medecine*, **205(3),** pp. 219-226 (2021)

14. A. Cherepovitsyn, A. Lebedev, Drill Cuttings Disposal Efficiency in Offshore Oil Drilling, *Journal of Marine Science and Engineering,* **11(2)**, 317, doi:10.3390/jmse11020317 (2023)
15. P.S. Tcvetkov, A.E. Cherepovitsyn and A.B. Makhovikov, Economic assessment of heat and power generation from small-scale liquefied natural gas, *Russia Energy Reports*, **6**, pp. 391-402 (2020)
16. I.G. Gerasimova, I.S. Oblova, E.I. Golovina, The Demographic Factor Impact on the Economics of the Arctic Region, *Resources*, **10**, p. 117, doi: 10.3390/ resources10110117 (2021)
17. K. V. Matrokhina., V.Y. Trofimets, E.B. Mazakov, A.B. Makhovikov and M.M. Khaykin, Development of methodology for scenario analysis of investment projects of enterprises of the mineral resource complex, *Journal of Mining Institute*, **259**, pp. 112-124, doi: 10.31897/PMI.2023.3 (2023)

Variability of the ice area and the possibility of achieving an ice-free regime in the Barents Sea

Eugene Istomin[1], *Yaroslav* Petrov[2,*], *Irma* Martyn[2], Valery Abramov[2], *Natalia* Yagotinceva[2]

[1]Doctor of technical sciences, Professor, Department of Applied Computer Science, Russian State Hydrometeorological University, 192007, Saint Petersburg, Russian Federation
[2]Candidate of technical sciences, Associate professor, Department of Applied Computer Science, Russian State Hydrometeorological University, 192007, Saint Petersburg, Russian Federation

Abstract. The results of the analysis of the interannual variability of the Barents Sea ice area, including trends in the area of sea ice distribution for the period 1979-2018 based on remote sensing data, are presented. The analysis of trends is based on a dimensionless trend index, which allows comparing different areas of sea ice distribution. Estimates of the Barents Sea ice-free regime in January and August are presented on the basis of extrapolation of trends for the period 1979-2018. The ice-free regime of the Barents Sea in January is expected in 2055, in August - 2026.

1 Introduction

In modern conditions, changes in climatic characteristics occur sequentially, changes in one of the components of the ocean-atmosphere system through feedback leads to changes in others. According to sources, as a result of intensifying polar phenomena, the area of Arctic sea ice is rapidly decreasing. Since the Arctic is a region for the development of natural resources, a long-term assessment of the forecast of climatic parameters is required for their development [1]. One of the most important parameters affecting the work in the Arctic region is sea ice. Currently, there are complex models for sea ice forecasting, but it is necessary to develop alternative approaches using statistical methods [2].

The purpose of this work is to estimate the variability of the area of distribution of the sea ice area of the Barents Sea and to forecast its ice-free regime in winter and summer using January and August as examples [2].

2 Methods and materials

Remote measurement methods are the main source of information about sea ice in polar regions. Measurements in the microwave range, compared to measurements in other ranges, provide the most comprehensive data set, since it works well in cloudy and bad weather and at night, which is typical for polar regions. Additional advantages are the availability and continuity of data observations [3, 4].

Monthly data on the area of sea ice distribution since 1979 are freely available on the website https://neptune.gsfc.nasa.gov. This work uses data to estimate the maximum (January) sea ice distribution and minimum (August) sea ice distribution for the Barents Sea [5].

The interannual variability of the sea ice distribution in the Arctic is largely inertial in nature, as a result, linear trends describe most of the dispersion of the cover time series. In this case, a simple statistical trend method can be used to predict the area of ice. Its basis is an assumption of stationarity (immutability) of climatic conditions for the entire forecasting period. This method allows to calculate forecast estimates of the ice area for practically any given year, including the date when the sea enters the ice-free regime [6-9].

As a rule, a trend coefficient depending on the sea ice area is used to assess trends. But due to a large variability of the sea ice area, quantitative comparison of trends becomes impossible. To do this, a trend index is introduced, which represents the ratio of the trend spread to its mean value and is expressed in conditional units [10, 11]

$$I = 100 a_1 n / \overline{X} \quad (1)$$

Here a_1 is the coefficient of the linear trend equation, n is the length of the series.

3 Results

Statistical estimates of sea ice distribution trends in the Barents Sea for January and August for the period 1979-2018 are presented in Figure 1-3 shows the temporal variability of sea ice distribution area in January and August for the period 1979-2018 for the Barents Sea with expressed linear and nonlinear trends, as well as trend equations. According to linear and nonlinear trends, the year of ice-free sea regime for January and

* Corresponding author: yaroslav.petrov025@gmail.com

 https://doi.org/10.1051/e3sconf/202337805002

August is calculated, the results of calculations are listed in Table 1, extrapolation of the trend is shown in Figure 3-4.

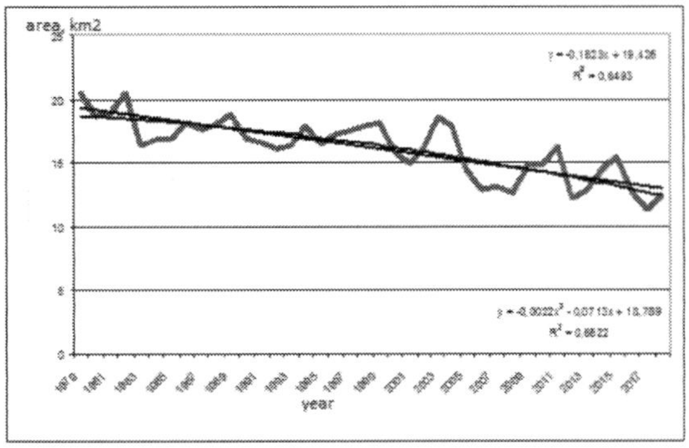

Fig.1. Temporal variability of the sea ice distribution area in January in the Barents Sea for the period 1979-2018 with linear and nonlinear trends

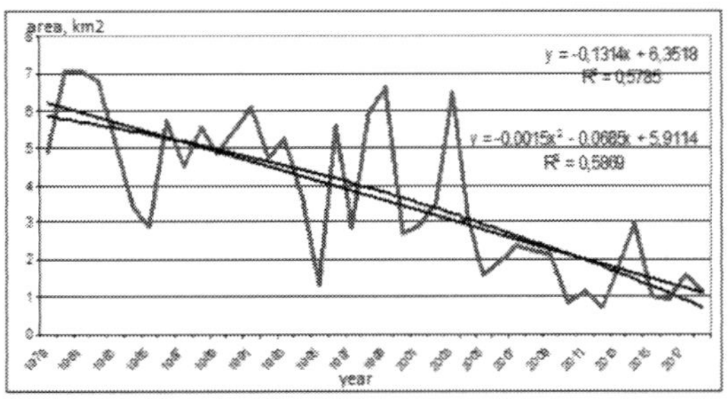

Fig. 2. Temporal variability of the sea ice distribution area in August in the Barents Sea for the period 1979-2018 with linear and nonlinear trends

Table 1. Characteristics of the linear trend of the sea ice distribution area (10^5 km2/year) for the Barents Sea in January and August for the period 1979-2018.

year	January				August			
	Average $\bar{x}$	R^2	a_1	trend index	Average $\bar{x}$	R^2	a_1	trend index
1979-2010	16.78	0.52	-0.15	-28.61	4.22	0.4	-0.124	-94.03
1979-2011	16.76	0.49	-0.14	-27.57	4.13	0.44	-0.129	-103.08
1979-2012	16.63	0.54	-0.15	-30.67	4.03	0.48	-0.136	-114.74
1979-2013	16.51	0.57	-0.158	-33.49	3.96	0.5	-0.135	-119.32
1979-2014	16.45	0.58	-0.15	-32.83	3.93	0.49	-0.128	-117.25
1979-2015	16.43	0.56	-0.15	-33.78	3.86	0.52	-0.131	-125.57
1979-2016	16.32	0.59	-0.15	-34.93	3.78	0.55	-0.133	-133.7
1979-2017	16.19	0.62	-0.159	-38.3	3.72	0.56	-0.131	-137.34
1979-2018	16.09	0.65	-0.162	-40.27	3.66	0.57	-0.131	-143.17

The interannual variability is expressed by the sum of the trend, cyclical and random components. The random component is predicted for a small number of steps, which is not suitable for the task. The cyclic component is expressed by the sum of significant harmonics and their calculation for a long time.

The results of the search for significant harmonics of the Barents Sea in January are listed in Table 2, in August in Table 3. The results of extrapolation of the sea ice distribution area in January and August for the Barents Sea are shown in Figure 3-4.

Table 2. Estimates of the contribution to the variance of the time series of January harmonics and trend values for the sea ice distribution area of the Barents Sea for the period 1979-2018.

Barents Sea, January	
Harmonic period, years	Contribution, %
5,7	21
4	11
Trend	64
Sum	96

Table 3. Estimates of the contribution to the variance of the time series of the August harmonics and trend values for the sea ice distribution area of the Barents Sea for the period 1979-2018

Barents Sea, August	
Harmonic period, years	Contribution, %
5,7	12
Trend	58
Sum	70

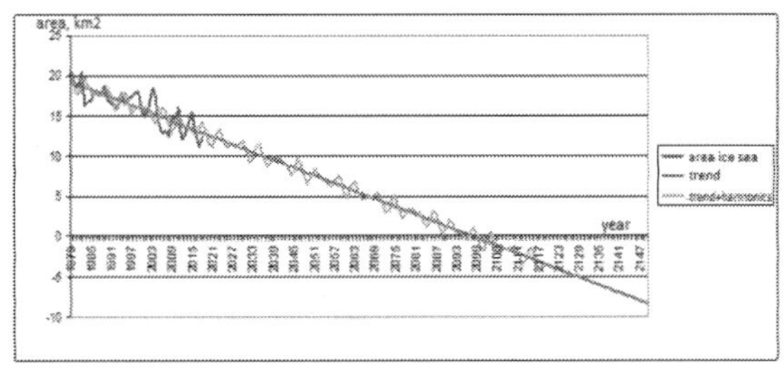

Fig. 3. Extrapolation of the January sea ice distribution area for the Barents Sea

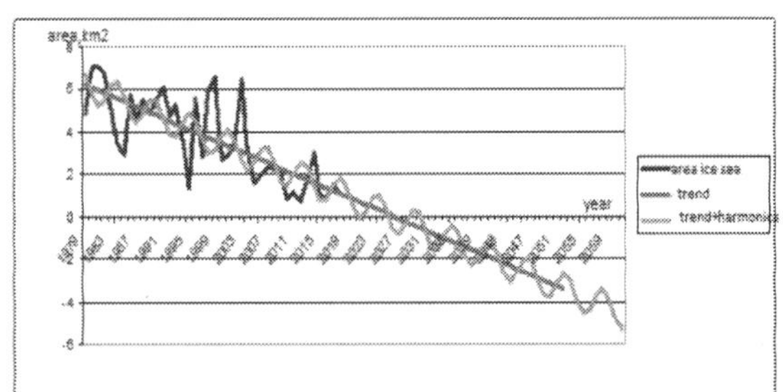

Fig. 4. Extrapolation of the August sea ice distribution area for the Barents Sea

4 Discussion

For the area of sea ice distribution in January for the period 1979-2018, linear and nonlinear trends were constructed and tested for significance according to the

Student's criterion. For a linear trend t*=14.5>tcrit=2.02 - the trend is significant, for a nonlinear trend t*=15.8>tcrit=2.02 - the trend is significant. So there is a tendency to decrease the area of ice by 0,16km2 /year. For August for the period 1979-2018, according to the Student's criterion, trends were checked for significance, for a linear trend t*=11.4>tcr=2.02 - the trend is significant, for a nonlinear trend t*=12.1>tcr=2.02 - the trend is significant. So there is a tendency to decrease the ice area by 0.13 km2/year.

According to linear and non-linear trends, the sea is calculated to go ice-free, which in January with a linear trend will occur in 2097, and with a non-linear trend faster in 2055. In August, the ice-free regime with a linear trend will occur in 2026, and with a non-linear trend in the current year 2021.

To assess the stability of the trend, the trend index is calculated. In Table 1, the maximum and minimum trend values are selected for January, the maximum for 2018 and the minimum for 2011. According to them, the output to the ice-free mode was calculated, so with the minimum trend, the output to the ice-free mode will occur in 2114, and with the maximum trend in 2097. For August, the maximum and minimum trend values are a1, the minimum is 2010 and the maximum is 2012. According to them, the ice-free regime was calculated in August, so with a minimum trend, the ice-free regime will occur in 2028, and with a maximum trend in 2025. The difference in these estimates does not go beyond the trend errors, which indicates a good quality forecast.

Next, a harmonic analysis was performed, significant harmonics for January were identified. Harmonics contribute about 32% to the variance of the series, together with the trend they describe 96% of the original series. The curve of the sum of significant harmonics is extrapolated to the year of ice-free operation, the results are shown in Figure 3. The ice-free regime for January, taking into account harmonics, will occur in 2099. For the values of the August ice area, significant harmonics have been identified, which together with the trend describe 70% of the variance of the original series. The curve of the sum of significant harmonics is extrapolated to the year of ice-free operation, the results are shown in Figure 4. The ice-free regime will take place in 2025.

5 Conclusion

Linear and nonlinear trends are significant for the area of sea ice distribution in the Barents Sea in January, with a trend value of 0,16 km2/year. The ice-free regime in January with a linear trend will occur in 2097, with a non-linear one - in 2055. With the minimum trend, the ice-free regime will occur in 2114, and with the maximum in 2097, that is, with a difference of almost 20 years. Taking into account the harmonics, the ice-free mode will be reached in 2099.

For August, trends are also significant, with a trend value of 0.13 km2/year. With a linear trend, the ice-free regime will occur in 2026, with a non-linear one – in 2021. With the minimum trend, the ice-free regime will occur in 2028, and with the maximum in 2025, with a difference of 3 years. Taking into account the harmonics, the output will occur in 2025.

In modern conditions of the development and development of the Arctic, further study of the interannual variability of ice is of great importance in a wide range of tasks for the development of the Arctic zone. In the future, the introduction of the described statistical methods into a geographic information system to automate calculations and apply greater variability in the selection of heterogeneous data sets [12, 13, 14, 15].

References

1. V.F. Dubravin and H.H. Nagornova, On estimates of water exchange through the Baltic Strait [in Russian – Ob ocenkax vodoobmena cherez Baltijskij proliv], *Innovations in science and education-2007: SB. Tr. Vol. Conf.,* KSTU, **Part 1** (2007)
2. E. Istomin et al. , *IOP Conf. Ser.: Earth Environ. Sci.*, **316**, 012011 (2019)
3. Y. Petrov et al, Development of a conceptual GIS model to support management decision making *IOP Conf. Ser.: Earth Environ. Sci.*, **574(1)**, 012062 (2020)
4. *The Hydrometeorological regime of the Vistula lagoon* Eds N N Lazarenko and A Mayevsky (Leningrad: Gidrometeoizdat, 1971), p. 6.
5. A.P. Tolstosheev, A metyod of estimation of the results of reconstruction of the trajectories of drifting buoys, *Physical Oceanography*, **19(6)**, pp. 358-365 (2010)
6. G.K. Korotaev et al, Seasonal, interannual, and mesoscale variability of the Black Sea upper layer circulation derived from altimeterdata, *J Geophys Res,* **108(4)**, pp. 22-31 (2003)
7. A. Gill, Dynamics of atmosphere and ocean, *Mir*, **1**, pp. 130-190 (1986)
8. B. Chubarenko and A. Babakov, *Sediment transport near the Vistula spit (Baltic sea) Atlantic Branch of P P Shirshov Institute of Oceanology of Russina Academy of Sciences* (Kaliningrad, 2016)
9. J.E. Ullgren et al, Observations of strong ocean current event offshore Pemba, *Northern Mozambique Operational Oceanography,* 9, pp. 55-66 (2016)
10. C. Hays Graeme, Ocean current and marine life, *Current Biology,* **27**, pp. 470-473 (2017)
11. J.B. Klemp, On the dynamics of gravity current in a strait, *J Fluid Mechanics*, **269**, pp. 169-198 (1994)
12. E. Istomin et al., About the methodology of geo-risk management in forestry, *IOP Conf. Ser.: Earth Environ. Sci.*, **507(1)**, 012006 (2020)
13. E. Istomin et al., Study of intra-day dynamics of currents in the area of the navigable strait of Baltiysk to adjust the movement of water transport *IOP Conf. Ser.: Mater. Sci. Eng.*, **817(1),** 012013 (2020)

14. E. Istomin et al., Application of Kalman-Bucy filter for vessel traffic control systems in the northern sea route *IOP Conf. Ser.: Mater. Sci. Eng.*, **817(1)**, 012012 (2020)
15. I. Martyn et al., Spatial and temporal variability of Bering sea ice cover *IOP Conf. Ser.: Mater. Sci. Eng.*, 539(1) 012198 (2020)

 https://doi.org/10.1051/e3sconf/202337805003

Assessment of hunting pressure on Arctic-nesting shorebirds: first results from the Northeast of Russia

Konstantin Klokov [1, 2,*], *Yuri* Gerasimov[3,4] and *Eugene* Syroechkovskiy[5,6]

[1]Doctor of Geography, Professor, Department of Regional Policy and Political Geography, 199034, Saint Petersburg State University Saint Petersburg, Russian Federation
[2]Senior Researcher, 199034, Peter the Great Museum of Anthropology and Ethnography (Kunstkamera), Russian Academy of Sciences, Saint Petersburg, Russian Federation
[3]Candidate in Biology, 683000, Kamchatka Branch of Pacific Institute of Geography of Far-eastern Branch of Russian Academy of Science, Petropavlovsk-Kamchatsky, Russian Federation
[4]Candidate in Biology, 121099, BirdsRussia (MOO ROSIP), Moscow, Russian Federation
[5]Candidate of Geographical Sciences, 121099, BirdsRussia (MOO ROSIP), Moscow, Russian Federation
[6]Candidate of Geographical Sciences, 117628, All-Russian Research Institute of Environmental Protection (FSBI"VNII Ekologiya"), Moscow, Russian Federation

Abstract. Conservation of Arctic migratory birds is based on a holistic approach that considers all habitats of a species within its annual life cycle. Hunting for Arctic-nesting shorebirds in the Northeast of Russia can negatively impact Arctic shorebird populations, especially endangered species. The Arctic Migratory Bird Initiative (AMBI) program was initiated by CAFF in 2015 to improve the conservation status of declining of Arctic migratory bird populations. BirdsRussia began a project to assess hunting pressure on the Arctic shorebirds nesting in the East Asian-Australasian Flyway in 2019 in Kamchatka. This is the first project focused on estimating hunting pressure on Arctic shorebirds in Russia. Its methodology is based on an anonymous survey of hunters. The result showed that about 45,000 shorebirds were hunted per year in Kamchatka, of which 37,000 are Whimbrel, about 1,600 of large and medium-sized shorebirds other than Whimbrel, and about 6,000 small shorebirds of different species. Hunters often do not distinguish between different shorebird species, and by mistake they shoot many birds of protected species; in addition, they often shoot mixed flocks. Such shooting threatens the endangered Spoon-billed Sandpiper and other protected shorebirds, such as the Far-Eastern Curlew, Bar-tailed Godwit and others.

1 Introduction

Arctic-nesting migratory birds are a significant component of both Arctic culture and ecosystem health. Many of them require special attention of researchers as they are rare and endangered species or game resources. The protection of Arctic migratory birds, as well as the issues of sustainable use of their resources (concerning game species) is complicated by the fact that Arctic birds use various flyways to move from Arctic breeding grounds to wintering grounds or stopover sites in lower latitudes. Flyways from breeding areas in the eastern part of Russian Arctic to wintering and stopover sites run mainly along the west coast of the Pacific Ocean and cross 22 Asian countries. The aggregate of these routes is defined as the East Asian-Australasian Flyway (EAAF). The EAAF is the most species-rich of the world's nine major flyways. In addition, the EAAF accounts for the highest proportion of wetland bird populations, which are declining at an unprecedented rate especially shorebirds. Of the 63 populations of 52 shorebird species migrating along this flyway, 20 populations of 17 species may become extinct or approach extinction in the near future if no measures are taken [1, 2, 3].

Over the past 30 years, this problem has attracted the attention of many ornithologists and conservationists. The main reason for the decline in EAAF shorebird populations is the reduction of the main habitats: coastal and inland wetlands [4, 5]. Hunting is also perceived as a medium or high-level threat, but it is difficult to estimate its scale due to poor knowledge of the issue [6, 7].

The Arctic Migratory Bird Initiative (AMBI) was launched by CAFF in 2015 and aims to improve the conservation status and secure the long-term sustainability of declining in Arctic-breeding migratory bird numbers through flyway-level cooperation with Arctic and non-Arctic countries and partners. the AMBI works across the four major flyways to engage global partners to address bird and habitat conservation, including mitigation of harmful anthropogenic actions and hunting. AMBI Work Plan Goal 3 is to prevent illegal hunting and regulate the unsustainable legal harvest of Arctic migratory birds along the EAAF, with a focus on the Spoon-billed Sandpiper, Lesser White-fronted Goose, Bar-tailed Godwit, and other priority

* Corresponding author: k.b.klokov@gmail.com

species. Action 3.1 initiates surveys of hunting pressure on Arctic-nesting shorebirds at stopover areas in northeastern Russia, including Chukotka, Kamchatka, Sakhalin and the mainland coasts of the Sea of Okhotsk [8]. Implementation of these activities started in 2019 by BirdsRussia in cooperation with the Northern Eurasia Wader Working Group. It is the first project aimed at the assessment of hunting pressure on Arctic shorebirds in Russia. The main aim is to identify the territories of the largest hunting pressure on shorebirds and to clarify the timing (season) of hunting them, primarily for the priority species of shorebirds (waders) of the East Asia-Australasia Flyway Partnership: Curlew Sandpiper, Red Knot, Great Knot, Far Eastern Curlew, Black-tailed Godwit and Spoon-billed Sandpiper. Special attention is paid to the Whimbrel, which is the most popular of wader for legal hunting in the Russian Far East, as well as to the most popular endangered species: the Far Eastern Curlew, Godwit, Spoon-billed Sandpiper, etc.

This paper focuses on the results of the first stage of this project in 2019 in Kamchatka. The results of the next stages – Sakhalin in 2020 and Khabarovsk Krai in 2021 – will be published later.

2 Methodology

The methodology is based on the experience of estimating waterfowl hunting in the eastern part of the Russian Arctic. It was developed by E.E. Syroechkovskiy and K.B. Klokov and was used in 1999–2006 to estimate bird harvests in 22 villages of Chukotka and northern Yakutia near the sea coast [9]. We slightly changed this method in connection to the fact that, unlike waterfowl, shorebirds are not the main object of hunting.

According to our methodology, the survey of each village includes two stages. First, an in-depth interview is conducted with 2–3 experts to identify on a qualitative level the general picture of how shorebird hunting occurs at this place and how important it is for local hunters. As a result of the interviews, we learned about the general picture of shorebird hunting in the village.

The second step was the survey using anonymous questionnaires that were filled out by the hunters themselves. The questionnaire was made as short as possible, because every extra question increases the probability that the hunter would find the questionnaire too complicated and would not want to waste time completing it.

The majority of hunters do not distinguish species of shorebirds. Therefore, in anonymous questionnaires, we mostly do not use the names of shorebird species, but ask hunters to divide shot shorebirds into the following groups:

Whimbrel, which is very popular and well known to Kamchatka hunters

other large and medium shorebirds with the exception of Whimbrel

small shorebirds.

Next, we consider results related to Whimbrel and these two groups of species. In addition, we asked to list the species of shorebirds shot if the hunter knew them, but only 6% of respondents did so.

We visited all towns and villages of Kamchatka connected with Petropavlovsk-Kamchatsky by car, except for a few villages located in the central part of the peninsula, where it is not possible to hunt shorebirds. It was much more difficult to survey remote villages, which have no automobile connection with capital of the region, as they can only be reached by plane or helicopter. The flights are rare and very expensive. But we still managed to visit some of them.

Sampling. Since it was not possible to create a random sample, we used two methods:

Asked hunters to fill out a questionnaire at the time when they visited the office of the hunting society in order to return their seasonal hunting permit. This method was mainly used in Petropavlovsk-Kamchatsky with the help of employees of the hunting society office.

Snowball Method, where each hunter, when filling out a questionnaire, gave the contact details of one or more other hunters. This method gives good results in small villages, where we received help from representatives of the hunting society in each village.

In total, after discarding incorrectly filled out forms, we got 402 forms suitable for processing.

Extrapolation. The number of shot shorebirds was calculated separately for:

whimbrel

all other big and medium shorebirds except Whimbrel

all small shorebirds.

First, the average hunting bag (the number of birds shot by a hunter) for each administrative district of Kamchatka was calculated. Then the districts were unified into five groups based on their geographical features and level of participation of local residents in shorebird hunting. Then the average hunting bag in each area group was then multiplied by the number of hunters, who had received hunting permits in each of the five area groups in the previous year.

3 Main results

	Area of Kamchatka					
	North	West	East	Centre	South	Total
Number of anonymous questionnaires	107	60	55	14	166	402
% surveyed hunters	39.0	17.0	11.0	2.0	5.0	8.0
% of hunters who shot Whimbrels	21.2	96.6	100.0	0	51.2	55.1
% of hunters who shot big and medium shorebirds (except for Whimbrel)	9.6	27.6	3.7	0	4.9	9.2
% of hunters who shot small shorebirds	11.5	34.5	22.2	0	7.3	14.3

Fig. 1. Number of surveyed hunters and % of hunters who shot shorebirds

	Area of Kamchatka					
	North	**West**	**East**	**Centre**	**South**	**Total**
Number of hunters	275	343	481	572	3244	4915
Average number of shorebirds shot by one hunter per year						
Whimbrel	2,06	23,21	15,85	0	6,45	8,86
Big and medium species other than Whimbrel	0,52	1,93	0,04	0	0,24	0,53
Small species	1,62	10,07	0,44	0	0,61	2,23
Total number of shorebirds harvested per year by all hunters						
Whimbrel	566	7960	7625	0	20928	37078
Big and medium species other than Whimbrel	143	662	18	0	791	1614
Small species	444	3454	214	0	1978	6090
Total	**1153**	**1207 6**	**7857**	**0**	**23697**	**44782**

Fig 2. Estimation of number of shot shorebirds

Below are the comments by species groups.

3.1. Whimbrel

Whimbrel is a very popular bird for hunting in Kamchatka among all waterfowl and shorebird species. According to our data, 55 % of hunters harvest Whimbrel in Kamchatka in 2019. The average bag of one hunter (total number of harvested birds divided on total number of hunters who received permissions) across Kamchatka made 8.9 Whimbrels.

The greatest number of Whimbrels is shot by hunters on the western coast of Kamchatka in the Tigilsky and Sobolevsky districts. The average hunting bag (birds per one hunter) for the year was 23.2 birds, and the total number of birds shot was about 8,000 Whimbrels. In the Tigilskiy district, located to the north of Sobolevsky, the average catch was slightly lower (18.4 vs 26.6, respectively). On the eastern coast of the peninsula (Ust-Kamchatsky district), the average bag per one hunter was 15.9 birds, while the northern districts shot fewer Whimbrels – an average of 2.0 birds per year.

In the central districts, Bystrinskiy and Mil'kovskiy, the Whimbrel, as well as other shorebird species, do not stop on their southward migration, so nobody hunts them.

The main part of Whimbrels is harvested in the south and southwest of Kamchatka (Elizovskiy and Ust-Bolsheretskiy districts). More than 90 % of the population of Kamchatka is concentrated in southeast of Kamchatka in the city of Petropavlovsk-Kamchatsky, towns of Yelizovo, Vilyuchinsk and nearby settlements. The total number of hunters (with the seasonal permits) in Petropavlovsk-Kamchatsky and nearby towns is 3,244 people, which is 66% of all hunters in Kamchatka. For urban hunters, the best places for hunting Whimbrels are during their southward migration. The greatest number of their stopovers are located near the town of Ust-Bolsheretsk, in the southwest of the Kamchatka Peninsula. This area is connected with Petropavlovsk-Kamchatsky by a fairly good road. The trip by car from Ust-Bolsheretsk to the regional center takes three to four hours. The average number of Whimbrel taken by one city hunter is not so high – 6.5 birds, because many hunters living in the city go hunting rarely. But since there are quite a lot of hunters in the city, about 21,000 Whimbrels are taken there, or more than half of the total number of Whimbrels taken in Kamchatka.

Our results show that the average hunter's bag mainly depends on their ability to reach out to Whimbrel autumn concentrations. The total number of the birds taken depends on the number of hunters as well. In the southwest of the Kamchatka Peninsula, both parameters coincide. In Sobolevsky and Tigil'sky districts (western coast of Kamchatka) there are a lot of Whimbrels, but there are comparatively few hunters. In the east coast, in the Ust-Kamchatka area, there are more hunters but there are fewer Whimbrel than on the western coast, and the average number of shot Whimbrels is rather high. And further North, in Penzhinsky and Olyutorsky districts, there are no good places to hunt Whimbrels, as there are few hunters either (Fig. 2).

As our calculations have shown, the total number of Whimbrels shot in Kamchatka is about 37 thousand per year. This is a very large number compared to published estimates [1, 2], where the total EAAF population of the Whimbrel is supposed to be about 55,000 birds. Considering that besides Kamchaka Whimbrels are also harvested in other regions, our results suggest that the flyway population estimate of 55,000 birds for Whimbrel is significantly lower than the reality. Published observations of Whimbrel concentrations in the several staging places [10, 11, 12, 13] and our long-term observations [14, 15] in Kamchatka also confirm this. We believe that hunters of Kamchatka have the greatest impact on the Whimbrel population on the flyway. But to verify this conclusion, we need to survey other parts of the flyway.

3.2 Far Eastern Curlew and Godwits.

According to the anonymous questionnaire survey, in 2019, 9.2% of hunters shot big and medium shorebirds in Kamchatka. The average of shorebirds taken by one hunter in Kamchatka was 0.5 birds per year. the total number of harvested shorebirds is estimated at 1,600 individuals. Their spatial distribution mostly overlaps with the spatial distribution of the Whimbrel bag. On average, according to anonymous questionnaires, there was 1 large shorebird of other species for each 22 Whimbrels.

However, interviews show that hunters often shoot large shorebirds, including Bar-tailed and Black-tailed Godwits when hunting Whimbrel. Some hunters do not know the correct name for these species, though many are aware of them. Officially, the Black-tailed Godwit is a game bird, while Bar-tailed Godwit used to be a game bird until 2018, and from 2019, it was included in the Kamchatka Red Book, also in part thanks to the efforts of the authors of this report. However, it is still being shot – 6 % of anonymous questionnaires reported cases of Godwit harvesting (usually hunters do not distinguish between two species of Godwits). Information on protected status does not reach hunters.

According to the interviews and other our studies in recent years, the Far-Eastern Curlew is also harvested regularly, though it has been included in the Red Data Book of Russia for a long time.

Here we should distinguish between "proper poaching", when hunters do it on purpose – they shoot all large waders that have come close to them, and "accidental poaching", when hunters shoot Far Eastern Curlews by mistake. The main reason for the error is the fact that young Curlews have a shorter bill than adults (similar to the Whimbrel's bill), and they confuse these two species.

Black-tailed Godwit has rather many significant staging places on west coast of Kamchatka. We have the data, and we ourselves have observed illegal hunting on it before the beginning of the hunting season.

Bar-tailed Godwit makes very long non-stop flights during seasonal migrations. We assume that only a relatively small part of the population of Bar-tailed Godwits visits Kamchatka during migration, and therefore hunting on this species here may not have a significant impact on the size of the whole population.

3.3 Other big and medium shorebirds

Other big species include Grey Plover and Pacific Golden Plover. Both this species migrate late and do not make any concentrations during their southward migration. However, there is no doubt that they are shot by hunters in Kamchatka. Our expert estimates are that from tens to 2–3 hundred individuals of each of these species are shot per season.

3.4 Small shorebirds

Small shorebirds are not actually an object of purposeful hunting. Many of them are non-game species in Russia, which are not supposed to be shot. According to informal interviews, hunters do not shoot them very often.the main reason of hunting small shorebirds is often the absence of other game. It is easy to shoot some small shorebirds in a flock, when the birds are sitting at the edge of the water or fly by.

According to the anonymous questionnaires in 2019, 14.3 % of hunters shot small shorebirds. The average harvesting across Kamchatka was 2.2 birds per hunter per year. The total number of birds taken calculated by extrapolation was over 6,000. More than half of the birds were taken in the western regions (Sobolevsky and Tigil'sky districts), where the average hunter's harvesting per year was 10 small shorebirds. In the other regions it was less than 2 birds per year.

When reporting the number of small shorebirds shot in anonymous questionnaires, hunters never indicate species. In interviews, hunters noted that they shoot mainly flocks of small shorebirds. They never think about the species, which they shot. With such type of shooting, it is evident that for every bird shot, there may be an equal number of wounded birds. These birds will die later, and such birds are not mentioned in either hunter questionnaires or in our calculations. So, the number of small shorebirds shot can be estimated nearly two-fold as compared with the data of the questionnaires, and the total number of small shorebirds shot, both legally and illegally, can be estimated to be at least 10,000 individuals.

When we studied the migration of birds in Kamchatka in the late 1970s and during 1980s, we noted hunting of small shorebird as a common occurrence. Now it had decreased, as probably ammunition has become more expensive. This hypothesis still needs to be tested. In the 2010s, most legal hunters with licenses reported that they rarely hunt small shorebirds, or even medium shorebirds, preferring only Whimbrels. But our knowledge on poachers, particularly in commercial salmon fishing camps is still very limited. And in Kamchatka there are hundreds of salmon fishing stations with thousands of people exactly in the period of late summer and autumn migration. The study of this type of hunting should be the subject of a special additional research. They can shoot whatever they see, including critically endangered species such as Spoon-billed Sandpiper (SBS).

As this species is very rare, the probability of this fact is not great, though the number of shot small shorebirds is rather significant. It is very difficult to estimate it. We can talk about handfuls of Spoon-billed Sandpiper shot every year. The probability of SBS shooting in the northern part of the Kamchatka west coast is supported by the fact that signal transmission from three of the nine birds with radio transmitters stopped sending signal from this area [16].

4 Conclusions and recommendations

Arctic nesting migratory bird populations are declining significantly in several regions and a conservation approach at the level of major flyways is needed to improve the status of their populations.

Our study of Arctic-nesting shorebirds hunting is the first significant research in this direction. It allowed to get the first insight into shorebird hunting in Kamchatka. To obtain a complete picture along the northern end of the EAAF, it is necessary to conduct similar studies in all other regions of the Russian Far East and Russian Arctic. Just as well, hunting assessment and monitoring in Kamchatka should be continued in order to be able to track the hunting process and obtain more detailed data.

The results of our study showed that hunting shorebirds in Northeast of Russia (the northernmost part of the EAAF) can negatively impact Arctic shorebird populations, especially endangered species. Hunting of protected shorebirds is often the reason of low level of awareness of hunters and lack of ability to distinguish shorebird species. Awareness and education must be improved.

Hunting for small shorebirds in Kamchatka has declined significantly over the past 40 years, however it still remains an additional threat to the conservation of the rare protected species, including Spoon-billed Sandpiper. In some villages it could be high.

Although our study in Kamchatka filled a significant gap, the available data on shorebird hunting in the Russian Northeast is still quite insufficient to propose and substantiate an effective conservation program for Arctic shorebirds. Also, more research needed to expand knowledge of Kamchatka, particularly of the northern areas.

The methodology we used in Kamchatka gave acceptable results. It can be used in other regions of the Russian Far East and Arctic. However, for quantitative assessment of catch of rare species (for example, the Far Eastern Curlew), it must be supplemented with other methods.

Hunting control in Kamchatka, as in the entire Northeast of Russia, is weak. For shorebird conservation, special education efforts are needed among hunters and game managers. Since hunters often shoot protected waders due to their inability to identify species and lack of knowledge about the protection status. In addition, hunting inspectors cannot control this process because they also cannot identify many species of birds.

Mainstream Kamchatka hunters are really interested in knowing the species they harvest. But there is no place where they can learn it. The publication of a field guide of waders of the Russian Far East is essential for improving the level of knowledge of hunters and hunting officers.

Various educational activities to increase the level of knowledge of hunters and game managers can also be useful. These activities can be conducted together with hunting societies and regional hunting agencies. As we found out in the course of our research, hunting societies in Kamchatka are very open for such activities, but they have limited resources.

To effectively conserve shorebirds, we need significantly more data on hunting bag. First of all, we need information on other regions to get a complete picture along the entire flyway. Secondly, it is necessary to move from a one-time survey to year-to-year harvest monitoring.

5 Acknowledgments

The project was supported by Karl Kaus Fondation, the East Asian – Australasian Flyway Partnership, Australian Government and UNEP/CMS with the participation of the Russian Science Foundation (project no. 18-18-00309). The Russian Bird Ringing Centre provided data on shorebirds ring recoveries. The Forestry and Wildlife Protection Agency of Kamchatka provided logistics help and data on the number of hunters and hunting permits issued. Kamchatka Hunter Society provided support in establishing contacts with hunters and distributing anonymous questionnaires. Local communities in Il'pyrskiy, Korf, Manily, Oktyabrsky, Paren', Sobolevo, Ustyevoe, Ust-Bolsheretsk, and Ust-Khairyuzovo Kamchatka villages were cooperative in estimation of their hunting efforts. We are grateful to all these organizations and Kamchatka hunters who completed questionnaires and answered interview questions to support our study.

Special thanks for contributing to our research to Alexander Matsyna, Evgeniy Lobkov, Nikolay Gerasimov, Sergei Kharitonov, Sergei Vakhrin, Andrey Gorovenko, and Julia Zavgarova.

References

1. M. Bamford et al., Migratory shorebirds of the East Asian-Australasian Flyway: Population estimates and internationally important sites (Canberra: Wetlands International – Oceania , 2008)
2. J.R. Conklin et al., Prioritizing migratory shorebirds for conservation action on the East Asian-Australasian Flyway (Hong Kong, 2014)
3. J.W. Pearce-Higgins et al., A global threats overview for Numeniini populations: synthesizing expert knowledge for a group of declining migratory birds, *Bird Conservation International*, **27(1)**, pp. 6–34 (2017)
4. D.S. Melville et al., Shorebirds along the Yellow Sea coast of China face an uncertain future: A review of threats Emu, **11**, 100–110 (2016)
5. C.E. Studds et al., Rapid population decline in migratory shorebirds relying on Yellow Sea tidal mudflats as stopover sites, *Nature Communications*, **8**, 14895 (2017)
6. L.C. Naves et al., Shorebird subsistence harvest and indigenous knowledge in Alaska: Informing harvest management and engaging users in shorebird conservation The Condor, *Ornotholodgical applications*, **121**, pp. 1–19 (2019)
7. E.J.W. Cajiao et al., Extent and potential impact of hunting on migratory shorebirds in the Asia-Pacific, *Biological Conservation*, **246**, pp. 1–12 (2020)
8. CAFF Congress MB5: Worldwide partnerships to conserve migratory birds: The Arctic Migratory Bird Initiative Available from: https://www.caff.is/arctic-migratory-birds-initiative-ambi [Accessed 8th October 2021] (2018)
9. E.E. Syroetchkovskiy and K.B. Klokov, Using questionnaire method to study the impact of hunting on waterfowl in the Russian Arctic Cazarka, **13**, pp. 76-103 (2010)
10. Yu.N. Gerasimov, Spring migration of waders in the mouth of Bolshaya River (West Kamchatka), *Ornithology*, **28**, 222 (1998)
11. F. Huettmann and Yu. Gerasimov, Using Sampling to obtain density estimates for Whimbrels (Numenius phaeopus) and other birds in the coastal tundra of the Moroshechnaya River Spit, Sea of Okhotsk, during fall migration, *Avian Ecology and Behavior*, **8**, pp. 49–69 (2002)
12. R.F Schuckard. et al. Shorebird and Gull Census at Moroshechnaya Estuary, Kamchatka, Far East Russia, During August 2004 Stilt, **50**, pp. 34-46 (2006)

13. Yu.N. Gerasimov et al. Some results of studying the summer-autumn migration of waders on the Bolshaya Vorovskaya River Lagoon, Western Kamchatka Topical issues of studying waders of Northern Eurasia, *Materials of the XI International conference (Minsk)*, pp. 15-19 (2019)
14. Yu.N. Gerasimov and N.N. Gerasimov, Whimbrel Numenius phaeopus on Kamchatka, *Russia Stilt*, **41**, pp. 48-54 (2002)
15. N.N. Gerasimov and Yu.N. Gerasimov, Study of migration of waterfowl and shorebirds of Kamchatka Materials International scientific conference "Areas, migrations and other movements wild animals" (Vladivostok, Russia, 2014), pp. 52-61.
16. Qing Ch et al., Post-breeding migration of adult Spoon-billed Sandpiper, *Wader study*, 127(3), pp. 200-209 (2020)

https://doi.org/10.1051/e3sconf/202337805004

Indigenous hunting in the Russian Arctic: toward sustainable wildlife resource management

Konstantin Klokov [1, 2, *]

[1]Doctor of Geography, Professor, Department of Regional Policy and Political Geography, 199034, Saint Petersburg State University Saint Petersburg, Russian Federation
[2]Senior Researcher, 199034, Peter the Great Museum of Anthropology and Ethnography (Kunstkamera), Russian Academy of Sciences, Saint Petersburg, Russian Federation

Abstract. Arctic wildlife resource management must secure biodiversity without infringing on indigenous hunting rights. The article focuses on the bottlenecks in resource management of game species hunted by Arctic indigenous population. It is based on data collected by the author on subsistence and commercial hunting by indigenous peoples in several Russian Arctic regions and on the results of CAFF Arctic Migratory Birds Initiative (AMBI) projects on the assessment of hunting impact on Arctic migratory birds. Three main problems of Arctic game resource management are discussed: gapes in legislation related to the contradiction between the right of Arctic indigenous peoples to hunt without permission and the need to restrict hunting in order to maintain the balance of ecosystems; lack of reliable information on hunting resources, including both population numbers and harvest size; the practically total removal of hunters from the management of hunting resources. The Russian game resource management system in its current state does not meet the Arctic conditions. It needs to be revised towards official liberalization of indigenous hunting and more flexible game resource regulation considering interests of different groups of hunters.

1 Introduction

The sustainable management of wildlife resources in the Arctic aims, on the one hand, to minimize risks to biodiversity and, on the other hand, to ensure indigenous peoples' harvesting rights. However, while fully recognizing the importance of traditional hunting practices and the political dimension of indigenous rights, it should be noted that traditional harvesting is not necessarily synonymous with its sustainability [1]. The indigenous hunters' interests do not always coincide with the conservation goals, and competition for resources may arise between indigenous and non-indigenous hunters. This result is a rather complex set of Arctic wildlife resource management problems that require an interdisciplinary approach to analyze and address.

In Alaska and Canada, these issues have been actively developed for more than half a century. classic works are H. Huntington's monograph [2], in which he describes and evaluates various federal, state, local, and cooperative wildlife management regimes and discusses international treaties governing the migration of waterfowl, whales, polar bears, and caribou, and publications by F. Berkes [3], who outlines the role of traditional knowledge in wildlife resource management. In Canada and Alaska there are a large number of publications with statistical data on indigenous subsistence hunting (e.g. [4]), as well as on the management of the most important game populations for indigenous peoples (primarily, caribou populations).

In Russia, there are no scientific publications specifically devoted to Arctic aboriginal hunting in general, but there are many works on the management of specific populations: wild reindeer, sable and others. Below we will focus on some of them.

This article discusses the resource management bottlenecks of the most important large and small game species hunted by the indigenous peoples of the Russian Arctic. It is based on data collected by the author in the course of many years of research on indigenous hunting, both subsistence and commercial, in various regions of the Russian Arctic from the Kola Peninsula to Chukotka. These studies used methods of informal and formal interviews, anonymous questionnaires, participatory mapping, content analysis of documents, and examination of statistical data. Of particular importance among them were the international project "Tundra. Drivers of Change in Tundra Ecosystems" (Norwegian Research Council Project, 2010-2014), which included three case studies in Kola Peninsula, Yamal and Taimyr, and a series of projects aimed at assessing the impact of hunting on populations of migratory Arctic birds (including geese, ducks, waders, seabirds, etc.) based on an anonymous survey of hunters. These studies were conducted jointly with Dr. Evgeny E. Syroechkovskiy in Chukotka, northern Yakutia, and the Nenets Autonomous District [5].

* Corresponding author: k.b.klokov@gmail.com

These projects were followed by a study of hunting pressure on Arctic-nesting bird populations on their migration routes, which was launched in the Northeast Russia in 2019 as part of the Arctic Migratory Birds Initiative (AMBI) of the Arctic Council Working Group on the Conservation of Arctic Flora and Fauna (CAFF) [6]. All these materials, as well as data from the official website of the Federal Center for the Development of Hunting Economy [7] and a number of scientific publications have been analyzed and summarized as part of the project of the Russian Science Foundation (No. 18-18-00309) and discussed in this paper.

2 Game resources of the Russian Arctic

The traditional use of wildlife resources is the basis of the distinctive cultural values of the indigenous peoples of the Russian Arctic. Game is an important resource for them. The list of the main game species hunted by the Arctic indigenous population includes two groups of birds: migratory waterfowl (geese and ducks) and sedentary grouse (ptarmigan, grouse, capercaillie, black grouse); fur-bearing animals (arctic fox in the tundra, sable and squirrel in the taiga, as well as several widespread species – fox, wolverine, ermine, muskrat, etc.); ungulates (wild reindeer, elk, snow sheep, musk ox), large predators (brown and polar bears and wolf), and the Arctic hare.

Over the past decades, a decline in the purchase price of fur-bearing animal skins, combined with an increase in the price for cartridges, hunting weapon and equipment, has made commercial hunting of most of these species unprofitable. Now, only wild reindeer and sable are essential for commercial hunting. For subsistence hunting, primarily feathered game (ducks, geese, ptarmigan) and ungulates (wild reindeer and elk) are important. Indigenous peoples hunt bears and wolves mainly for their own safety and to protect herds of domestic reindeer. Bears and ungulates are important for trophy hunting, however, indigenous people do not participate in it.

The game resource state in the Russian Arctic is generally satisfactory, but not for all species. Over the past two decades, there has been a sharp decline in the tundra populations of wild reindeer, especially in northern Yakutia, and in the Taimyr population of reindeer, the largest in Eurasia. The state of migratory birds resources (ducks and geese) varies in different sectors of the Russian Arctic. In the western sector, populations of Arctic-nesting birds migrating along the East Atlantic Flyway are mostly stable. On the contrary, in the eastern Arctic, many populations of waterfowl and waders following the East Asian-Australasian Flyway are depressed and continue to decline in numbers [8]. This is due to anthropogenic reduction of wetlands and strong hunting pressure at wintering areas and migration stopovers in the countries of East Asia. Hunting at migration stopovers in Kamchatka and Sakhalin can also affect the stability of some Arctic-nesting waterfowl and wader populations.

Some Arctic animals species, on the contrary, are increasing in their numbers [7]. This is, for example, the brown bear and, in some regions, the wolf. The increase in the number of predators significantly complicates the traditional reindeer husbandry of indigenous peoples. In addition, every year there are cases of human deaths from bear aggression.

3 The main features of indigenous hunting in the Russian Arctic

Unlike the central regions of Russia, where hunting is mainly a recreational activity, in the Arctic it is an important part of the subsistence economy. For the indigenous population of the Arctic, hunting is primarily a source of meat to feed the family and maintain ethnic traditions. Among the indigenous people, there are also those who is engaged in professional hunting, receiving their main income from it. These are mainly sable hunters and trappers in the northern taiga of Eastern Siberia, and wild reindeer hunters in Taimyr and Yakutia. In contrast, for most non-indigenous hunters in the Arctic, hunting is most often a hobby, and obtaining food products from hunting is secondary.

The traditional use of natural resources by the indigenous population is considered environmentally balanced. This is noted even in the federal law of the Russian Federation [9]. It is believed that the members of the traditional society when using natural resources follow the imperative: do not take more from nature than it is necessary to satisfy your essential needs. This imperative stems from the desire to minimize labor costs in traditional communities [10]. It took root as a tradition among all northern peoples and has enabled them to survive in the extreme conditions of the Arctic. In should be noted that, following it, humans can upset the environmental balance, if it is necessary to provide themselves with food and other necessary products of hunting. As Igor Krupnik showed [11], this imperative did not exclude local overhunting, sometimes on a large scale. The fact that the indigenous population considers hunting primarily as a source of food was confirmed by the results of anonymous questionnaire survey of indigenous hunters, which we conducted together with Dr. Evgeniy Syroechkovskiy in Chukotka, Yakutia and the Nenets Autonomous District. Answering the question, what the main importance of bird hunting for your family is, most of the indigenous hunters chose the answer "food supply".

Modern (non-traditional) society is widely known to be guided by the rule not to take from the ecosystem more than it can give without compromising resource reproduction. People of modern society are ready to limit their needs in order to preserve the balance of the ecosystem (at least in theory). it is implicitly assumed that such a restriction does not threaten their life and health. However, from the point of view of Arctic indigenous people, the need to limit their hunting needs for reasons of environmental ethics is not always obvious. Although there are no longer people in the Russian Arctic whose survival is completely tied to the results of hunting, there are indigenous communities where families' budgets depend mainly on the amount of game they caught. Therefore, it is not surprising that in response to the "provocative" question in our anonymous questionnaire, "Do you think it is acceptable to shoot red-listed species if it is necessary family sustenance?" we received up to 15% answers stating that these species can be shot for sustenance.

4 Hunting regulation

The regulation of hunting in the Russian Arctic is based on federal and regional laws and by-laws issued by federal and regional agencies. Federal legal acts apply to the entire territory of Russia and are focused primarily on the population of central regions, where the majority of hunters live. Although hunting legal acts have provisions specifically related to indigenous peoples of the North, in general they do not meet the specific conditions of the Arctic. A number of problematic situations are associated with this.

Firstly, according to the federal law "On Hunting" [12], hunting grounds are divided into two categories: public and assigned. Assigned lands are considered to be those which are leased to an organization or an individual entrepreneur for a certain period of time under a special agreement. Only the person who won first place in the auction can conclude such a contract. The contract obliges the lessee to ensure the rational use and protection of game at its own expense. The assigned hunting grounds may occupy up to 80% of the area of each region, so that the public lands are no less than 20%. However, in the Russian Arctic, the area of assigned hunting lands is still insignificant, and almost the entire territory belongs to public lands managed by public services. These services should ensure the rational use of game resources, but they have neither the financial nor the human resources to do so.

Secondly, according to the federal law "On Hunting", a hunter must be a person registered in the state hunting register. The hunter must have a hunting ticket and a permit for storing and carrying of hunting weapons, which is issued in accordance with the Federal Law "On Weapons" [13]. To fulfill all the formal requirements of the law on firearms is really difficult for indigenous hunters, especially for those who lead a nomadic lifestyle and rarely visit villages. Here is just one example. According to the existing rules, firearms must be stored in a locked metal box. However, a nomadic reindeer herder roaming the tundra cannot carry such a box on a reindeer sled. Without a rifle or a carbine, he cannot protect his reindeer herd from wolves and bears. Besides, he needs a hunting gun to provide his family with food during nomadic migrations.

In addition, there are gaps in the laws related indigenous hunting rights. Because of this, poachers can often take advantage of indigenous hunters, and indigenous hunters cannot always get sufficient access to game resources.

The Russian Law "On Hunting" (Article 12) [12] states the three most important types of hunting for Arctic residents:

- Commercial hunting: this is hunting for income, and the products of hunting may be sold
- Amateur and sport hunting: this is recreational hunting, the products of which must be used only for personal consumption or as a trophy
- Hunting in order to ensure the traditional way of life and traditional economic activities of indigenous small-numbered peoples of the North, Siberia and the Far East of the Russian Federation (refer to it hereafter as "traditional hunting").

According to Article 19 of the Federal Law "On Hinting" [12], the products of traditional hunting may both be used for personal consumption and sold to commercial companies that purchase hunting products. Traditional hunting is allowed only to indigenous peoples, and they have the right to hunt without permits, which means that in fact they may shoot or trap unlimited amounts of game. Only in some cases their hunting is restricted by special decisions of the regional authorities. These are very good rules from the point of view of protecting the rights of indigenous peoples. At the same time, they create problems, since many non-indigenous hunters shoot too much game, and in case of inspection they claim that the game was shot by their neighbors or friends from among the indigenous people. This is exactly what happens when sable furs and wild reindeer are taken in excess of the established limits.

5 Regulation of the use of hunting resource

The Russian system of game resource management system includes: limitation of hunting periods, restriction of hunting territories, and regulation of game resource harvesting. Let us consider the latter in more detail. Two approaches are used for this purpose: limits (quotas) and daily or seasonal norms of harvesting.

The limits are set by regional authorities, but for the most important species they must be agreed upon at the federal level (i.e., in Moscow). This list includes several Arctic species: wild reindeer, elk, musk ox, snow sheep, lynx and sable. The law states that the limits for them in each region are approved annually on the basis of information on the population size and the norms for withdrawal rate (potential biological removal) for each species. However, for this approach to work, firstly, regular population surveys must be conducted. Secondly, hunters must regularly report on how much game they have shot or trapped without significantly understating their hunting bags. Unfortunately, this is not always the case.

Let's consider, for example, how this approach works in managing the resources of the two most important commercial species for the indigenous population of the North of Russia: sable and wild reindeer.

At first glance, when one analyzes the data presented on the official website of the Federal Center for the Development of Hunting [7], the number of harvested animals everywhere meets the standards of game taking, and there are no problems in the management of game resources. However, the results of expert assessments show that these data are far from reality. Of course, it is difficult to check all estimates, but just in the case of sable, there is a good opportunity to do so. The main part of the sable skins shot and trapped by hunters is sold at international fur auctions, and there are official and accurate data on the number of sable skins sold annually. A comparison of these data showed that the number of skins sold at auctions is 1.5-2 times higher than the number of officially shot or trapped sables [14]. In addition, it should be noted that not all skins are sold at international auctions, since a significant part of them is sold inside the country. The reason for this discrepancy is that hunters have to pay for each license and they are interested to reduce the data on the number of harvested sables. This example shows that official records produce severely underreported data if hunters are not motivated to tell the truth.

Another example is the management of tundra wild reindeer populations. The largest Taimyr reindeer population is best studied. It grew steadily from the beginning of the 1960s, when the first aerial survey of its numbers was carried out, until 1985, and in 1985-1990, it stabilized at 590-620,000 head. In the 1970s and 1980s, intensive commercial shooting of reindeer began in the Taimyr, as well as in the neighboring regions of Evenkia and Yakutia, where reindeer annually went for wintering. For commercial shooting of reindeer, a special state enterprise was created – Taimyr Gospromkhoz (state farm). During this period, about 80,000 animals were shot annually, and young reindeer dominated among the harvested animals [15]. Both indigenous and non-indigenous hunters participated in hunting. Hunting was controlled by state agencies. The population size and number of shot reindeer were monitored. Aerial surveys were conducted annually [16, 17].

In the 1990s, during the crisis connected with market reforms, monitoring stopped. The shooting of reindeer was almost halved, and the herd increased, presumably, to 1 million heads by 2000. Later, the shooting increased again to a level of about 80,000 reindeer per year, of which half were shot without permits [15]. The reindeer population number and hunting were no more monitored. The 2017 aerial counts showed that the population had decreased to 384,000 heads. According to expert opinions [15, 17], there were several reasons for this. Firstly, uncontrolled shooting became selective – mainly large animals were killed. Secondly, since 2010, a mass harvesting of velvet antlers (young non-ossified reindeer antlers) began, which hunters cut off from the largest males, approaching the river-crossing animals in boats. After that, about 70% of the males died, and the rest were not capable of normal reproduction that year. The selective shooting and cutting of velvet antlers severely disrupted the structure of the population and considerably reduced its productivity. Only in 2019, the cutting of antlers from live reindeer was prohibited and measures were taken to monitor compliance with this ban. Third, climate change has played a role. The rivers began to clear from the ice earlier, and female reindeer are now unable to reach calving areas in time. So, many calves die.

According to experts [15, 17], in order to stop the reduction of the reindeer population, the structure of which is severely disrupted, it is necessary to reduce harvesting to 3% of the population (i.e., to about 11,000 animals). However, the authorities have so far ignored this opinion. In the hunting season of 2020–2021, the Executive Order of the governor of the Krasnoyarsk Krai allocated limits of 30,405 reindeer to the hunting farms of Taimyr and Evenkia [18]. In addition, according to the Resolution of Krasnoyarsk Krai Government, each member of the indigenous population has the right to take 8 reindeer per year in Taimyr and 7 reindeer per year in Evenkia as part of traditional hunting, while the number of animals shot by indigenous hunters is not registered. In addition, a significant portion of Taimyr reindeer migrate for wintering to Yakutia, where they are shot by local hunters according to the limits set by the Government of the Republic of Yakutia. According to experts, such a strategy will soon result in the collapse of the Taimyr population [15, 17].

Thus, the hunting of Taimyr reindeer is a case of a common-pool resources drama [19, 20]. Reindeer are hunted by several groups of hunters from three different regions (Taimyr, Evenkia, and Yakutia). Some of them shoot animals for commercial purposes and some for subsistence. Unless their actions are appropriately coordinated, the depletion of the overall resource pool is inevitable.

To manage the resources of game birds in Russia, a different approach is used: the maximum possible hunting bag per one hunter per day or per hunting season is established. Rare bird species are completely prohibited from hunting. In theory, this approach allows to reduce the hunting pressure on the population, if there is evidence that its numbers are decreasing. However, in order to use it efficiently, three conditions should be met:

- Hunters must know and comply with the established norms and rules of hunting, including even the places where permanent control over them is impossible (and this is a large part of the Arctic territory).
- Hunters should be able to distinguish one species from another, before shooting.
- Trends in the population number and in harvesting of each species should be known to the managing agencies.

Surveys in Kamchatka and Sakhalin have shown that hunters, as a rule, are well aware of mass species of Arctic birds, but do not distinguish rare ones [6]. Because of this, they often accidentally shoot endangered species. For example, along with the widespread and allowed to hunt white-fronted goose, they often shoot the lesser white-fronted goose, which looks similar to the first one but is included in the Red Data Book of the Russian Federation. Virtually all hunters are unable to distinguish between species of waders, with the exception of a few of the largest. Waders in the Arctic are not prime targets for hunting and are harvested in small numbers. However, there are many rare and critically endangered species among small and medium-sized waders. For example, the spoon-billed sandpiper which is endemic to Chukotka. Its abundance in the world is only about 200 pairs and continues to decline [21]. In addition, hunters in the Arctic shoot banned species of waterfowl because they are unaware of the ban on hunting.

hunting authorities in the Arctic regions do not have reliable information on both game population number and the number of birds shot by hunters. Here are some examples of the severe underestimation of the Arctic bird resources harvesting. According to the calculations of V. Krivenko and V. Vinogradov [22], based on official data from state hunting agencies, the annual number of harvested ducks throughout Chukotka was estimated at 2,600, and geese at 1,600 birds (it should be noted that the authors themselves indicated that this assessment was greatly underestimated). According to one of our surveys in Chukotka in 2005, in one village of this region alone (we do not disclose the name of the village for privacy reasons) hunters shot more than 2,800 geese and about 5,000 ducks (including almost 400 Spectacled and Steller's Eiders which were prohibited from hunting).

A survey using an anonymous questionnaire of hunters in 2019 in Kamchatka [6] revealed that the total number of waders taken there by legal hunters was about 45,000 birds. Most of these birds were Arctic-nesting Whimbrel, which fly south through Kamchatka in the fall, but with them hunters also killed a significant number of rare species of Arctic waders. However, according to the Kamchatka Forestry and Wildlife Protection Agency, only 2,345 waders were shot in this region in the fall of 2019, and 2,510 waders in the fall of 2018. Many Arctic waders fly further south through Sakhalin Island. According to a similar survey conducted in 2020 on Sakhalin, at least 50,000 waders were taken on the island during the year, while according to the official data, only 302 waders were taken there. Thus, the information available to the agencies that regulate and control hunting is very far from reality.

6 Conclusion

We can conclude that the Russian game resource management system does not match the conditions of the Arctic. The bottlenecks discussed above in the management of Arctic game resources can be reduced to three main problems. The first is gapes in legislation related to the contradiction between the right of Arctic indigenous people to hunt without permission, and the need to restrict hunting in order to maintain the balance in ecosystems. The second is the lack of reliable information about game resources, including both population numbers and number of shot animals. The third is the practically total exclusion of hunters from the management of hunting resources.

All of the above problems can be solved through adaptive management of game resources and co-management with the full participation of indigenous hunters. Indeed, despite the fact that the restriction of traditional hunting by the authorities contradicts the Russian federal law, the indigenous hunters themselves have the right to limit their harvesting and they are really interested in this, so that not to undermine the base of game resources. However, to do this, they must be allowed to participate in the process of managing Arctic game resources.

Thus, in the Russian Arctic, there is a need to review the methods of game resource management in the direction of the official liberalization of indigenous hunting, more flexible and wise regulation of game resources, taking into accounts the interests of different groups of hunters.

Indigenous communities perceive themselves as guardians of traditional values, and wild animals are included in various contexts of traditional culture: myths, beliefs, rituals, etc. Thus, game species, as well as the natural environment as a whole, are an integral part of the ethno-cultural heritage, and representatives of traditional culture are the keepers of this heritage. In this context, traditional and modern ideas about the sustainability of game populations could complement, rather than contradict each other. To achieve this, it is important to adapt the mechanism of game resource management to the mentality of the indigenous peoples representing traditional cultures. Special multidisciplinary research integrating the approaches of applied environmental studies, game management, sociology and ethnology can contribute to solving this problem.

The research was supported by the Russian Science Foundation (project No. 18-18-00309).

References

1. CAFF 2013 Arctic Biodiversity Assessment. Status and trends in Arctic biodiversity. Conservation of Arctic Flora and Fauna, Akureyri
2. H.P. Huntington, *Wildlife Management and Subsistence Hunting in Alaska* (London, Belhaven Press, 1992)
3. F. Berkes, *Sacred ecology, traditional ecological knowledge and resource management* (Philadelphia, Pennsylvania, USA: Taylor and Francis, 1999)
4. C.L. Brown et al., Harvests and Uses of Wild Resources in 4 Interior Alaska Communities and 3 Arctic Alaska Communities, 2014, *Technical Paper No. 426. Alaska Department of Fish and Game, Fairbanks* (2016)
5. E.E. Syroetchkovskiy and K.B. Klokov, Using questionnaire method to study the impact of hunting on waterfowl in the Russian Arctic Kazarka, *Zoologicheskiĭ Zhurnal*, **13**, pp. 76-103 (2010)
6. K. Khlokov et al., First attempt to evaluate hunting Pressure on Shorebirds in Kamchatka: Progress

Report Spoon-billed Sandpiper Task Force, *News Bull*, **22**, pp. 31-34 (2020)

7. Federal Center for the Development of Hunting Available online: http://www.ohotcontrol.ru (2021)
8. CAFF Congress 2018 MB5: Worldwide partnerships to conserve migratory birds: The Arctic Migratory Bird Initiative 2018 Available online: https://www.caff.is/arctic-migratory-birds-initiative-ambi (2018)
9. Federal Law "On Territories of Traditional Nature Use by Indigenous Small-numbered Peoples of the North, Siberia and the Far East of the Russian Federation" of 07.05.2001 N 49-FZ Available online: http://www.consultant.ru/document/cons_doc_LAW_31497/ (2001)
10. A.N. Yamskov, *Ecologically Significant Cultural Archetypes of Human Behaviour Ethnoecological Aspects of Spiritual Culture* (Moscow, IAEA RAS, 2005), pp. 266-296
11. Krupnik I Arctic Adaptations: Native Whalers and Reindeer Herders of Northern Eurasia (Hanover and London, University Press of New England, 1993)
12. Federal Law "On Hunting and Conservation of Hunting Resources and on Amendments to Certain Legislative Acts of the Russian Federation" of 24.07.2009 N 209-FZ Available online: http://www.consultant.ru/document/cons_doc_LAW_89923/ (2009)
13. Federal law "On weapons" of 13.12.1996 N 150-FZ Available online: http://www.consultant.ru/document/cons_doc_LAW_12679/ (1996)
14. A. L. Waisman, You are my sables, *Russian Hunting Journal*, **8**, pp. 10-20 (2015) (in Russian)
15. V. Mikhailov and L. Kolpaschikov Population dynamics of the Taimyr wild reindeer herd, *Int. J. Environ. Stud.*, **74(5)**, pp. 862–883 (2017)
16. K.B. Klokov, Northern Reindeer of Taymyr Okrug as the Focus of Economic Activity: Contemporary Problems of Reindeer Husbandry and the Wild Reindeer Hunt, *Polar Geography*, **21(4)**, pp. 233-271 (1997)
17. L.A. Kolpaschikov et al. The modern history of the Taimyr wild reindeer population: dynamics, management, threats and ways of conservation, *Proceedings of the Karelian Research Centre of the Russian Ac. Sci.*, **11**, pp. 5-20 (2019) (in Russian)
18. Krasnoyarsk Territory Governor Decree "On Approving the Limits of Hunting Resources Production in Krasnoyarsk Territory during the 2020-2021 Hunting Season" of 30.07.2020-205-ug. Krasnoyarsk Territory. Official Internet portal of legal information. Available online: http://zakon.krskstate.ru/0/doc/68473 (2021)
19. T. Dietz et al., *The Drama of the Commons* (Washington, DC, National Academy Press, 2002)
20. T. Dietz et al., The Struggle to Govern the Commons, *Science*, **302**, pp. 1907–1912 (2003)
21. Ch. Zockler et al., The Spoon-billed sandpiper Calidris pygmaea conservation project in 2019 and 2020: population trends continue to be negative BirdingASIA, **33**, pp. 51–56 (2020)
22. V.G. Krivenko and V.G. Vinogradov, BIODAT. Waterfowl resources of Russia. Available online: http://biodat.ru/doc/ducks/index.html (2001) (in Russian)

E3S Web of Conferences **378**, 05005 (2023) https://doi.org/10.1051/e3sconf/202337805005
IIRPCMIA 2021

Efficiency of sorption-biological treatment of oil-polluted coastal areas in the Euro-Arctic region

Vladimir Myazin[1,*]

[1] Institute of North Industrial Ecology Problems – Subdivision of the Federal Research Centre "Kola Science Centre of the Russian Academy of Science", 184209 Apatity, Russian Federation

Abstract. A biotechnology based on sorption-biological treatment for the restoration of the Arctic coastal areas after oil spills was proposed. A hydrocarbon-oxidizing microbial preparation based on microfungi (Penicillium janthinellum, Penicillium simplicissimum) and bacteria (Pseudomonas fluorescens, Pseudomonas putida) was created. Granular activated carbon, thermoactivated vermiculite, and peat were used as sorbents. The sorbents are concentration centers for microorganisms, prevent the spread of hydrocarbons, and improve the condition of the polluted substrate by optimizing the water-air regime. After treatment, sorbents are not removed from the soil. As a result of laboratory studies at +10°C, the hydrocarbon content in the polluted coastal sand was reduced by 44-60%, and in the coastal soil by 25-68% after three months. This technology can be used at the final stage of coastal areas cleaning, as well as the only method of cleaning in highly sensitive areas. The use of aboriginal hydrocarbon-oxidizing microorganisms, typical of most coastal substrates of the Euro-Arctic region, excludes changes in the structure of the microbial community in these areas. The adaptation of these microorganisms to extreme climatic conditions allows biological treatment to be carried out with high efficiency in the Subarctic and Arctic.

1 Introduction

In recent years, the Arctic region has become the center of intersection of national interests of the Arctic states, which leads to the formation of a new system of global and regional security. Reserves of natural resources in the Arctic, primarily hydrocarbons of the continental shelf, cause competition for their development and exploitation, including control over the shelf zone of the Arctic seas [1,2].

At the same time, the Arctic is an area of valuable biodiversity. The growth and expansion of infrastructure, industrial activities and resource exploitation can damage vulnerable ecosystems of the Arctic [3]. The development of the Northern Sea Route is also a negative factor for oil pollution in coastal areas of the Arctic seas [4, 5].

Restoration of coastal ecosystems after oil cleanup can take a long time. A promising method of restoration of oil-polluted coastal areas is biotechnological, based on acceleration of natural processes in the environment [6-7], including the use of sorbents [8-9].

The purpose of this study is evaluation of the effectiveness of the sorption-biological method for cleaning and remediation of oil-polluted coastal areas in the Euro-Arctic region.

2 Materials and methods

2.1 Substrates

Coastal soil and sand from the coast of the Pechenga Bay of the Barents Sea were taken as substrates for laboratory studies (Figure 1).

Fig 1. Areas of substrate selection for laboratory experiments

* Corresponding author: myazinv@mail.ru

Table 1. Characteristics of the substrates for the laboratory experiment

Type of substrate	TOC (%)	pH	Dehydrogenase activity (mg TTC/10g)	Total petroleum hydrocarbon (mg/kg)	Description of the site
Soil	4.74	5.11	1.79	97	salt marsh: 0-2 cm – organogenic, deeper than 2 cm – medium loam
Sand	0.10	7.19	0.01	46	sandy beach

2.2 Oil-water emulsion

The oil-water emulsion was prepared from three types of oil: light, medium and heavy. The ratio of oil: water = 1 : 10. Before preparing the oil-water emulsion, volatile components were removed from the oil by heating it. After this treatment, the oil corresponded in its characteristics corresponded to the oil which had been on the water surface for 1-2 months. sea water of 22‰ salinity was used to prepare the oil-water emulsion.

2.3 Sorbents

We used as sorbents:
- granular activated carbon (GAC) with the size of 2-3 mm granule. GAC consists of 87-97% carbon; its sorption capacity of hydrocarbons is 200-980 g/kg.
- thermoactivated vermiculite (VER) is a mineral of hydromica group with a layered structure. When heated, vermiculite swells and increases in volume several times; it has a high absorption coefficient.
- Milled peat with a low degree of decomposition (no more than 35%).

2.4 Microbial preparation

On the basis of microorganisms isolated from seawater and soil of the Barents Sea coast, a microbial preparation of hydrocarbon-oxidizing action was developed, which includes strains of microscopic fungi (*Penicillium janthinellum, Penicillium simplicissimum*) and bacteria (*Pseudomonas fluorescens, Pseudomonas putida*).

2.5 Design of the laboratory experiment

oil-water emulsion of different types of oil was introduced into the substrates (soil and sand) and mixed. Polluted substrates were placed into containers in an amount of 200 g, and then microbial preparation, sorbents and mineral fertilizers were added (table 2). The studies were carried out at a constant temperature of +10°C. The soil and sand in the containers were periodically moistened and stirred 2 times a month. The experiment lasted for 3 months. Soil samples for analysis were taken once a month.

Table 2. Design of the laboratory experiment

Type of oil	Variant of treatment	Sorbents (g)			Mineral fertilizers	Microbial preparation
		Peat	GAC	VER		
All types (light, medium and heavy)	no treatment	-	-	-	-	-
	Peat	10	-	-	+	+
	GAC	-	2	-	+	+
	VER	-	-	1	+	+

2.6 Methods of analysis

The content of total petroleum hydrocarbons (TPH) in the soil was determined by infrared spectrometry using the AN-2 analyzer [10]. This method is based on extraction of TPH from soil by tetrachlorocarbon, separation of oil products from the polar hydrocarbons in a column filled with aluminum oxide, and further spectrophotometric identification of hydrocarbon content by the intensity of infrared radiation absorption.

The activity of soil dehydrogenase was determined by the colorimetric method (540 nm wavelength) based on the reduction of the colorless salt 2,3,5-triphenyltetrazolium chloride to red triphenylformazan [11].

2.7 Statistical processing

Statistical analysis of the data was carried out using Microsoft Excel 2007.

3 Results and discussion

3.1 Bioremediation of polluted sand

TPH content in the sand before its pollution with oil-water emulsion was 46 mg/kg, and after the pollution – from 2831 to 4677 mg/kg for different types of oil.

Three months after the start of bioremediation, the TPH content decreased by 44-60%. The maximum decrease was noted in the sand polluted with light oil. When polluted with medium and heavy oil emulsion of, the degree of oil products destruction was lower (Figure 2).

Type of oil	Variant of treatment	TPH (mg/kg)				Decrease of TPH (%)
		Initial	1 month	2 months	3 months	
Light	no treatment	4677	4494	4343	4080	13
	Peat		2091	2342	2293	51
	GAC		1903	2014	1881	**60**
	VER		2395	2305	2497	47
Medium	no treatment	2879	2585	2457	2213	23
	Peat		1540	1855	1416	**51**
	GAC		1742	1937	1408	**51**
	VER		1606	1736	1477	49
Heavy	no treatment	2831	2372	2313	2035	28
	Peat		1662	1823	1544	45
	GAC		1767	1781	1285	**55**
	VER		1837	1831	1595	44

Fig. 2. Dynamics of the TPH content in polluted sand

During the first month, biodegradation in variants with light oil proceeds more intensively than with heavier types of oil. There is a tendency of more intensive biodegradation of hydrocarbons with the use of GAC, but no statistically significant difference between types of sorbents was found.

In the first two months after the beginning of bioremediation, the greatest increase in the activity of soil dehydrogenase was noted in the variants with the addition of peat, but after three months the maximum activity was in the variants with the use of GAC (Figure 3). In general, the activity of the enzyme during the observation period increased by 10-12 times when using different sorbents, which indicates an increase in the processes of microbiological destruction of hydrocarbons.

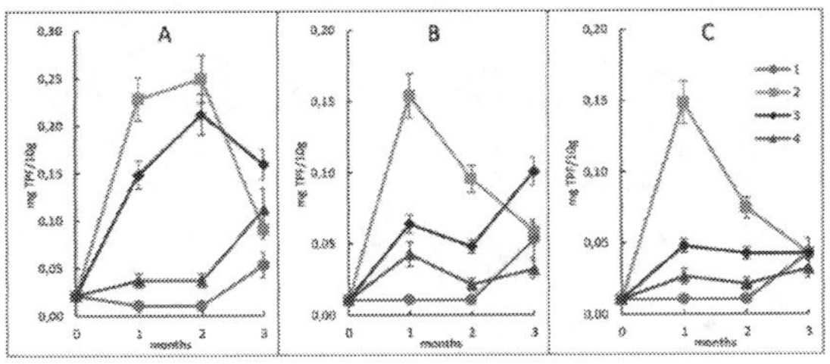

Fig. 3. dynamics of dehydrogenase activity in sand after pollution with light (A), medium (B) and heavy (C) oil emulsions. 1 - no treatment; 2 - peat; 3 - GAC; 4 - VER.

The most effective method proved to be sorption-biological treatment of sandy substrates using granular activated carbon. this method of treatment reduces the content of petroleum hydrocarbons, as well as high-molecular-weight polar organic compounds – products of hydrocarbon transformation.

3.2 Bioremediation of polluted soil

The TPH content in the sand before its pollution with oil-water emulsion was 97 mg/kg, and after the pollution it ranged from 2159 to 4981 mg/kg for different types of oil. The petroleum hydrocarbons predominate in the organic matter of polluted soil.

Three months after the beginning of bioremediation, the TPH content decreased by 25-68%. A significant decrease in the TPH content when using sorbents was noted only in variants with light oil emulsion. The use of granular activated carbon is the most effective among the sorbents (Fig. 4).

Type of oil	Variant of treatment	TPH (mg/kg)				Decrease of TPH (%)
		Initial	1 month	2 months	3 months	
Light	no treatment	4981	4544	4296	3690	26
	Peat		2122	2316	2137	57
	GAC		1684	2147	1601	**68**
	VER		1806	2467	2067	59
Medium	no treatment	2193	1921	1963	1390	37
	Peat		1548	1505	1362	38
	GAC		1548	1528	1438	34
	VER		1759	1776	1376	37
Heavy	no treatment	2159	1609	1690	1331	38
	Peat		1873	1526	1469	32
	GAC		1816	1855	1484	31
	VER		1862	1916	1610	25

Fig 4. The dynamic of the TPH content in polluted soil

Bioremediation with sorbents also reduced the content of high-molecular-weight polar organic compounds by 54-62% for variants with light oil and by 34-39% for variants with medium oil.

dehydrogenase activity increased immediately after contamination to 2.85-4.03 mg TPF/10g. three months later, the enzyme activity decreased 2-2.5 times to values comparable to uncontaminated soil. The use of sorbents with microbial preparation did not significantly effect on the activity of the enzyme (Figure 5).

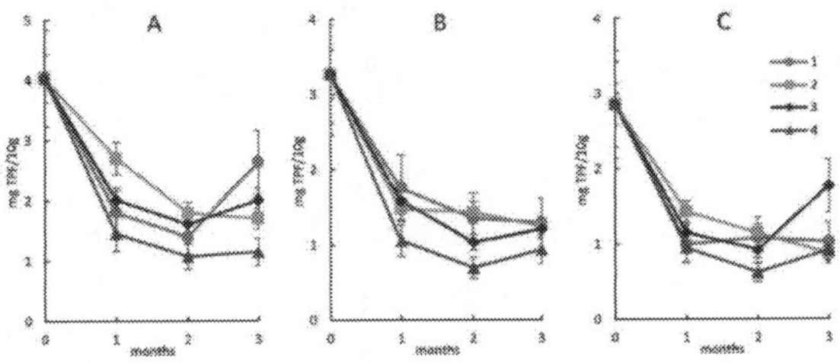

Fig.5. dynamics of dehydrogenase activity in soil after pollution with light (A), medium (B) and heavy (C) oil emulsions. 1 - no treatment; 2 - peat; 3 - GAC; 4 - VER.

An increase in dehydrogenase activity immediately after pollution of coastal soil indicates the presence of an active microbial pool in it, capable of using oil components as a source. As the available components for microbial oxidation decrease, dehydrogenase activity decreases.

3.3 Efficiency of remediation

Based on the results of the analysis of TPH content in the soil, we can calculate the time of their almost complete decomposition, i.e., at 99% (T99). The formulas for the calculation are as follows [12]:

$$T99 = \ln 100/k;$$
$$k = \ln(\frac{y0}{y})/t,$$

where k is the rate constant of hydrocarbon decomposition, y is the residual hydrocarbon content at the time t; y0 is the initial content of hydrocarbon.

The use of sorption-biological method accelerated the cleaning of coastal substrates (Figure 6).

Type of oil	Variant of treatment	The rate constant of hydrocarbon decomposition	T99 (days)
sand			
Light	no treatment	0.00152	3033
	Peat	0.00792	581
	GAC	0.01012	455
	VER	0.00697	660
Medium	no treatment	0.00292	1575
	Peat	0.00788	584
	GAC	0.00795	579
	VER	0.00742	621
Heavy	no treatment	0.00367	1256
	Peat	0.00673	684
	GAC	0.00877	525
	VER	0.00637	723
soil			
Light	no treatment	0.00333	1381
	Peat	0.00940	490
	GAC	0.01261	365
	VER	0.00977	471

Fig. 6. The time of almost complete decomposition (T99) of petroleum hydrocarbon

The sorption-biological method will reduce the cleaning time of sandy beaches by 2.0-5.5 times when polluted with different types of oil, and of coastal soils by 3.2 times when polluted with light oil. The obtained results confirm the conclusions about the effectiveness of the sorption-biological method of bioremediation of oil-polluted soils in the Subarctic that we made earlier on the basis of laboratory [13] and field [14] studies.

4 Conclusion

Studies have shown that the sorption-biological method of bioremediation can reduce the sand cleaning time by 2-5.5 times. Biodegradation in variants with light oil in the first month proceeds more intensively than heavier types of oil. The maximum decrease in the amount of hydrocarbons in the sand and the increase in the activity of dehydrogenase were noted when using GAC and peat (together with a microbial preparation and mineral fertilizers).

The efficiency of sorption-biological treatment of coastal soil was lower. The maximum effect was noted when GAC was added to the polluted soil. The use of this method is justified only when coastal soils are polluted with light oil. An increase in the activity of soil dehydrogenase immediately after pollution indicates the presence of a pool of microorganisms capable of utilizing oil components. Under these conditions, biostimulation, based on the application of mineral fertilizers to the polluted soil, will be an effective method of cleaning.

There is no significant difference between the sorbents used, therefore, each of them can be used depending on the availability and economic feasibility. The sorption-biological method of bioremediation allows cleaning with high efficiency in the Arctic and Subarctic.

The research was supported by International Project «Arctic Coast Bioremediation KO 1001» within the framework of the Kolarctic CBC 2014-2020 Programme and State scientific program FMEZ-2022-0011.

References

1. E. Samylovskaya, et al., Digital Technologies in Arctic Oil and Gas Resources Extraction: Global Trends and Russian Experience, *Resources*, **11**, 29 doi: 10.3390/resources11030029 (2022)
2. Yu. N. Gladkiy et al., History and perspectives for the expansion of the Russian Arctic o Ecumene, *IOP Conf. Series: Earth and Environmental Science*, **434**, 012005 (2020)
3. O.I. Kazanin and V.V. Yaroshenko, Decrease in coal losses during mining of contiguous seams in the near-bottom part at Vorkuta deposit, *Journal of Mining Institute*, **244(4)**, pp. 395–401 (2020)
4. E.V. Travkina et al. Northern Sea Route: Formation of Russian Transport Policy in the Arctic, *IOP Conf. Series: Earth and Environmental Science*, 302, 012088 (2019)
5. Y.F. Lukin, International sea routes of cargo transportation in the Arctic, Arktika i Sever, **40**, pp. 225-253 (2020)
6. I.A. Nechaeva, et al., Stimulation of microbial destruction of oil by introduction of bacteria associations and mineral fertilizers in soil under laboratory and field conditions, *Biotechnology in Russia*, **1**, pp. 90-99 (2009)
7. A.H. Molla and A. Fakhru'l-Razi, Mycoremediation – a prospective environmental friendly technique of bioseparation and dewatering of domestic wastewater sludge, *Environ. Sci. Pollut. Res. Int.*, **19** pp. 1612-1619 (2012)
8. L. Bandura, et al., Application of mineral sorbents for removal of petroleum substances: a review, Minerals, **7**, pp. 37-43 (2017)
9. G. Vasilyeva, et al., Adsorptive bioremediation of soil highly contaminated with crude oil, *Science of The Total Environment,* **706**, 135739 (2020)
10. PND F 16.1:2.2.22-98 Method for measuring of the total hydrocarbon content in soil and sediments by IR-spectrometry (Moscow, 1998), p. 17
11. Determination of dehydrogenize activity Practical work on Agrochemistry (Moscow, Moscow State University Press, 2001), p. 689.
12. V.N. Bashkin and R.V. Galiulin, Geoecological Risk Management in Polar Areas (Switzerland, Springer Verlag, 2019), p. 156
13. V.A. Myazin, et al., Influence of granular activated carbon on the rate of bioremediation of soils in the Murmansk region, historically contaminated with oil products, *Problemy regional'noy ekologii*, **2**, pp. 20-26 (2020)
14. V.A. Myazin, et al., The effectiveness of biostimulation, bioaugmentation and sorption-biological treatment of soil contaminated with petroleum products in the Russian Subarctic, *Microorganisms*, **9**, p 1722 (2021)

Problems of Environmental and Technogenic Safety Regulation in the Arctic Region: Oil Spills

Sergey Yakovlev[1,*] and *Alexey* Shemyakin[2,*]

[1] Senior Researcher, Putilov Institute for Informatics and Mathematical Modeling — Subdivision of the Federal Research Centre «Kola Science Centre of the Russian Academy of Sciences» (IIMM KSC RAS), 184209, Apatity, Russian Federation
[2] Junior Researcher, Putilov Institute for Informatics and Mathematical Modeling — Subdivision of the Federal Research Centre «Kola Science Centre of the Russian Academy of Sciences» (IIMM KSC RAS), 184209, Apatity, Russian Federation

Abstract. The article analyzes the changes in the regulatory and legal framework related to the development of the main document on oil spill response planning. the key points of the regulatory and methodological framework applied until January 1, 2021, which should have been given special attention, are highlighted. The second part of the article analyzes the updated regulatory and legal framework for preventing and responding to spills. the third part provides a comparative analysis of the old and new legislation governing the development of oil and petroleum product spill prevention and response plan, and also discusses some methodological aspects. The results of the analysis are formulated in the conclusion. It is noted that despite innovations, scientific and methodological support of development remains insufficient.

1 Introduction

Accidents and emergencies (ES) associated with oil and petroleum product spills (hereinafter - ES(O)) continue to be topical for the Arctic zone of the Russian Federation (AZRF). Thus, according to the report of the Ministry of Natural Resources and Environment of the Russian Federation [1], 2,629 oil spills were registered in federal districts, which include Arctic regions, in 2018, while the total number of registered pollutions in the Russian Federation is 3,053. The total area of these pollutions was 58 hectares, and the volume of oil products that entered the environment was 1,054 m3. More than half of the reported accidents were caused by corrosion of pipeline metal. One of the dangerous consequences of oil spills can be contamination of nearby water bodies. These consequences can be much more dangerous if oil products get into rivers. For example, in 2019, about 193,000 tons of oil products flowed into rivers in the Arctic Ocean basin [1].

It is worth mentioning the largest emergency of 2020 in the Arctic zone of the Russian Federation - at one of production facilities of PAONorilsk Nickel, more than 20 thousand tons of oil products spilled into water bodies and rivers. Rosprirodnadzor estimated the damage caused to the environment at 146 billion rubles. These funds were recovered from Norilsk Nickel through court proceedings [2].

It is obvious that the issues of emergency risk management do not lose their relevance. Since January 1, 2021 there have been significant changes in the legislation in this area. These changes are considered on the example of the key document – the plan for the prevention of and response to oil and petroleum product spills (OSRP).

This article reviews the methodologies for developing OSRP used before January 1, 2021; the main normative acts regulating this process; and mentions the key points that need to be given special attention to when developing OSRP. The second part of the article analyzes the updated regulatory framework. The third part presents a comparative analysis of the old and new legislation regulating the development of OSRP. the conclusion is based on the comparison of the old and new regulatory and methodological framework.

The purpose of the work is to analyze changes in the legislation regulating oil spill prevention and response, and to assess the impact of the new requirements on the effectiveness of ES(O) risk management.

2 Development of the OSRP by January 1, 2021

The previous (until 2021) regulatory and methodological support for the development of OSRP is considered in [3, 4]. Let's touch on some points.

The main regulatory document was the Order of the EMERCOM [5], which approves the Rules for development and approval of OSRP on the territory of the Russian Federation. The rules contain definitions of the terms used. The notion of OSRP coverage area was introduced, which became null and void in 2012. The main planning tasks were named, including:

* Corresponding author: yakovlev@iimm.ru, shemyakin@iimm.ru

substantiation of the level of a possible emergency situation

definition of the interaction procedure

substantiation of the sufficient number of forces and means for emergency liquidation

drawing up a situational schedule of the emergency response operation.

The OSRP level was determined in accordance with the Government decree [6] and the Order of the Ministry of Natural Resources of Russia [7]. Spills not covered by the ES(O) classification were identified, for which internal regulations should be developed in accordance with [5].

The process of approval and concordance depends on the level of the OSRP. In this case, the whole OSRP and the Schedule have separate approval and concordance sheets.

Appendix 1 to the Regulation [5] contains a rather detailed example of the OSRP structure described in [3, 4]. The OSRP includes, in particular, the following elements:

description of technological processes with oil products

volumes and areas of spills, boundaries of emergency zones, risk assessment

scenarios of possible ESs at different levels

determination of sufficient composition of forces and means for emergency response, as well as firefighting units (in case of fire)

levels of response, areas of responsibility.

The operational part of the OSRP includes, among other things, algorithms of operations and technologies used in emergency response, firefighting action planning, technologies for collecting spilled oil products, and waste storage.

Appendix No. 2 to the Rules [5] contains a sample Schedule of operational activities for the events of threat and emergency.

An OSRP examination was a complex and costly stage, usually requiring several iterations, the order of which was determined by the level of the OSRP. However, in 2012, the examination was canceled.

The OSRP is valid for three to five years, depending on the level. the OSRP is subject to adjustment (revision) upon expiry of its validity or in case of significant (affecting the level) change in the initial data.

There are annexes to the OSRP - mandatory and recommended. The mandatory ones include:

diagram of the facility with the boundaries high-risk zones and priority protection areas

maps, scenarios and consequences of emergencies at different levels

schedules and regulations

calculation of sufficiency of forces and means

contracts, certificates, licenses.

The Decree of the Government [6] approved the Basic Requirements for the development of the OSRP. In particular, for different objects (vessel, terminal, tank, train, pipeline, stationary tank) the maximum possible spill volumes are defined. So, for stationary facilities, this is 100% of the volume of the largest reservoir (tank). Possible spill volumes and areas define emergency categories - from local to federal (as amended by the Decree [8]).

The Order of the Ministry of Natural Resources of Russia [7] approved the Instructions on determining the lower level of a spill to classify a spill as an emergency situation. Thus, onshore, the minimum values are specified depending on the source and type of the spill, as well as the type of territory. For example, for large stationary storage facilities, for light oil products and an industrial site without cover, the lower level of a spill is 7 tons.

Resolution [8] approved the Rules for the organization of the events provided by the OSRP. In particular, the maximum permissible time of localization is s specified: 4 hours - for water area, 6 hours - for territory.

The Acts [5-8] provide the basis for the development of the OSRP. However, in fact, dozens and hundreds of documents and materials of various levels are needed to fill all sections of the OSRP. The Ministry of Emergency Situations of Russia is the main agency that supervises the development and approval of the OSRP. Let us note insufficient methodological support of the development. Not a single methodological manual has appeared at the federal level.

In conclusion, the OSRP was supposed to provide an analysis of the risks both oil spills and explosions and\or fires. Depending on the level, the OSRP was subject to coordination and approval by various departments. sections of the OSRP were interconnected, duplicative, and required annexes and copies of various valid certificates, contracts, etc. As a result, the OSRP was comparable to a security declaration. numerous legal requirements were difficult to comply with in practice. The region might have lacked resources for the collection and disposal of oil waste, emergency rescue teams (ERTs) with the necessary scope of certification and equipment, which made it impossible to organize the fight against the maximum category of emergency at the site, and involvement of ERTs from other regions did not ensure the containment of the spill within the required time, etc.

3 Updated requirements for the development of the OSCR

Let's give a brief description of the changes.

The Decree of the Government [9] canceled the Order of the EMERCOM [5]. The Decree of the Government [10] declared documents [6, 8] null and void from January 1, 2021. By the Resolution of the Government Resolution [11], the document [7] was abolished.

New the requirements for the development of the OSRP are contained in [10, 12].

The Resolution [10] approved the Rules, which contain, among other things, definitions of the concepts of containment and response to a spill, as well as requirements for the content of an OSRP. The rules comply with the Federal Law [12] (Article 46).

The criterion for identifying facilities (for which, among other things, the development of an OSRP is required) on land is the maximum estimated volume of spills - 3 tons or more.

An OSRP should contain the following data:

- general information about operations with petroleum products
- possible initiating events of spills
- volumes of maximum spills
- possible areas of spills and consequences for the environment, population and life support systems (taking into account the characteristics of the area);
- priority actions of personnel on spill events
- estimated time to contain and eliminate the maximum spill
- calculation of sufficiency of forces and means to eliminate the maximum spill, taking into account technologies, as well as calculation of spill containment time from the moment of detection or receipt of information (in case of a spill on water bodies within 4 hours, on land - within 6 hours)
- composition and procedure of forces action and use of equipment
- organization of additional resources attraction in case of spills exceeding the maximum ones
- control and communication schemes
- organization of temporary storage, transportation and disposal of the collected oil;
- cost of the OSRP operations (own expenses)
- schedule for response to maximum spills (containment, spill collection, disposal of collected oil).

Copies of documents on emergency services and groups, as well as license copies of contractors' licenses for waste transportation are attached to the OSCR.

The maximum estimated spill volumes for different objects (ships, barges, wells, train, terminals, pipelines, warehouses) are specified. So, for oil product warehouses this is 100 percent of the volume of the largest container.

Further [10] describes the procedure for conducting integrated exercises, the procedure for issuing an opinion on the organization's readiness for containment and response to spills (based on the results of the exercises), the procedure for notification on approval of the OSRP, the procedure for reporting a spill, and the procedure for attracting additional forces and means.

Article 46 of the Federal Law [12] contains, in particular, information about an OSRP. It is noted that it is necessary to conduct an OSRP when carrying out activities related to hydrocarbons. The OSRP is approved by the organization if there is an opinion on readiness and coordination with the environmental supervision. The procedure for amending the OSRP is described. The operating organization is obliged to comply with the OSRP and have adequate financial support. In the event of a spill, the organization shall be obliged to provide notification of the fact of the spill, ensure conducting works in accordance with the OSRP, apply in in accordance with the established procedure for engaging additional forces and means, carry out reclamation and other recovery works after the spill liquidation, fully compensate the damage and additional costs.

Now let's compare the OSRP requirements before and after the changes in the legislation.

4 Comparative analysis and discussion of changes in the regulatory and methodological framework

First, we note the main similar components, elements of an OSRP developed by the operating organization:

- general information about the organization, basic operations with oil products
- possible sources of spills
- maximum volumes of spills
- areas of spills, possible consequences for the environment, population, life support

 https://doi.org/10.1051/e3sconf/202337805006

systems (taking into account the characteristics of surrounding area)

- priority actions of personnel in the event of a spill
- terms for containment and elimination of the maximum spill, in particular, the time for containment of a spill on land should not exceed 6 hours
- calculation of forces and means sufficient to eliminate the maximum spill
- composition and procedure for the actions of forces and the use of equipment
- schemes of control and communication
- information about oil waste storage, transportation and disposal
- schedules for maximum-size of spill
- contracts with emergency rescue teams, certification, licenses of contractors.

- Now, let's list the significant differences between the new legislation and the previous one:
- regulatory framework - a number of documents have been abolished, instead of them it is required to be guided by the Rules [10];
- there is no concept of the level and term of validity of OSCR. However, the Rules [10] have a clear validity period - from January 1, 2021 to January 1, 2027. This can lead to the fact that in 2027 there will be a need to update the existing OSRP, quite possibly - according to the updated rules
- spills with a volume of less than 3 tons are not considered
- there is no indication of the need to assess risk (probability)
- there is no mention of such accident scenarios as fires and explosions (in the old OSRP, significant attention was paid to possible fires);
- possible accident scenarios are not mentioned
- there are no levels of response and issues of transfer of control
- there is no mention of an emergency committee
- the time of spill containment from the moment of detection or information receipt in the event of a spill on water bodies (including their water protection zones - previously this note did not exist) - within 4 hours
- the procedure for approval of an OSRP has been changed - one of the most significant changes in the new rules for developing plans;
- the concept of a spill that exceeds than the maximum design spill ("beyond the design basis") and the associated involvement of additional (not provided for in the OSRP) forces and resources has appeared
- information about the scope and cost of work (own costs) is required, in the previous OSRP there was a concept of reserves of material and financial resources
- Rosprirodnadzor appears as the supervising agency coordinating the OSRP, while previously the EMERCOM of Russia acted in this role.

5 Conclusion

Climate changes, aging of equipment, and other factors increase the urgency of oil spill prevention and response issues in the Arctic zone of the Russian Federation. An important aspect of combating ES(O) is the legislative framework, which has undergone significant changes at the turn of 2020-2021. Therefore, a comparative analysis of the introduced changes is important and timely.

In general, at first glance, the procedure for the development and approval of OSRP has been significantly simplified. Instead of a detailed (albeit approximate) table of contents, general guidelines are provided. A number of concepts and provisions are excluded.

The process of preventing oil spills has been substantially revised: whereas previously accidental spill drills were conducted in parallel with the development and approval of an OSRP and, in general, did not depend on each other, from 2021 onwards successful exercises have become a key stage in the development of the oil spill response plan. Moreover, if the exercises reveal that the organization is not prepared to deal with an emergency, approval of the OSRP by the supervisory authorities becomes impossible.

On the one hand, simplification of the process of developing and approving a OSRP is a positive thing: the administrative burden on the operating organizations is reduced, resources for the development and approval of planning documents are saved. On the other hand, the main focus is on assessing the readiness of the organization to respond to an ES(O) This approach, undoubtedly, allows to minimize damage from emergencies, but it overlooks the prevention of oil spills as such.

At the same time, the phenomenon of possible oil spills and related accidents and emergencies has not become simpler, protected and secured. Scientific and methodological support for the development of an OSRP is still insufficient. This raises serious concerns about the risk of an ES.

As part of further research it is planned to analyze the practice of applying the new regulatory framework for the regulation of EC(O) in the Arctic regions of the Russian Federation.

The results were obtained within implementation of the state task of the IIMM KSC RAS (No. 122022800547-3).

References

1. State report "On the state and protection of the environment of the Russian Federation", Ministry of Natural Resources and Ecology of the Russian Federation Available online: http://www.mnr.gov.ru/upload/iblock/cf1/07_09_2020_M_P_O%20(1).pdf (2019)
2. Norilsk Nickel fully paid for the damage caused to environment in connection with the fuel spill at CHPP-3. Available online: https://www.nornickel.ru/news-and-media/press-releases-and-news/nornikel-polnostyu-oplatil-ushcherb-nanesennyy-okruzhayushchey-srede-v-svyazi-s-razlivom-topliva-na-tets-3/type = releases
3. S.Yu. Yakovlev, et al., Development of plans for prevention and response of oil spills for territorial objects, *Modeling and analysis of safety and risk in complex systems: Proceedings of the International Scientific School MA BR-2010,* St. Petersburg, July 6-10, 2010, (St. Petersburg, SUAI), pp.447 – 452 (2010)
4. S.Yu. Yakovlev, A.S. Shemyakin, and A.G. Oleinik, Information and analytical support for the safety of the Arctic industrial and natural complexes (in press).
5. Order of EMERCOM dated 28.12.2004 No. 621 "On Approval of the Rules for the Development and Coordination of Plans for the Prevention and Response of Oil and Oil Products Spills in the Territory of the Russian Federation (2004)
6. Decree of the Russian Federation Government dated August 21, 2000 No. 613 "On urgent measures to prevent and eliminate emergency oil and oil products spills" (2000)
7. Order of the Ministry of Natural Resources of Russia dated 03.03.2003 No. 156 "On approval of the Instructions for determining the lower level of an oil and oil product spill for classifying an accidental spill as an emergency"(2003)
8. Decree of the Government of the Russian Federation dated 15.04.2002 No. 240 "On the procedure for organizing measures to prevent and eliminate emergency oil and oil products spills on the territory of the Russian Federation"(2002)
9. Decree of the Government of the Russian Federation dated 11.07.2020 No. 1034 "On invalidating regulatory legal acts and certain provisions of regulatory legal acts of the Russian Federation, on the abolition of acts of federal executive bodies containing mandatory requirements, compliance with which is assessed when carrying out control measures for implementation of federal state fire supervision and licensing control in the field of fire safety, federal state supervision in the field of protecting the population and territories from natural and man-made emergencies, state supervision over the use of small boats, bases (structures) for their parking in internal waters and the territorial sea Russian Federation" (2020)
10. Decree of the Government of the Russian Federation dated December 31, 2020 No. 2451 "On approval of the Rules for organizing measures to prevent and respond to oil and oil products spills on the territory of the Russian Federation, with the exception of the internal sea waters of the Russian Federation and the territorial sea of the Russian Federation, as well as on the recognition as invalid some acts of the Government of the Russian Federation " (2020)
11. Decree of the Government of the Russian Federation dated August 06, 2020 No. 1192 "On recognizing as invalid some normative legal acts and certain provisions of normative legal acts of the Government of the Russian Federation, on the abolition of some normative legal acts and certain provisions of normative legal acts of federal executive bodies containing mandatory the requirements, compliance with which is assessed when carrying out control measures in the implementation of federal state supervision in the field of industrial safety and state mining supervision, and the recognition of the Instruction on observation of deformations of sides, slopes of benches and dumps in open pits and the development of measures to ensure that their stability, approved by the State Committee for Supervision of Safe Operations in Industry and Mining Supervision under the Council of Ministers of the USSR on July 21, 1970 " (2020)
12. Federal Law No. 7-FZ dated 10.01.2002 "On Environmental Protection" (as amended on 30.12.2020) (2020)

Oil spill response technology for ice-covered waters in the Arctic

Askar Mingazhev[1,*], *Andrey* Shchipachev[2,*] and *Alisa* Mingazheva[3]

[1] Candidate of Technical Sciences, Associate Professor, Ufa State Petroleum Technical University, 450064, Ufa, Kosmonavtov 1, Russian Federation.
[2] Doctor of Technical Sciences, Professor, Head of the Department of Transport and Storage of Oil and Gas, Saint Petersburg Mining University, 199106, Saint Petersburg, Russian Federation
[3] Assistant, Ufa State Aviation Technical University, 450008, Ufa, K. Marx 12, Russian Federation

Abstract. A technology for removing oil spills from under the ice, based on the use of the vortex funnel effect, is proposed. A comparative analysis of methods oil pollution removal from under the ice cover is carried out. Models of the processes of separation of the "ice-oil-water" system occurring in the vortex funnel are presented, as well as, based on the experiment, photo illustrations of the stages of formation of the vortex funnel and removal of oil from under-ice. The schemes of deployment and operation of mobile equipment for removing oil spills from under the ice are proposed. It is shown that the proposed approach to liquidation of emergency oil spills in the Arctic can be used to create means and technologies of liquidation of more large-scale disasters of this kind.

1 Introduction

The Arctic, with its enormous reserves of hydrocarbons, the demand for which grows every year, is becoming a hostage to the energy problems of mankind. The stimulating factor for industrial development of the Arctic is also the decrease of ice cover, which facilitates the development of new oil fields and its transportation.

The harsh natural and climatic conditions of the Arctic imply a significantly greater likelihood of occurrence and response to oil spills, although even in more "favorable" areas, such as the Gulf of Mexico, in 2010 it was not possible to prevent an accidental oil spill and promptly eliminate its consequences.

When eliminating oil pollution in the Arctic, it is necessary to take into account the following factors. The main factor complicating the removal of pollution is ice cover. The presence of ice cover, especially thick ice, being a natural obstacle, greatly complicates the detection of oil spills and their elimination. Other factors include harsh climatic conditions, in particular, low air and water temperatures, long polar nights, and remoteness from land.

At the same time, there are many causes of oil spills, which create their own specific features of their detection, localization and removal. Oil spills can occur due to leaks from onshore and offshore wells, accidents on pipelines, oil storage facilities and oil tankers.

Despite the recommendations of various world environmental protection organizations [1], exploration and development of new hydrocarbon deposits in the Arctic region continues. In this regard, ensuring the environmental safety of oil production requires the development of new effective ways to eliminate its accidental spills in the Arctic, especially during the exploitation of sub-ice fields.

2 Analysis of existing methods of oil pollution removal from under the ice

Despite some success in the development of new technologies for oil spill response in the Arctic [2], they are usually not completed, in particular, due to insufficient funding and lack of interest from oil companies.

The main methods for removing oil include such as mechanical technologies for collecting oil, oil combustion, the use of sorbents and dispersants. However, these methods of oil removal have shown some efficiency in open water, but not in the presence of ice cover [3, 4].

At the same time, of all the listed technological directions for removing oil barriers from under the ice, the mechanical methods are the most environmentally safe, which determines the relevance of their application.

However, the disadvantage of many mechanical methods for removing oil pollution is their limited coverage of processing areas or use in flowing water bodies covered with ice. For example, such technical solutions as those listed in the patents, which consist in the use of devices that provide for the for the arrangement of an ice hole, installation of a winch on the ice, placement of an inflatable boom along the contour of a rectangle under the ice, winding it on the winch drum, pulling oil to the ice hole and pumping it [5

* Corresponding author: mad-20007@yandex.ru, schipachev_am@pers.spmi.ru

, 6] refer to preventive measures to protect water bodies from oil pollution, which are mounted in advance over possible accident zones, which is extremely costly and not always acceptable.

In real emergency situations, it is difficult to determine in advance the area of the oil spill. Therefore, the requirements for technology and equipment for liquidation of oil spills under the ice cover should provide for the following conditions.

1. Fast detection and response to an emergency
2. Rapid deployment of emergency response tools
3. Preliminary measures for containment of oil spills
4. Localization and reduction of the spill area
5. Pumping the main volume of oil pollution into reservoirs, their transportation and regeneration
6. Elimination of residual pollution.

One of the important tasks, effective elimination of emergency oil spills, is its rapid detection, the most rational use of underwater vehicles [2]. Quite a lot of developments are devoted to the problem of detecting petroleum products [7, 8].

Despite the intensification of oil production in the Arctic regions, mainly in the sea areas of the Arctic Ocean, there are still no sufficiently reliable methods and devices for collecting oil and oil product spills from under the ice cover [1].

Noteworthy is the technology of oil spill response using the idea of drilling wells in ice and installing booms to collect oil from under the ice cover of a reservoir [9]. Technical solutions in this direction provide for the following operations: drilling a well in the ice, installing under the ice cover a hollow shell equipped with an inflatable fence along the contour, pumping air into its cavities, and injecting a working agent into the contact zone of the shell with the polluted area of ice, removing the working agent from under the ice cover, pollutants from the cavities of the shell and the inflatable air barrier, and removing the shelland the inflatable barrier from the well: drilling another well in the ice cover with operations including removing pollutants from under the ice cover. When drilling wells in the ice, the installation of the shell with an inflatable fence is carried out directly in the oil spill zone, the cavities of the shell and the inflatable fence are filled with hot air.

However, such technologies have low productivity, high labor intensity, limited possibilities of appon, while oil is collected from a small area of the water area.

A method for removing oil from under the ice through an ice-hole made in the ice. The device for implementing the method includes arrows passing under the ice in two different directions from the hole so as to form an angle of less than 180 degrees to each other. The method involves the use of a device for collecting oil, placed in the ice hole and an oil water flow device placed at the level of the lower surface of the ice to ensure flow direction to the hole between the arrows [10].

Methods for removing oil pollutants from the surface of water using oil arrows and skimmers equipped with rotating brushes have become widespread. Oil is also collected using a long rope made of hairy yarn that can absorb oil. It moves along the surface of the water collecting oil, then it is taken out of the water and squeezed out, after which the cycle is repeated. These methods, which have proven themselves quite effectively in open water, turned out to be ineffective for extracting oil from an environment containing ice or from under the ice cover [11, 12].

The Oil Spill Cleaner [13] is a multihull vessel containing the middle and side hulls. The vessel is equipped with a separator detaching the ice blocks by moving them between the middle hull and the side hulls from bow to stem (Fig. 1).

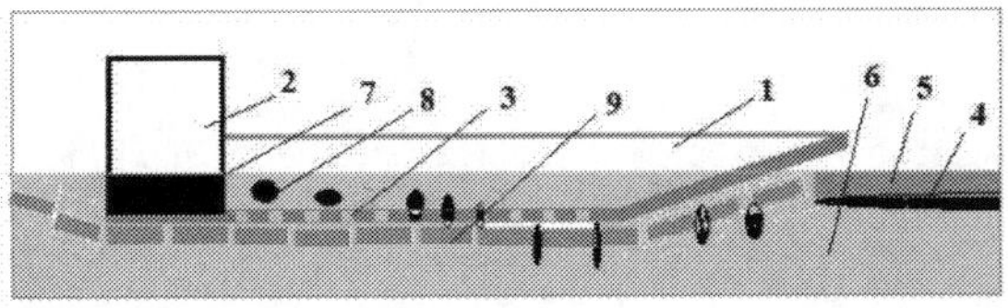

Fig.1. A vessel for removing oil contaminants in ice cover conditions (1 - icebreaker vessel, 2 - oil collection tank, 3 - separator, 4 - oil spill under the ice, 5 - ice cover, 6 - water, 7 - collected oil, 8 - fragments of oil in water, 9 - cracked ice) [13]

The separator (Fig. 1) is a grid that separates ice from oil. The vessel pushes the ice down through the separator, and while the water and oil pass through the separator, oil is separated from the water and collected in a container. The oil can be separated from the water using a collection device, which can be, for example, a brush skimmer.

3 Technology and equipment for oil removal from under the ice

The purpose of the technology for removing oil pollution from under the ice cover, proposed in this article, was designed to ensure the above requirements and conditions for effective elimination of the consequences of an accidental oil spill.

The proposed technology [14,15] consists in localization of an oil or oil product slick and subsequent removal of oil or oil product by pumping it into an oil receiver. At the same time, a well is drilled in the area of localization of the oil or oil product slick in the ice cover; a swirler with an evacuation device is submerged into the slick area, the swirler rotation creates a vortex funnel in the water under the ice (Fig. 2), in which oil is collected and it is pumped out from the vortex funnel formed (Fig. 3).

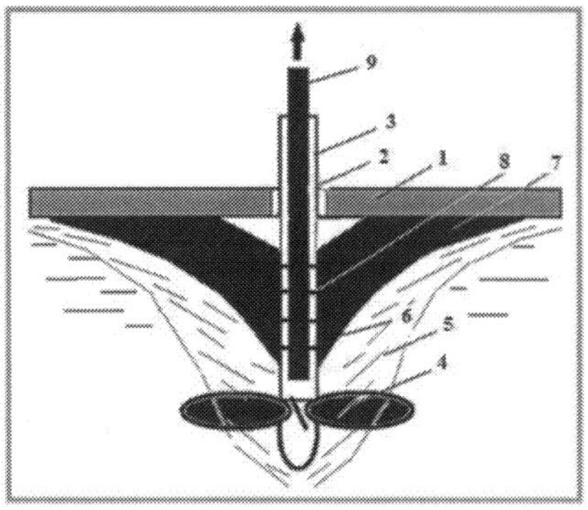

Fig. 2. Scheme of the formation of a vortex funnel under the ice cover (1 - ice cover, 2 - well, 3 - hollow swirler shaft, 4 - swirler, 5 - vortex funnel, 6 - oil in a vortex funnel, 7 - oil spill under the ice 8 - perforations in the hollow shaft, 9 - pumping pipe oil)

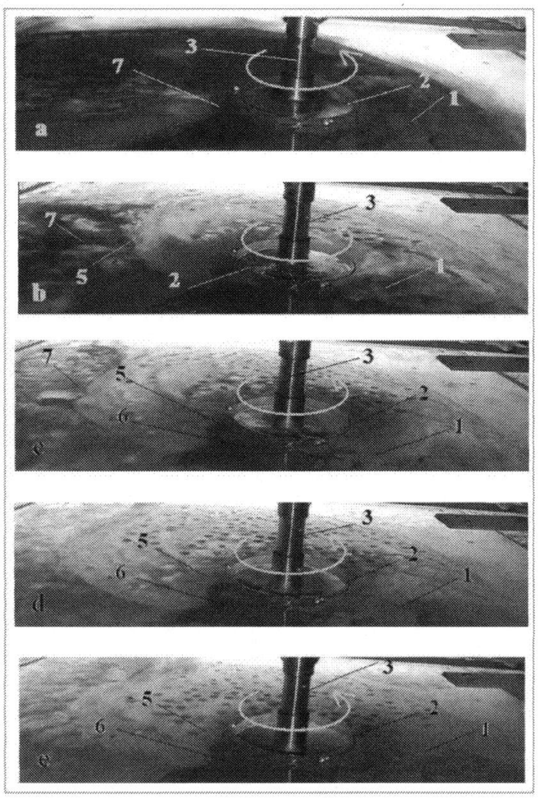

Fig. 3. Stages of vortex formation under the ice cover (a - initial state; b, c, d - the process of oil pulling into the vortex funnel; e - the end of the oil removal process. 1 - ice cover, 2 - well, 3 - hollow swirler shaft, 5 - vortex funnel, 6 - oil in vortex funnel, 7 - oil spill under the ice)

To remove oil pollution, the oil slick is first localized, which can be done, for example, by placing the boom 10 under the ice cover 1 (Fig. 4a) by transporting it by underwater drones or submersibles [2].

Above the area of the localized oil slick 7, a mobile installation for drilling and pumping oil 11 is placed (Fig.4a) and a well 2 is drilled in the ice cover 1, a swirler 4 with a pumping device is immersed through it into the slick area (Fig.4b).

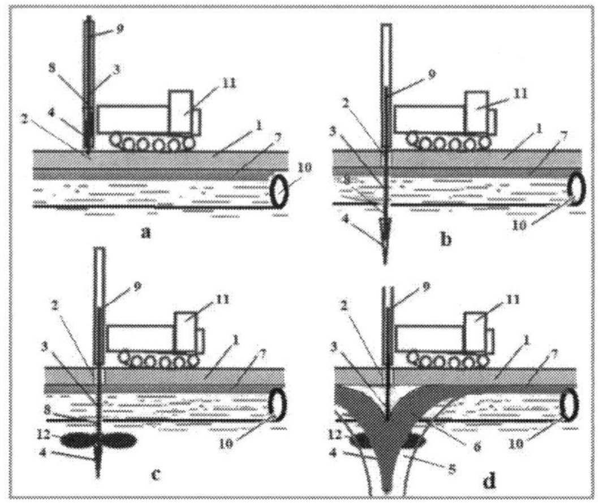

Fig. 4. Scheme of deployment and operation of equipment to remove oil spills from under the ice (a - preparation for drilling a well, b - drilling a well, c - deploying a swirler, d - pumping oil from under the ice cover. 1- ice cover, 2 - well, 3 - hollow swirler shaft, 4 - swirler, 5 - vortex funnel, 6 - oil in the vortex funnel, 7 - oil spill under the ice, 8 - perforations in the hollow shaft, 9 - oil pumping pipe, 10 - booms, 11 - mobile rig for drilling and pumping oil, 12 - blades)

After immersion of the swirler 4 into the water to the required depth, the blades 12 of the swirler 4 (Fig. 4c) open and the rotation of the swirler 4 creates a vortex funnel 5 in the water under the ice (Fig. 4c), which collects oil 7 into it, by capturing it into a vortex funnel 5.6. Then, oil 6 is pumped out of the vortex funnel 5 (Fig. 4c). When a vortex funnel 5 forms in the water (Fig. 2 and Fig. 4), due to the difference in the densities of water and oil, lighter oil 6 is collected in the central part of the vortex funnel 5. Further, through perforations 8 in the hollow shaft of the swirler 4, oil 6 through a pipe for pumping oil 9 is pumped out into a tank for collecting oil products (Fig. 5).

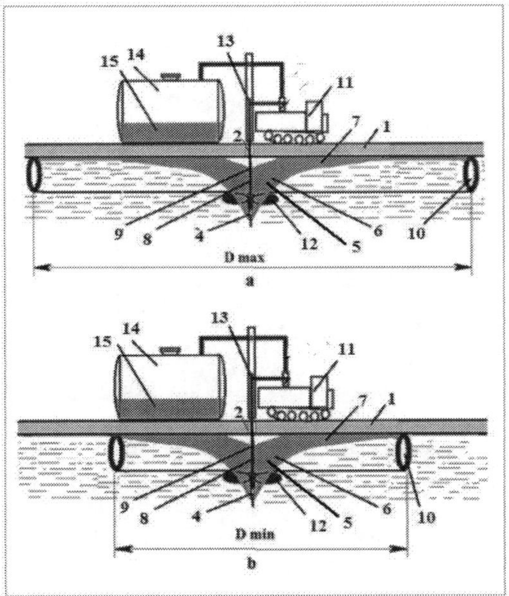

Fig. 5. Schematic diagram of a mobile installation for removing oil spills from under the ice. (a - the beginning of oil spill response, b - oil removal with a gradual decrease in the area covered by booms. 1 - ice cover, 2 - well, 3 - hollow swirler shaft, 4 - swirler, 5 - vortex funnel, 6 - oil in a vortex funnel, 7 - oil spill under the ice, 8 - perforations in the hollow shaft, 9 - pipe for pumping oil, 10 - booms, 11 - mobile rig for drilling and pumping oil, 12 - swirler blades, 13 - pipeline for pumping oil into tank, 14 - tank for oil, 15 - pumped out oil)

Further, through the perforations 8 in the hollow shaft of the swirler 4, the oil is pumped through the oil transfer pipeline 13 to the oil collection tank 14 (Fig. 5). Then, after removing the oil, the blades 12 of the swirler 4 are folded and the swirler 4 is removed from under the ice. The mobile unit is folded down and, if necessary, moved to another site.

At formation of two-phase ("water-oil") vortex funnel (Fig. 6 and Fig. 7), the following factors act. When the swirler blades rotate from the underlying water layers to the upper water layers and from them to the oil layer, the rotation of each layer (conventionally adopted in this model) is transmitted by the friction force between the layers in question (Fig. 6). Since energy losses occur in this case, the angular velocity of each successive layer decreases when approaching the ice cover. At the same time, the friction force between the water layers is much greater than at the "water-oil" interface, which leads to a sharp lag in the angular velocity of the oil layer rotation from the water layers.

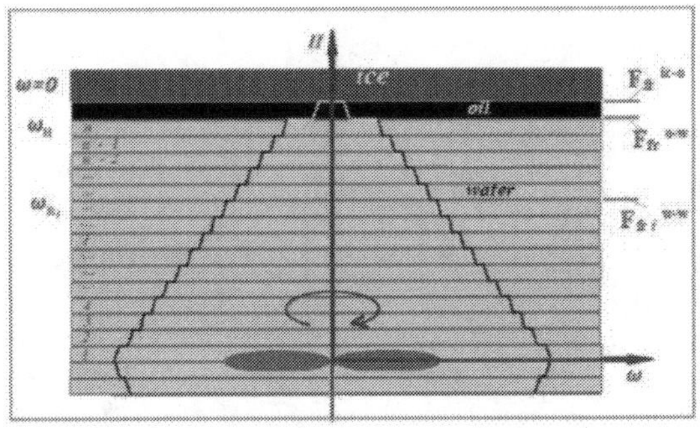

Fig. 6. Distribution of angular velocities at the initial stage of vortex vortex formation. ($F_{fr}^{ic\text{-}o}$ is the friction force between ice and oil, $F_{fr}^{o\text{-}w}$ is the friction force between oil and water, $F_{fr\,i}^{w\text{-}w}$ is the friction force between layers and water, ω is the angular velocity of rotation of the swirler, ω_n is the angular velocity of rotation of the layer oil, $\omega_w^{\,i}$ - angular velocity of rotation of the *i*-th layer of water)

As a result of this, the formation of a vortex water funnel occurs in advance. In addition, a larger specific gravity of water (Pw) creates an advantage in the centrifugal force of each layer of water (Fcf w), which affects the resulting strength of the i-th layer of water (Rin), which significantly exceeds the resulting force of the i-level of oil (Rn) and, being the dominating force, collects the vortex funnel of oil into the inner cavity of the vortex funnel. Moreover, the lower specific gravity of oil also contributes to its accumulation above the water funnel.

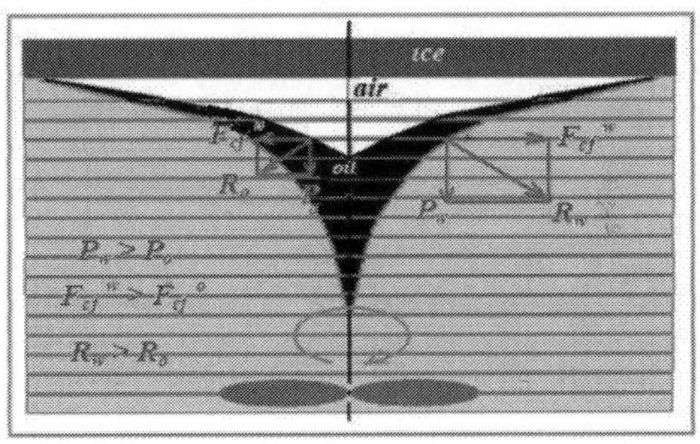

Figure 7. Factors influencing the formation of a vortex funnel. ($F_{cf}^{\,o}$ - centrifugal force acting at the *i* - level of the oil funnel, $F_{cf}^{\,w}$ - centrifugal force acting at the *i* - level of the water funnel, P_o - gravity acting at the *i* - level of the oil funnel, P_w - gravity at the i - level funnel, R_w is the resultant force acting at the *i* - level of the water funnel, R_o is the resultant force acting at the *i* - level of the oil funnel)

The forces acting in the binary vortex funnel of the "water-oil" system can be represented as the following expressions:

$$Pw > Po, \qquad (1)$$

where: Pw - the force of gravity acting at i - the level of the water funnel;

Po - the force of gravity acting at the i - level of the oil funnel.

Fcf w > Fcf o, (2)

where: Fcf w - centrifugal force acting at i - water funnel level;

Fcf o - centrifugal force acting at the i - level of the oil funnel.

Rw > Ro, (3)

where: Rw - the resulting force acting at i - the level of the water funnel;

Ro - the resulting force acting at the i - level of the oil crater.

Further, since the molecular interaction forces of the "oil-oil" system are greater than those of the "ice-oil" system, when oil is pumped into the interior of the vortex funnel (as a result of the effect of the vortex funnel effect), oil formed in the vortex funnel removes oil layers from the "ice-oil" boundary, cleaning its surface, which is replaced by the "ice-water" boundary. The latter effect is also confirmed by experiments (example.g., the series of photographs in Fig. 3), where gradual removal of oil from the "ice-oil" boundary is observed.

4 Conclusion

Despite some success in the development of new technologies for oil spill response in the Arctic, they are usually not completed due to insufficient funding and lack of interest from oil companies.

A new method of oil pollution removal has been proposed and investigated, which consists in localizing an oil slick and then removing it from under the ice using the vortex funnel effect and pumping it into an oil receiver.

The authors are grateful to the Technopark of Aviation Technologies (Ufa) for providing the opportunity to carry out experiments to remove oil contaminants from under the ice.

References

1. Oil spills. Problems associated with the elimination of the consequences of oil spills in the Arctic seas. Available online: https://wwf.ru/upload/iblock/011/arctic_oil.pdf (2011)
2. A.A. Toropov, et al., Elimination of oil spills under the ice in the remote Arctic waters of JSC "Central Design Bureau of Marine Engineering" Rubin", *Arctic: ecology and economics*, **32(4)** (2018)
3. Arctic Marine Oil Spill Prevention and Response and Emergency Preparedness / ExxonMobil. Available online: https://cdn.exxonmobil.com/~/ media / russia / fi les / arctic / arctic-osr_russian-fi nal.pdf (2021)
4. V.I. Pavlenko, et al., Actual problems of oil spill prevention and response in the Arctic and methods for assessing environmental damage to coastal territories, *Arctic: ecology and economics*, 19(3), pp. 4-11 (2015)
5. Auth. USSR No. 1765292, ICl E02B 15/04, Device for collecting oil from under the ice cover of the reservoir. Bul. No. 36, 1992
6. S.N. Popov, et al., Device for collecting oil under water. Patent RU 2471925, ICl E02B 15/04, 10.01.2013
7. V.M. Orlov, et al., *Remote control of the upper layer of the ocean* (Novosibirsk, Science. Siberian Branch 1991) p. 49
8. M.L. Belov, et al., Laser method for monitoring thin films of oil products on the water surface, based on the measurement of the first and second derivatives of the reflection coefficient, *Optics of the atmosphere and ocean*, **24(7)**, pp. 568-571 (2011)
9. RF Patent No. 2533920, ICl E02B 15/04, Method and device for collecting oil from under the ice cover of a reservoir. Bulletin No. 33, 2014.
10. US Patent No. 9982407. Method and a system for removal of oil under ice. 2018 Nov.
11. RF patent No. 2134739. A device for removing oil and oil products from the water surface Application: 97111715/13, 08.07.1997. 1999
12. RF Patent No. 2109876. A device for collecting oil products from the water surface. Application: 96111211/13, 06.06.1996. 1998
13. US Application No. 20100252508. Oil Combatting Vessel. 2010
14. A.D. Mingazhev, et al., RF patent No. 2572765. ICl E02B 15/04 Method of collecting oil from under the ice cover of a reservoir. Bull. No. 2, 2016.
15. A.D. Mingazhev, et al., RF patent No. 2683496. ICl E02B 15/04 Method for collecting oil spills from under the ice cover of a reservoir. Bull. No. 10, 2019

Permafrost degradation and devastating floods of the 20th – 21st centuries for the Livelihoods of Yakut Arctic villages

Alexander Suleymanov[1,*]

[1]Candidate of Historical Sciences, Senior Researcher, Siberian Branch of the Russian Academy of Sciences, Institute for Humanities Research and Indigenous Studies of the North, 677027, Yakutsk, Russian Federation

Abstract. The article uses the example of the Argakhtakh village of the Srednekolymsky ulus of the Republic of Sakha (Yakutia) located in the middle course of the Alazeya river basin to analyse the consequences of floods during the period of a relatively stable state of the environment and in recent decades, characterized by its imbalance due to global climate transformation, based on analysis of archival data, published literature and own field materials. Due to this, the causes, history and consequences of h floods of 1977-1978 were examined. It is shown that it caused significant damage to agriculture, specifically to the fodder supply of bovine cattle and horses. At the same time the materials found confirm that the territory of the village itself was not significantly affected. A different situation occurred during the floods of 1997, 2007-2008 and 2017-2018. In the first case, not only the farmlands were flooded, but the new streambed of the Alazea River appeared next to Argakhtakh and the local farm zoo was deprived of its stable land connection to the village. The flood rises of the 21st century, apart from disastrous consequences for cattle ranges and meadows, caused damage to the housing stock of Argakhtakh, household outbuildings and infrastructure objects. all this led to a radical reduction in the scale of involvement of the local population in traditional sectors of economic activity (cattle and horse breeding), the elimination of fur farming, the degradation of transport accessibility of the village and the costs of the population to restore the economy.

1 Introduction

Water is life. During the whole history of human evolution, the presence of a river or other large water reservoir has been the most important condition for life. The largest subject of the Russian Federation, which is the Republic of Sakha (Yakutia), has about 800 thousand lakes and about half a million large and small rivers, the total length of which is over one-and-a-half million kilometers. In Yakutia, rivers are not only benefactors, but actually transport carcasses, connecting everyone living along them. If during the hot summer month, ships, boats and other floating crafts can move along the rivers, then the long winter turns streambeds into winter motor roads allowing not only passenger, but also cargo transportation. The republic has entire districts that can only be reached by automobile transport by winter roads. Currently winter roads take more than 60% of the total length of roads in Yakutia, and about 80% of all cargo is transported on them. At the same time, of course, rivers are a source of not only additional opportunities, but significant risks which are associated primarily with destructive floods.

More and more often, one of the causes of such floods is the degradation of permafrost, within the zone of continuous spreading of which almost the whole territory of Yakutia is located, [1]. It is here that the frozen earth reaches its maximum capacity exceeding 1.3 km. At the same time, 31% of the total area of the region is covered by the wet permafrost rock massif, which is the most vulnerable to the consequences of global climate changes of the last decades, which clearly confirms the fact that permafrost, i.e., permanent freezing of the land, unfortunately, is not at all permanent [2].

About 15 years ago the problem with rising water levels of one of the northern rivers of Yakutia, which is the Alazea River, and flooding of settlements located on its banks (Andryushkino, Argakhtakh and Svatay) became so serious that in 2007-2008 a scientific group was created at first and then an expedition was organized in order to discover the reasons of these flood events according to special decrees of the Government of the Republic of Sakha (Yakutia). According to the results of these studies of 2018, the members of these examinations published a monograph in which they noted that the genesis of floods is based on the degradation of permanently frozen ground, which led to the growth of amount of supra permafrost groundwater due to an increase of the depth of seasonal melting and

* Corresponding author: alexas1306@gmail.com

degradation of a wet array of permafrost rock mass. The work also dealt with a range of consequences of 'big water' on the ecosystem of the Alazea River basin. However, the social consequences were not considered [3].

In general, the risk of devastating floods in the Arctic is now not prevalent in the scientific discourse aimed at analyzing natural disasters in this region. Floods are approximately on a par with coastal flooding, erosion and oil spills [4, 5]. In addition, authors who study the impact of climate change on northern communities pay some attention to the risk of flooding in the Arctic [6, 7, 8]. The case of Yakutia is represented mainly by studies of the consequences of floods for local communities in Central and Southern Yakutia [9, 10, 11].

The consequences of the floods of 1997 and 2007-2008 for livelihoods were briefly mentioned in a chapter on the migration of Yakutian population published in «Global Warming and the Human Natural Dimension in Northern Eurasia [12]. The possibilities of calculating the economic damage caused by floods on the Alazea River were analysed by O. Parfenova and D. Nogovitsyn in their article [13].

At the same time, the literature existing to date does not answer a range of important questions. Firstly, it is unclear if such floods have occurred earlier and to what consequences they led. Were there any differences from today's floods? Secondly, the certain impact of the recent floods on the Alazea River on the life support system formed here over decades has not been studied. In this article, the author will try to fill the existing historiographic gap. Its purpose is to analyze the history of the floods that took place during the 20th-21st centuries in the Alazea River basin and their influence on the life support system though the example of the village of Argakhtakh, as well as to identify the differences between the consequences of floods during the period of active climatic transformations and the time when perpetually frozen ground was, as it is believed relatively stable.

2 Materials and Methods

This study is based primarily on the results of field studies conducted in Srednekolymsky ulus in March-April 2021. In the course of these studies, 39 respondents were interviewed, the consequences of flooding (affected residential and household buildings) were recorded and one-time observations of the permafrost temperature regime in mine workings 9 meters deep (2 boreholes) were conducted. important materials were obtained while getting acquainted with the permanent exposition of the museum of of Argakhtakh village and studying materials of the Municipal Archive of the municipal district "Srednekolymsky ulus". This work was financially supported by the Russian Science Foundation, project number 19-78-10088. Interpretation of the obtained data, inclusion of accumulated materials in the modern concepts of analysis of destructive floods in the Republic of Sakha (Yakutia) and their impact on the existing life support systems of the local population was carried out with the financial support of FORMAS (project DNR: 2019-02332), RFBR (project No 20-55-71005) and JST (Grant No. JPMJBF2003) through the Belmont Forum Collaborative Research Action: Resilience in the Rapidly Changing Arctic, HYPE-ERAS project.

3 Results and Discussions

3.1 History of the village of Argakhtakh and the Floods on the Alazea River

The village of Argakhtakh is located in the middle reaches of the Alazea River and is administratively classified as a part of the Alazea nasleg of Srednekolymsky ulus of the Republic of Sakha (Yakutia) (See Fig. 1). According to some sources, the village was founded in 1946 [14]; according to others, it was founded two years earlier, as part of the Soviet policy of settlement formation (concentration of scattered clans, many of whom led nomadic lifestyle, in the established settlements) conducted in norther Yakutia from the 1930s through the 1950s [15]. The central base of the Alazeysky state farm (sovkhoz) was located in Argakhtakh. The base of the state farm activity included meat-and-milk cattle breeding, meat horse herd farming, fur farming (silver fox breeding), fishing and fur trade. The population of Argakhtakh is decreasing. according to the 2002 census of Russia, the population of the village was 581 people, while in 2021 it is 469 persons. The vast majority of the population of Argakhtakh is traditionally Yakuts.

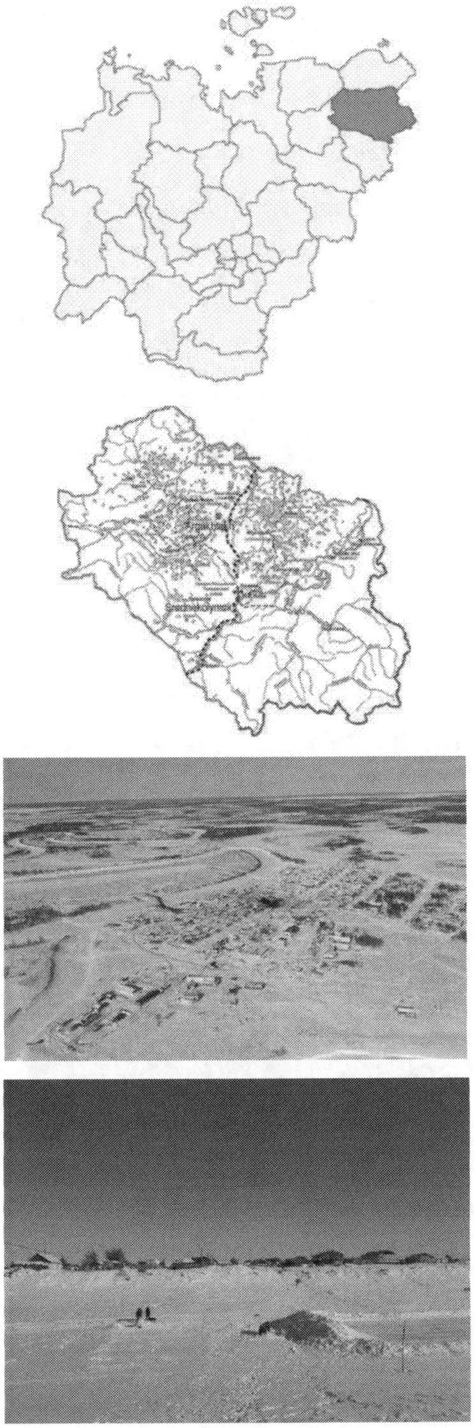

Fig. 1. Argakhtakh village, Srednekolymsky ulus of the Republic of Sakha (Yakutia).

As follows from the materials found in the administration of the Alazea nasleg and the archive of the Srednekolymsky district, as well as from a survey of the local population, from the 1950s through the 1980s, Argakhtakh suffered from floods 4 times: in 1958, 1968-1969, 1977-1978 and 1985. In 1969 and 1978, the water level reached 807 cm and in 1985 it reached 766 cm. [Archive of the administration of the Alazea nasleg].

At the same time, unlike the fast-passing underfloods lasting for a few days that are characteristic of most rivers in Yakutia, the Alazea River is characterized by a long rise in water level, counted in months due to a number of features. This explains that floods in Argakhtakh may last for two consecutive summer seasons in those cases when ice formation took place when the water levels were critically high. This explains that floods in the Argakhtaya can last for two consecutive summer seasons in those cases when ice formation occurred at critically high water levels.

In particular, this happened in 1977-1978, when one of the well-documented floods on the Alazea River occurred. These documents remained after the work of the district commission on the determination of the damage from this natural disaster. they show that the flood was caused by continuous heavy rainfall from July 20 to September 15, 1977. As a result, the precipitation norms were exceeded almost 3 times. According to the data from the Andryushkino gauging station, ice formation took place on October 08, 1977, when the water level was 878 cm, while the average value of this indicator at the moment was 696 cm. Moreover, by October 30, the water level reached 905 cm. As it was mentioned during the work of the commission, something similar was observed only once, in the autumn of 1968, when ice formation started at a water level of 892 cm. However, that year, on the contrary, it decreased by the end of October.

As a result, by the winter of 1977-1978, 150 tons of stored haystacks, 300 tons of silage and 104 tons of mowing were found in the water at the Alazea department of Alazeysky state farm. During the summer of 1977, the Alazeysky state farm fulfilled less than 70% of the hay harvesting plan. The situation was aggravated by intensive snowfall, which started on September 14, 1977, and further increase of air temperature accompanied by new rains, which then were replaced by frosts. As a result, an ice crust formed on the winter pastures cattle and animals could not reach the grass. Such situation, when cattle ranges were flooded and then frozen, was noted for the first time in the history of observations on the Alazea River (which, however, was not that long at that time). Besides, due to water flooding of the areas, the depth of seasonal melting of perpetually frozen ground increased, which contributed to the formation of swamps at the hayfields.

The situation continued to develop negatively in the spring and summer of 1978, when melt waters began to flow into the Alazea River as well. By the end of June, around 80% of all haylfields and cattle ranges of the Alazea department of Alazeysky state farm were flooded. As a result, the state farm lost 156 heads of bovine cattle and 186 horses in 5 months of 1978 alone. Therefore, the main consequence of the floods of 1977-1978 for the life-support system of the village of Argakhtakh was the damage caused to traditional economic activity. Besides, the documents mention that the take-off grounds were affected by the underfloods. Argakhtakh is not included in the list of named affected

https://doi.org/10.1051/e3sconf/202337805008

settlements, but the large number of helicopter flights to transport workers conducted at the expense of the Alazeysky state farm in 1977-1978, suggests that the situation contributed to the deterioration of the transport accessibility of the village. It should also be noted that among the types of damage caused by the 'big water' to the state farm listed in the documents not a single mention of any construction within the territory of the central estate were damaged was found. Besides, the cattle numbers exceeded the 1976 data in 1979 (548 and 569 correspondingly) [Municipal Archive of Srednekolymsky ulus municipal district].

3.2. Floods of the end of the 20th century and the first decades of the 21st century

The 1990s and early 2000s in the history of Yakutia were marked by a number of devastating floods. The most famous example is the spring flood on the Lena River, which is the largest in the region. In 1992, due to the superposition of two waves of flood (due to a rapid increase of air temperature, the snow in the river heads melted earlier than normal and the main ice motion was followed by the so-called black water), for instance, in Olekminsk, where the critical water level is 930 cm, it reached the indicator of 1,109 cm. In 1998, as a result of the spring flood on the Lena River, 172 settlements were affected, 160 bridges, 133 dams and 760 km of roads were destroyed. 97 thousand residents were in the flooded area, and 15 people died. Another natural disaster happened in the spring of the following year. However, the most infamous event is the 2001flood, when the water level at the source of the Lena River exceeded the critical mark by 6.5 meters and almost washed away the city of Lensk [9].

Alazea was no exception. In 1991, the water level here reached 767 cm. In 1997, it exceeded the critical (for Argakhtakh) mark of 782 cm and reached 834 cm. Three years later, these numbers were 835 cm; in 2002 it was 802 cm; in 2003 – 762 cm; in 2007-2008 – 771 cm, and in 2011 – 737 cm. The most disastrous were the floods of 2007 and 2017-2018. In the first case, the water level reached 869 cm, and in the second – 897 cm. No wonder that the floods of those years, as well as the flood of 1997, are most vivid in the memory of the local population [Archive of the administration of the Alazea nasleg].

In 1997, the elements flooded the agricultural lands in the same way they had two decades earlier. The high water level lasted from July to September. In the following years, no fodder grasses grew on these lands, because everything was covered with sludge. As a result, all the lands in the vicinity of the village were cycled out. The requirement to use distant lands with the active use of transport naturally led to an increase in the economic load on economic subjects. It should be noted that if during the 1977-1978 flood the administration of Alazeysky state farm had enough resources to practically continuously use helicopters to transport brigades and required cargoes, then after the collapse of socialist economy even the extensive use of ground freight transport became difficult due to the high cost of fuel in the Arctic districts of Yakutia.

Probably the main reason why the 1997 flood was well-remembered by the local population was that the Alazea River changed its streambed next Argakhtakh, washed away the road and blocked the farm zoo from the rest of the village. Our respondents associate the change of in the nature of the following devastating floods with the fact that the Alazea River has become closer to the residential buildings as a result of previous cataclysms.

For instance, the flood of 2007-2008 not only flooded the agricultural lands, but also directly affected Argakhtakh, where the elements affected 29 households, and residents travelled along the village streets by boat. Ten years later, the flood affected 85% of agricultural lands and 50% of the stored-up hay. 22 residential houses were under water [12] (See Fig. 2). [12] (see Fig. 2).

Fig. 2. Argakhtakh village, flood of 2007 (photo from the funds of the Argakhtakh museum)

As the study has shown, natural disasters had the most negative consequences for the life support system of the Argakhtakh residents. First of all, enormous damage was caused to animal breeding: the lack of fodder led to the forced butchering of animals or further abandonment of its maintenance due to significant expenses and labour costs. Particularly, if in 2015 the population of Argakhtakh kept 139 heads of cattle, in 2019 – 87. The number of horses during the same period decreased from 477 heads to 325. Of course, this trend was influenced not only by the 'big water', but also by the general social and economic situation, as well and the aforementioned high cost of animal keeping.

However, the fact that the most significant changes in livestock numbers took place in 2017 indicates that floods made the main contribution. Besides, between 2015 and 2019, another branch of traditional economic activity, which is fur farming, went down from 45 heads to nought [Archive of the administration of the Alazey nasleg]. In this case, according to local residents, the situation was influenced not only by general trends in the development of agriculture in Yakutia and the lack of demand for the products during the post-Soviet period, but also by the consequences of the 1997 flood, which made it much more difficult to reach the farm zoo (See Fig. 3).

Fig. 3. Abandoned farm zoo in Argakhtakh village, March 2021 (Author's photo)

The increased transport isolation of Argakhtakh was also an important consequence of the 2017 flood (See Fig. 4). traditional, the village had a stable land connection with the administrative centre of the ulus, which is Srednekolymsk, only by winter road the "Arktika" (See Fig. 5). In summer, there were flights to Argakhtakh. In 2017, the take-off site was flooded and so damaged that an expensive helicopter communication became the only option. Helicopters fly to the village twice a month [Author's field materials, March 2021].

Fig. 4. Argakhtakh village, flood of 2017 (photo from the funds of the Argakhtakh museum)

"Big water" also led to the limited use of ice boxes – "glaciers", which are special constructions buried in the frozen ground, which provide for storing frozen products all year round (See Fig. 5, Fig. 6) [author's field materials, March 2021]. This option is an important component of the traditional farming of the rural population of the Arctic districts of Yakutia [16]. However, in Argakhtakh, as a result, many ice boxes were flooded and abandoned. Besides, the widespread distribution of freezer chests and ground thawing affect the reduction of their used [17, 18]. According to the data we obtained during the research at the end of March 2021, the temperature of the grounds in Argakhtakh at a depth of 3 m was only -2.65°C [Author's field materials, March 2021].

Fig. 5. Winter road "Arktika", March 2021 (photo of the author)

Fig. 6. One of the surviving ice boxes in Argakhtakh, March 2021 (photo of the author)

The floods seriously impacted the housing stock of Argakhtakh, as well as the well-being of local population; some residents had to build new homes to replace those that had been damaged by the disaster. In particular, one of the retirement-age families we interviewed had to abandon the house they had lived lived in for four decades and build a new.

Traditional collectivism and mutual assistance, characteristic of the rural residents residents of Yakutia, especially in its Arctic part, are an important adaptation mechanism that makes it possible to neutralize at least partially, the resulting costs. Because of this, residents of flood-stricken territories could wait for months for the end of the natural disaster in the houses of their relatives and households located in the higher part of the village.

4 Conclusions

Therefore, the above data confirm that Argakhtakh has been prone to flooding since the first decades of its existence. At the same time, if from the 1960s through the 1980s the main strike was mainly on agricultural land, and agriculture recovered fairly quickly on the basis of available resources, then at the turn of the 21st century floods began to occur more frequently and, most importantly, on a larger scale. At the same time, agriculture no longer had the same stability as before. As a result, the combination of new social and economic conditions and powerful floods led to a radical sequestration of traditional sectors of economic activity, which the Yakut population had been involved in for ages (animal and horse breeding), as well as to the actual liquidation of the fur trade, successfully practiced during the Soviet period. The main difference between the floods of recent years, which occurred during the time of forced degradation of the permafrost, and the floods that had taken place earlier, is their disastrous conditions directly for the village of Argakhtakh, its residential and household buildings, as well as infrastructure facilities.

This work was supported by the Russian Science Foundation within the framework of the Project No. 19-78-10088 "'Cryoanthropology: natural low temperatures in the life support system of rural communities of Yakutia (traditional practices, modern challenges and adaptation strategies)" and with financial support from FORMAS (project DNR: 2019-02332), RFBR (project No 20-55-71005 "Hydrology, Permafrost and resilience in Eastern Russian Arctic and Subarctic") and JST (Grant No. JPMJBF2003) through the Belmont Forum Collaborative Research Action: Resilience in the Rapidly Changing Arctic, HYPE-ERAS project.

References

1. R.V. Desyatkin and A.R. Desyatkin, Effect of Increasing the Depth of the Active Layer on Changes in the Water Budget in the Cryolithic zone, *Eurasian Soil Science,* **52**, p. 1447–1455 (2019)
2. V. Lytkin, et al., 2021 Influence of Permafrost Landscapes Degradation on Livelihoods of Sakha Republic (Yakutia) Rural Communities, Land, **10(2)**, p. 101
3. S.P. Gotovtsev, et al. *Cryoecosystems in the Alazeya River Basin* (Academic Publishing House "Geo", 2018)
4. IRDR (Institute for Risk and Disaster Reduction) 2014 Arctic Risk: A Discussion of the Possible Outcomes of Two Disaster Scenarios (IRDR Report 2014-02). Available online: https://www.ucl.ac.uk/rdr/publications/irdrspecial-reports/irdr-special-report-2014-01 (2014)
5. B. Rossi, Summary Report and Recommendations on the Prevention of Marine Oil Pollution in the Arctic (EPPR Summary Report and Recommendations on the Prevention of Marine Oil Pollution in the Arctic, 2013). Available online: https://oaarchive.arctic-council.org/handle/11374/106 (2013)
6. R. Brunner, et al., An Arctic Disaster and Its Policy Implications, *Arctic*, **57(4)**, pp. 336–346. (2004)
7. J.P. Clement et al., Managing for the Future in a Rapidly Changing Arctic. A report to the president. Available online: http://www.afsc.noaa.gov/publications/misc_pdf/iamreport.pdf (2013)

8. I. Kelman, et al., Climate Change's Role in Disaster Risk Reduction's Future: Beyond Vulnerability and Resilience, *Int. J. Disaster Risk Sci.,* **6**, p. 21–27 (2015)
9. L.I. Vinokurova, et al. Waiting for the ice-break: social aspects of devastating floods in rural Yakutia (late 20th – early 21st century), *Humanitarian studies in Eastern Siberia and the Far East*, **35(1)**, 28–44 (2016)
10. Y.Y. Kontar, et al., Springtime Flood Risk Reduction in Rural Arctic: A Comparative Study of Interior Alaska, United States and Central Yakutia, Russia, *Geosciences*, **8(3)**, p. 90 (2018)
11. N.I. Tananaev, et al., Assessment of the community vulnerability to extreme spring floods: the case of the Amga River, central Yakutia, Siberia, *Hydrology Research*, **52(1)**, pp. 125-141 (2021)
12. Global Warming and the Human-Nature Dimension in Northern Eurasia, ed. by T. Hiyama, H. Takakura (Springer, 2018)
13. O. Parfenova and D. Nogovitsyn, Estimation of Economic Damage from Floods on the Alazeya River in the Republic of Sakha (Yakutia), *Problems of the modern economy*, **65(1)**, pp. 124–126 (2018)
14. S.P. Gotovtsev, "Ylyk" – a step into the future, *Science and technology in Yakutia*, 19(2), pp. 70–73 (2010)
15. V. Filippova, *Indigenous small peoples of the north of Yakutia in the changing space of life. Second half of the twentieth century* (Novosibirsk, Publishing House Nauka, 2007)
16. A.A. Suleymanov, Cryoanthropology: Influence of Natural Low Temperatures on the Yakut Traditional Life Sustenance System, 19th Century to the 1930s., *Oriental Studies*, **14(1)**, pp. 115–133 (2021)
17. L.V. Lytkin and I.I. Syromyatnikov, *IOP Conf. Ser.: Earth Environ. Sci.*, **666**, 062030 (2021)
18. N. Tananaev, On the collective monograph "Cryoecosystems of the Alazeya River Basin", *Science and technology in Yakutia*, **34(1)**, pp. 109–112 (2018)

Effectiveness of the right to health and favorable environment in the Arctic regions

Jean-Paul Markus[1], *Uolan* Borisov[2*]

[1]Phd, public law professor, Université de Versailles Saint-Quentin-en-Yvelines, 78000, Paris, France
[2]Senior Lecturer of arctic law department of law faculty, North-Eastern Federal University, 677007, Yakutsk, Russian Federation

Abstract. The main problem in the realization of the rights to a favorable environment, including in the Arctic regions, is the difficulty of establishing and proving the causal link between economic activity and damage to health. Modern scientific achievements make it possible to establish the connection between the micro-environment and health and, therefore, to provide indemnification of damage. But the causal link between degradation of macro environment and human health is less studied, which leads to complications in judicial practice.

1 Introduction

While environmental protection was not yet seen as a purely social problem, it was mainly of interest to marginal social groups. In the 1990s, environmental protection took a new direction of development and new laws were adopted, establishing requirements for the protection of a favorable environment. This circumstance caused a number of favorable changes in environmental protection.

First of all, environmental protection became a hot political topic, which was widely discussed in various political debates, especially during elections.

Involvement of people in "environmental thinking" made it possible to issue laws focused at protecting human health and their rights to a favorable environment. One of the most significant law is the French Environmental charter of 2005 [1], whose first article reads as follows: "Everyone has the right to live in a favorable environment". As a result of these changes, the legal system has taken a "green" direction [2].

Secondly, the popularization of the environmental topic has served to develop the direction of environmental law in French jurisprudence since the 1980s. As result, new educational programs and a new discipline "santé environmentale" (which can be translated as "environmental health") emerged [3].

Thirdly, the World Health Organization has begun research in the new area of "environmental health" ("santé environmentale"), which includes the following aspects: human health, the quality of human life, which is determined by physical, chemical, biological, social, psychosocial and aesthetic factors of the environment. Research has been used to identify environmental factors that affect human health. Research is also used to study how to manage, control and prevent environmental changes that may affect the health of current and future generations. In this regard, there have been many scientific dissertations on this topic [4].

In France there is a scientific organization that deals with environmental health - the Francophone Society for Environmental Health [5], as well as a world association - the International Federation for Environmental Health [6]. According to some researchers, the discipline of "environmental health" may eventually replace the discipline "santé publique" (public health) [7].

The roots of this science go back to 19th-century public health, the goal of which was to combat epidemics caused, in particular, by urban development as a result of the Industrial Revolution. Many French laws stem from this trend (Law of May 13, 1850 on unsanitary housing, law of February 15, 1902 creating compulsory vaccination, and even before, an imperial decree of October 15, 1810 "relating to factories and workshops which spread an unhealthy or uncomfortable smell ")

Currently, "environmental health" ("santé environnementale") is a subject of legal research and jurisprudence. This discipline, however, is somewhat ambiguous: it focuses on the immediate human environment ("microenvironment"), which directly affects human health, but tends to move away from human concerns to consider the global environment ("macroenvironment"). General environmental degradation is likely to threaten human health, but so does humanity itself threatening global planetary health. Nevertheless, the more the law addresses micro environmental issues (I), the more difficult it is to resolve macroenvironmental disputes (II). This phenomenon is more pronounced in the Arctic regions.

* Corresponding author: jean-paul.markus@uvsq.fr, buv989@mail.ru

2 Indemnification of damage caused to individuals

2.1 Problematics

At first glance, one might assume that negative changes in the microenvironment do not affect Arctic residents as much as urban residents, such as those living in buildings between the highway and the airport. However, for indigenous peoples, the negative changes in the microenvironment caused by industrial development of their lands - oil drilling, mining of diamonds, gold and other minerals, construction of temporary workcamps in which migrant workers live, also cause enormous damage to their way of lifestyle and health. Roads, pipelines and other infrastructures affect the areas of traditional residence of indigenous peoples, destroying hunting and nomadic areas. Also, the health of the indigenous population is affected by newcomers, who may be carriers of diseases that are not typical for these regions.

Another threat to the health and lives of indigenous people is a change in their lifestyle, a change in food habits: the modern food industry is replacing traditional food products. This happens because indigenous people cease hunting and fishing due to changes in living conditions and changes in the environment. In addition, indigenous people are losing their hunting and fishing skills. These phenomena have been documented among island peoples, such as the Melanesians (French Polynesia) or the Kanaks (New Caledonia) and among the peoples living in the Arctic regions of Canada [8].

2.2 Scientifically proven and legally established causal relationships

If health risks are scientifically proven, the state must take measures to prevent them. The responsibility of the state to protect the health of the population is enshrined in the constitutions of virtually all states. In the event that the authorities are unable to protect public health, mechanisms for indemnification of harm are provided. This classic legal mechanism has become possible due to scientific progress in identifying threats to the environment associated with human activities. The gradual formation of indicators or aggregates, which are now reliable [9], makes it possible to determine more accurately the causal relationships between harms caused to one or more subjects and the environment. The establishment of scientific reasons on which the judgement is based, with the help of an expert, is a condition for the condemnation of the alleged causer of harm and the award of compensation to the victim.

In all legal systems, the complainant must prove at least three facts: the harm caused to them; the source, and the causal link between them. For example, a small settlement and an association of local residents were able to prove health damage as a result of environmental pollution caused by an incinerator that significantly exceeded European standards for emissions of dust and dioxins [10]. In this particular case, the existence of established standards of dust emissions, which were violated, made it easier to obtain compensation. Another case, much more prominent in the media, concerned the spread of toxic green algae on the Breton coast due to the excessive use of nitrogen fertilizers by farmers [11]. Several people were poisoned, and entire beaches are still closed for health reasons. France has been condemned both by the Court of Justice of the European Union [12] and by the national courts: in a decision of 12 July 2017, the Council of State overturned the government's refusal to take all necessary measures to ensure compliance with European Directive 2008/50/ CE of 21 May 2008 on atmospheric air quality [13].

If there were no standards for the use of nitrogen fertilizers, applicants would first have to prove: 1 / the risks of using nitrogen fertilizers; 2 / the obligation of the state to establish limited norms; 3 / the connection between the lack of that standard and the harm to the health of those affected, which would cause many obstacles, often insurmountable due to lack of scientific methods. Representatives of small communities have tried to solve the problem and have addressed the problem of collisions with electrical objects, but their requests were rejected [14].

This burden of proof, or its relaxation, can be only be reversed by law, as was done in France with regard to the consequences of nuclear tests for the inhabitants and the military in the Algerian Sahara or in French Polynesia. [15] Except in these cases, compensation is not possible without sufficient evidence. Many interpretations of the environment have been rejected by the courts because they were based on people's "anxiety" and on rumors attributing some death to some allegedly dangerous activity [16]. Moreover, in practice, it is not uncommon for health concerns to be used to overturn construction, that is actually less hazardous to human health than the value of the plaintiffs' real estate [17].

Thereby, the development of microenvironmental science provides detailed information about the effects of activities on the human environment, which forms the basis for the judge in the proof. The Court of cassation was able to overturn the decision of the court of Appeal, which, instead of verifying for each applicant the real fact of harm and its causal relationship with the alleged factual violation, made a formal decision [18]. These scientifically proven causal relationships are the basis for judicial decisions, as well as for legal acts. The regulation of activities is a kind of violation of freedom, which can only be justified by the protection of public health. A similar example is the administrative court, when it checks the legality of an order by examining the proportionality of sanctions [19].

Thus, a resident of the arctic region directly affected by pollution can count on proven legal instruments. The situation is different when it comes to indirect victims in the macro environment.

3 Compensation for damage to a collective

3.1 Problematics

The scientific concept of "human bioclimatology" was presented in the medical work of researcher J.P. Besansenot [20]. The relationship between climate and human health has been emphasized since the 18th century as two systems, the interchange between which is stable over time: climate regulates human life, while humans have no or insignificant influence on climate. nowadays, however, climate has dramatic changes that affect the individual throughout his or her life. It has been scientifically proven that nature and human communities adapt to climate change from generation to generation [22], but on the scale of a single human life, this adaptation is much more difficult. However, it is at this scale that the adverse health effects manifest themselves, because, on the one hand, climate change accelerates, and on the other hand, life expectancy increases.

Researches of the impact of the strong cold waves interspersed with floods and winter storms to the health of indigenous people of Quebec have been carried out [23]. The global warming contributes to the appearance of parasites that were previously absent in the Arctic regions (mosquitoes and other biting insect vector of diseases) and infections which imported from other regions. Due to global pollution, accumulations of plastic enter into food, in particular through fish, which is a staple food in the Arctic regions. Another threat to health is the risk of food security, lack of fishing resources, sources of protein.

In addition to these direct health effects, there are changes in social behavior that have health consequences, such as food changes caused by the scarcity of traditional foods. As noted in the Arctic regions, as well as in other regions of the world, the replacements of traditional foods with imported foods, which have much more sugar and carbohydrates, leads to cardiovascular problems, diabetes, vitamin deficiencies, anemia, dental problems and obesity with all the ensuing consequences. In addition, the indigenous population's resistance to infections is reduced. The current mortality and morbidity rates among Inuit are higher than in other places of Quebec, mainly due to their poor nutrition and smoking. As a result, the life expectancy of the elderly is reduced, well-being is reduced, and consequently mental health risks arise, all of which may also be linked to climate change. In these cases, it is necessary to determine the amount of harm to be compensated.

3.2 Scientifically unproven and legally difficult-to-prove harm

As the court practice shows, only the possibility of connecting these two phenomena by causal relationship allows the court to solve the issue of indemnification. If the pollution of the water network of the region resulting from human agricultural activities can be easily identified. How to quantify the relationship between the global oil industry and the harm it inflicts to the population of Arctic region?

An application to the court for indemnification for harm to health caused by environmental violations can be considered in favor of plaintiffs only if reliable indicators would be developed, the ones which will allow the judge to go beyond the consideration of cases which court usually do [24]. Only these objectively formulated indicators will make it possible to identify measurable causal relationships between a global phenomenon (pollution, global warming) and the consequences of which the indigenous population suffers. However, the larger the scope of the dispute due to the increase in the community of plaintiffs (for example, the "Arctic peoples"), then less indicators we have. Therefore, it is difficult to deviate from the classical model of litigation based on the consideration of damage to an individual or a small community of persons faced with polluting activities, which has been studied and understood. So we have to return to litigation on the microenvironment due to the lack of scientific data on the macro environmental causes of negative changes in the environment.

An example of such practice is a petition made in 2005 by Inuit in Canada and the United States, with legal support from the nongovernmental organization (NGO) Earth Justice, to the Inter-American Commission on Human Rights. The Commission decided that there was insufficient information to judge that emissions from American industry directly or indirectly affected the rights of indigenous peoples in the Arctic (in particular, the right to life and health) [25]

This does not exclude positive changes in the future. In addition, some courts take a flexible approach in establishing causation. For example, the American court has the following approach: "an indirect causal relationship will be sufficient if there is a sufficiently traceable relationship between the alleged actual damage and the alleged behavior of the defendant" [26]. There are other methods, such as an environmental impact assessment procedure prior to construction with an environmental impact - which has been compulsory in France since 1976. Environmental impact assessments are reviewed by courts. However, the courts do not have sufficient scientific competence to consider them objectively, especially when they are contradicted by other studies provided by environmental associations. Even experts, despite their competence, cannot unambiguously interpret the assessments. Environmental impact assessments are often criticized for being based on only one scientific discipline, whereas it would be more effective to supplement the assessments with knowledge from several scientific disciplines to highlight the real impact of industry and infrastructure on the environment. So the problem doesn't lies in the personality of the expert or his possible conflicts of interest, but in his specialty and knowledge that he must mobilize to assess the threat to the environment or health [27]. International agreements are concluded to involve different branches of knowledge, which involve the public in the study of potential environmental impacts prior to the construction of infrastructure [28].

Meanwhile, the causal link between pollutants, climate change and victims is too vague to trigger the classic compensation dispute, even on the basis of environmental damage. Indirect impacts on health and the environment, these "global crises" make the task much more difficult [29]. Despite the existence of legal acts in which assessments of the impact of industrial activities on the environment and health are mandatory, modern scientific methods cannot provide objective data to the court and their justification in terms of causation. The forensic tool is not effective. Therefore, we must envision other legal instruments: foresight, precaution, preliminary collective compensation for damage, in other words: a preventive approach aimed at preventing damage or mitigating the consequences.

4 Conclusion

Of course, it is possible to "objectify" [30] responsibility for the damage caused to the climate, and its consequences for the environment and health, allocate funds for this from the state budget, which in turn is also reimbursed from the tax on environmentally polluting activities. But this presupposes the conclusion of international agreements, which, as you know, cause certain difficulties in the conclusion. It is also possible to introduce at the national level binding treaties on liability for health and the environment, which should precede any activity that carries risks to the environment and health. There are many preventive and legally promising methods, but which involve transparency and integrity of all stakeholders.

This article was carried out with the financial support of the Russian Foundation for Basic Research (RFBR) in accordance with research project No. 21-510-22001 "State regulation of subsoil use and environmental protection in France and in the Arctic zone of the Russian Federation: comparative research, methodology and practice" RFBR and the Foundation "House of Human Sciences" of France

References

1. Constitutional law the French Republic n° 2005-205 of March 1, 2005 relating to the Environmental Charter (2005)
2. The environmentalization of law, Studies in honor of Sylvie Caudal, coed. Varenne / Francophone Institute for Justice and Democracy (2020)
3. A. Cicolella, Le trente-troisième anniversaire de la santé environnementale, *Les tribunes de la santé,* **54**, pp. 31-37 (2017)
4. C. Guilleux, L'institutionnalisation de la santé environnementale France, *Aix-Marseille University doctoral thesis in sociology* (2015)
5. The Francophone Society of Health and Environment Available online: http://www.sfse.org/FR/presentation/qui_sommes-nous.asp (2021)
6. The International Federation of Environmental Health website Available online: https://www.ifeh.org/index.html (2021)
7. J. Le Moal, D. Eilstein and G. Salines, La santé environnementale est-elle l'avenir de la santé publique?, *Revue Santé publique*, **22(3)**, p. 281 (2010)
8. G. Imbert, Vers une étude ethnoépidémiologique du diabète de type 2 et de ses complications, Santé publique, **20(2)**, p. 113 (2008)
9. M. Calvez, Les plaintes en santé environnementale et l'expertise épidémiologique, *Sciences sociales et santé*, **33(4)**, p. 7 (2015)
10. C. Lepage, note Oct. 11, 2019 CA Paris no. 18-04919, Maincy incinerator case, Rev. Energy Environment Infrastructures 2019/12, n° 60 (2019)
11. C. Etrillard, Le droit face aux marées vertes en Bretagne, *Rev. Energie Environnemen t–Infrastructures,* 04 study, 11 (2018)
12. ECJ, June 27, 2002, case C-258/00, Com. EC v/ France: Rec. ECJ 2002, I, p. 5959. – CJEU, June 13, 2013, case C-193/13, Com. EU vs. France (breaches of the Nitrates Directive). – ECJ, September 23, 2004, case C-280/02, Com. EC v/ France: Rec. ECJ 2004, I, p. 8573 (breaches of the urban waste water directive). – V. on all these cases: A. Van Lang, The administrative judge, the State and green algae: AJDA 2010, p. 900. – Ph. Billet, V. Themis Ulva spec., Legal variations on green algae: BDEI, 2010, suppl., p. 31. – N. Hervé-Fournereau, The Court of Justice of the European Union and the quality of water: jurisprudential reflections of the paradoxes of the Union's water policy, In case C-280/02 The legal books, 2010, **51(3-4)**, p. 947 (2002)
13. EC, ass., July 10. 2020, No. 428409, Assoc. Friends of the Earth - France and others. Also CAA Nantes, 1 Dec. 2009, No. 07NT03775, Min. ecology, energy, sustainable development and the sea c/ Association "Halte aux tides vertes" (2020)
14. Administrative Court of Appeal of Paris, 20/12/2007, n° 05PA02760 (2007)
15. Law n° 2010-2 of January 5, 2010, article 4: "the person concerned benefits from a presumption of causality", the conditions of which are set by decree n° 2014-1049 of September 15 2014 (2014)
16. M. Calvez, et al., Les plaintes en santé environnementale et l'expertise épidémiologique, *Revue Sciences sociales et santé*, **33(4)** (2015)
17. Case of appeals against the construction of wind farms, or water treatment plants. Court of Cassation, Criminal Division, May 9, 2007, 06-87.174 (2007)
18. Cour Administrative d'Appel de Marseille, Juge des référés, 17/05/2006, 06MA00755, à propos du réaménagement d'une zone industrielle particulièrement polluante (Fos-sur-Mer) (2006)
19. J.P. Besancenot, Climat et santé PUF (Presses Universitaires de France, 2001)
20. J.P. Besancenot, Climat et santé PUF (Presses Universitaires de France, 2001), p. 4
21. J.P. Besancenot, Climat et santé PUF (Presses Universitaires de France, 2001), p. 116

22. P. Gosselin and D. Bélanger, Research impacts and adaptations of public health to the new climate of Quebec, *Revue Santé publique,* **22(03)**, p. 291 (2010)
23. Etude menée par l'Institut national de santé publique du Québec en 2004 Available online: https://www.inspq.qc.ca/sante-des-autochtones/enquete-de-sante-aupres-des-inuits-du-nunavik-2004 (2004)
24. M. Calvez, V. Van Tilbeurgh and S. Ollitrault, Environmental health complaints and epidemiological expertise, *Revue Sciences sociales et santé*, **33(4)**, 7 (2015)
25. Osofsk H.M. view on this petition «The Inuit Petition as a Bridge? Beyond Dialectics of Climate Change and Indigenous Peoples' Rights », American Indian Law Review 2007 675 Available online: http://www.inuitcircumpolar.com/inuit-petition-inter-american-commission-on-human-rights-to-oppose-climate-change-caused-by-the-united-states-of-america.html. (2007)
26. Comer et al vs. Murphy Oil USA inc. et al 2009 9 Available online: http://www.ca5.uscourts.gov/opinions/pub/07/07-60756-CV0.wpd.pdf (2009)
27. K. Favro, et al., *L'expert dans tous ses états* (Dalloz, 1996), p. 97
28. The Aarhus Convention on Access to Information, Public Participation in Decision-making and Access to Justice in Environmental Matters, signed on June 25, 1998 in Denmark (2008)
29. W. Dab, Santé et environnement PUF, *Que Sais-je*, p. 58 (2020)
30. A.S. Tabau and C. Cournil, *Rev. juridique de l'environnement,* **40(9)**, p. 683 (2015)

Problems of entrepreneurship development in the Arctic: Russian and foreign experience of observing the principles of green economy

Artur Kuchumov[1,*], *Elena* Pecheritsa[2,*] and *Natalia* Blazhenkova[3,*]

[1] Candidate of Sciences in Economics, Associate Professor, St. Petersburg State University of Economics, 191023, St. Petersburg, Russian Federation
[2] Candidate of Sciences in Sociology, Associate Professor, St. Petersburg State University of Economics, 191023, St. Petersburg, Russian Federation
[3] Doctor of Economics, Professor, Ufa State Petroleum Technical University, 450000, Ufa, Russian Federation

Abstract. In recent years, interest in the Arctic regions has grown. Global corporations, policymakers, and research scientists became interested in the Arctic, as the climate in the Arctic regions changes and as the consequences in the global world directly or indirectly extend to all countries. The desire to understand and manage the new economic opportunities associated with the vast natural resources of the Arctic regions is common among all interested parties. However, the achievement of economic goals depends on many factors. It should be noted that among the factors facilitating economic development, we should consider important the factors of development of local communities and local entrepreneurship, and observation of the principles of green economy. In recent years, Russian and foreign Arctic research scientists have published a significant number of articles on the study of the Arctic regions and the development of entrepreneurship. The purpose of this work is to study entrepreneurship in the Arctic regions by analyzing Russian and foreign scientific publications. A common negative demographic factor for the Arctic regions is the change in the structure of the population, so the development of entrepreneurship can be considered important for the development of the Arctic regions, capable of providing livelihoods for local residents. The article considers the current conditions for entrepreneurial activity in the Arctic to determine the problems and prospects for the development of entrepreneurship in the Arctic regions from the point of view of green economy. In the article, the authors offer advice on engaging in entrepreneurial activities in the Arctic, following the principles of sustainable development and adhering to the principles of a green economy.

1 Introduction

Long distances, challenging climatic conditions, and long periods of darkness pose unique challenges when it comes to developing business activities in the Arctic [2,3]. Despite the fact that in recent years the fishing fleet, cruise tourism, and research initiative have significantly increased, and as the ice melts further, new shipping routes are being considered, to the point that expedition tours on small and medium-sized passenger ships carrying from 12 to 300 people are becoming increasingly popular, communication services, satellite coverage, radio communication in remote maritime areas with severe weather conditions due to the limited range of operation in marine areas in the Arctic territories leave much to be desired [5].

Studying Arctic entrepreneurship, the researchers face many paradoxes. The most important factors in the development of entrepreneurship in the Arctic, creating a mosaic picture of the situation in the Arctic territories, are: geographical location, island or quasi-continental position; presence of large resource corporations and the specific stage of their exploitation of the resource field; institutional structure of the local economy in terms of degree of nationalization. The combined impact of these factors creates a multifaceted picture, with the Murmansk region at one pole of extremely favorable conditions for business, and the Chukotka Autonomous Region at the other, characterized by extreme harshness [6].

The changing Arctic environment could lead to an "Arctic gold rush", where states compete with each other to develop oil and gas reserves and sought-after natural resources in marine areas by expanding the legal definition of the outer limits of their continental shelves. The Arctic contains about 90 billion barrels of oil and 1 669 trillion cubic feet of natural gas, accounting for 22 percent of the world's oil and natural gas reserves. In addition, as sea ice depletes, the Arctic could become an alternative corridor for international shipping. The Northeast Passage, a roundabout sea route through the Norwegian and Russian Arctic, is actually 37 percent

* Corresponding author: arturspb1@yandex.ru, helene8@yandex.ru, bnm61129@yandrx.ru

shorter than traditional routes through the Suez Canal. However, before it becomes possible to extract natural resources or use the Arctic sea routes, it is necessary to eliminate huge technical, environmental, and operational risks [4], leading to the increased negative anthropogenic impact on the fragile environment of the Arctic.

In July 2020, the Federal Law on State Support for Entrepreneurship in the Arctic Zone of Russia, signed by Russian President Vladimir Putin [1], came into force. The new law establishes a system of preferences for residents of the Arctic zone of Russia and contributes to the development of northern territories, creation of appropriate conditions for attracting investment, and stimulation of the opening of new companies, including small and medium businesses. Experts expect that the law will aid in facilitating conditions for creating new jobs, developing infrastructure and improving transport accessibility. In accordance with the newly adopted law, support will be provided to individual entrepreneurs or resident companies of the zone that plan to start new investment projects with a planned investment capacity of one million rubles and above, including the cost of new construction, technical re-equipment, modernization of the main production facilities, repair of premises, purchase of machinery and equipment. An investment project is recognized as new if at the time of application, the volume of actual capital investments does not exceed 25% of the total cost of the project. According to the new law, the resident status implies a number of tax benefits and non-tax preferences. For example, a taxpayer with resident status in the Arctic zone of Russia may reduce the amount of mineral extraction tax by implementing a tax credit for investments in infrastructure. The law also provides tax incentives for projects not related to the extraction of minerals: reduction of federal and regional income tax, insurance payments, VAT cancellations for the sale of goods subject to the procedure of the free customs zone and for transportation of export cargoes by sea and related icebreaking operations. In addition, residents of the Arctic zone are granted non-tax preferences, such as duty-free import of equipment and export of finished products, interest rate subsidies on investment loans, reimbursement of the coupon rate on bonds. Residents receive the right to purchase land plots owned by the state or local municipalities without going through a tender procedure. The set of laws sets the timing of the provision of municipal services in urban planning. For example, the deadline for submitting a town-planning plan for a land plot in the Arctic zone is ten working days. It is assumed that these changes will attract more investment in the Arctic regions and improve the quality of life in the Arctic. We should note the lack of an environmental component in the law under consideration.

2 Results

Over the past few decades, climate change and globalization have dramatically changed the Arctic. As a result of global warming, Arctic sea ice is rapidly melting, potentially facilitating access to natural resources and opening up new sea lanes in the region. These changes have drawn the attention of the world community to potential commercial opportunities in the region [7]. We will consider some of them.

1. Cargo transportation. Linking Asia, Europe and North America, sea routes have become the main axes of container traffic. Now most of the cargo traffic originates in Northern China, with seven Chinese ports among the ten largest container terminals in the world. The northern polar routes between Asia, North America, and Europe include three main ones: the Northwest Passage (NWP) through the Canadian Arctic, the Transpolar Route (TPM), and the Northern Sea Route (NSR) along the coast of Russia. The Northern Sea Route runs from the Atlantic to the Pacific Ocean along the Russian coast of Siberia, mainly through Russian Arctic waters [19]. The special importance of the Northern Sea Route is clear, given its association with trade between Europe and the ports of the Far East of Asia, such as Yokohama and Busan, but also including Shanghai and Kaohsiung. Both Shanghai and Kaohsiung are ports with significant trade volumes. According to the World Shipping Council [8], Shanghai's trade volume per million twenty-foot containers (TEUs) was the highest in the world at 36.54 in 2015, and Kaohsiung ranked 13th with 10.26 million TEUs. Moreover, although the trade route between Asia and North America had the largest volume of cross-trade at 23 125 000 million TEUs in 2013, the trade route between Asia and Northern Europe took a clear second place with 13 706 000 million TEUs [8]. In terms of potential benefits, compared to the Suez Canal, the NSR could reduce travel time by 40 percent [9,10] or approximately seven days of travel between Yokohama and Rotterdam. According to [11], this represents a reduction from 11 200 to 6 500 nautical miles. Regardless of the exact distance, it is clear that sailing from these ports across the NSR will lead to a huge reduction in the number of nautical miles, and this route is rightly described as the one that "will connect East Asia (Japan, South Korea, Taiwan and China) with the Northwest Europe through the Arctic Ocean". This reduction in distance will save fuel, reduce emissions and save time, making it more energy-efficient and giving significant business opportunities to many countries. As an example, in 2017, a Russian tanker transported a cargo of liquefied natural gas (LNG) from Hammerfest in Norway to Poryong in South Korea in 19 days, which, according to expert estimates, saved about 30 percent of the time, compared with using the usual southern sea route through the Suez Canal [12].

In addition to decreasing the time and costs, and also lowering the naturogenic impact on the environment, a positive aspect of using the NSR is that there is no the problem of piracy [13].

However, there are many problems associated with choosing the NSR, which can be outlined as follows: sea ice and, therefore, a higher level of danger; high navigation needs and high cost of navigation equipment; higher unit costs per distance traveled, given the need to move through the ice and the corresponding support of the icebreaker; higher costs in the construction of vessels: vessels should be ice-class, which are 10–30% more expensive to build; and the possible environmental impact of oil spills and increased air pollution, need for specific crew training, cargo protection, hull maintenance, and special ship equipment. In addition, navigating through sea ice means that it is impossible to accurately specify the time of the vessel arrival, which can pose a problem for navigation. It is also necessary to take into account the keel depth of ships, which may be limited on the NSR (especially when passing the Sannikov Strait, the depth of which is limited), which will reduce the possible volume of cargo transportation. Another specific problem is the lack of ready port infrastructure. All these factors are inextricably associated with the technical issues of costs, logistics, information, and anthropogenic impact.

2. Large oil and gas corporations. Localization of production technologies and development of technologies for offshore oil and gas production in the Arctic zone can be an incentive for further import substitution for Russia. Currently, the experience of Arctic defense enterprises in the production of equipment for the extraction and processing of oil and gas is becoming increasingly popular [14]. At the moment, the production of liquefied natural gas (LNG) is one of the most dynamically developing segments of the hydrocarbon market worldwide. An analysis of the general situation in the energy supply market shows that the high pace of development of LNG projects is likely to continue for a few decades. Russia is among the top ten largest LNG exporting countries, and it is expected to strengthen its position in this segment [20].

In the Arctic, the cost of developing oil and gas fields is extremely high, so large Arctic oil and natural gas fields become particularly important in terms of the future development of these hydrocarbons, since a large field is capable of offsetting the costs of infrastructure for the subsequent development of smaller fields.

The discovery of large Arctic oil and gas fields began in Russia with the discovery of the Tazovskoye field in 1962, and in the United States with the Prudhoe Bay field in Alaska in 1967 [21]. Oil and gas reserves in the Arctic territories are presented in Figure 1 and 2.

Country	At the end of 1987, billion barrels	At the end of 1997, billion barrels	At the end of 2006, billion barrels	At the end of 2007, billion barrels
United States	35.4	30.5	29.4	29.4
Canada	11.7	10.7	27.7	27.7
Russia	n/a	n/a	79.3	79.4
Norway	6.6	12	8.5	8.2
Arctic States	n/a	n/a	144.9	144.7
World	910.2	1 069.3	1 239.5	1 237.9

Fig. 1. Proven oil reserves in 1987–2007. Source: [21].

Country	At the end of 1987, trillion cubic meters	At the end of 1997, trillion cubic meters	At the end of 2006, trillion cubic meters	At the end of 2007, trillion cubic meters
United States	5.3	4.74	5.98	5.98
Canada	2.69	1.81	1.62	1.63
Russia	n/a	45.17	44.6	44.65
Norway	2.29	3.65	2.89	2.96
Arctic States	n/a	55.37	55.09	55.22
World	106.86	146.46	176.22	177.36

Fig. 2. Proven natural gas reserves of Arctic states and the world over time. Source: [21].

Large oil and gas corporations unequivocally have a negative impact on the environment, since their production always releases substances into the air, moreover, they produce hazardous chemicals that are washed away by wastewater.

Any production is designed for economic growth and capacity expansion, this is one of the problems: often at such enterprises environmental protection and the economics contradict each other, therefore, it is necessary to assess where the limit is for increasing production.

3. Tourism. In recent decades, tourism has become one of the main reasons for human presence in the Arctic, although the first tourists began to explore the Arctic more than 500 years ago [18].

The shrinking area of sea ice and the subsequent increase in accessibility of polar regions has opened up new routes for expeditionary cruise ships, which has led to changes in cruise travel patterns and in the types of passengers visiting the region. In addition, in recent decades, global changes in the polar environment have changed the focus of tourism marketing, and terms such as "last chance to visit" have been used to attract a new demographic tourists, some of whom are highly adventurous and interested in autonomous experiences. On average, cruise passengers tend to be wealthy and quite often elderly; industry professionals call the average age of cruise passengers "very high" and describe some cruises as "floating retirement homes" [23]. Some industry professionals note the high level of education and desire for knowledge of these elderly passengers, but recently, there has been an influx of younger, more adventurous visitors, seeking, among other things, the thrill of "adrenaline tours". In addition to the change in the demographic composition of passengers, the Arctic cruise traffic has increased dramatically in recent years, both in the high-altitude regions of the Arctic and in the lower and sub-Arctic regions. The total number of cruise passengers in the high-altitude Arctic destinations of Svalbard and Greenland, the Canadian Arctic and Franz Josef Land has grown from about 50,000 in 2005 to just under 80,000 in 2016, with significant growth in the smaller vessel segment of 100–200 passengers compared to more than 3,000 (with more adventurous routes) [24]. The Association of Arctic Expedition Cruise Operators (AECO), which represents the majority of Arctic expedition cruise industry operators, reviews the number of passengers carried annually by AECO members. In 2016, expeditionary cruise passengers accounted for a quarter of the total number of cruise passengers, and the number of passengers carried by AECO member vessels in key AECO areas (Svalbard, Jan Mayen, Greenland, Arctic Canada, Russian Arctic National Park, and Iceland) increased from 13,480 passengers in 2008 to 26,296 passengers in 2017 [22]. In recent decades, a significant number of publications have recognized the various changes that the Arctic is facing in terms of social, cultural, economic and environmental aspects. The Arctic is a vast region of diverse climates, communities, and landscapes. This region is also characterized by strong seasonality, offering a wide range of experiences that cannot be attained in a single trip. For example, the northern lights and the midnight sun are phenomena that can only be seen in winter and summer, respectively. Therefore, to fully explore the Arctic, you need several trips to different Arctic regions. There is no doubt that tourism in the Arctic is different from tourism in other areas [26]. The reason for this statement originates in the form of exclusivity, since "very few other geographical regions are designated as a concept in tourism" [25]. However, it is now being questioned whether in the future there will emerge factors that contribute to the development of, for example, ski tourism in the Arctic regions of different countries. For instance, in Sweden, Finland, Norway, and Russia, conflicts in the field of land use arise in the case of the development of large ski resorts in their Arctic zones, especially around the specially protected natural areas, which in themselves are not only important attractions for nature tourism, but also traditional sources of livelihood for the indigenous inhabitants of the Arctic territories [15,30,31].

4. Animal husbandry. The paper of Reinert E. [16] lists an example of a sharp drop in the income of Saami reindeer herders in Northern Norway between 1976 and 2000, despite an increase in government subsidies. Saami reindeer herders maintain a legitimate monopoly on the supply of venison, a traditional luxury product in Norway. The reasons for this fall lies in the interaction of a number of factors, including: cyclical climate change in Northern Norway; a system with fixed prices for deer meat, independent of supply fluctuations, growing influence of natural cycles; increasing strict sanitary regulations forcing Saami reindeer herders to abandon the slaughter and procurement of reindeer meat; the oligopolistic market forces of non-Saami entities that undertake slaughter and processing. The paper confirms that the drop in the incomes of Saami reindeer herders was due to the failure of the Norwegian Ministry of Agriculture to understand the key factors that distinguish subarctic pastoralism from intensive farming. Sanitary requirements and the government's desire to achieve economies of scale in processing contributed to the fact that the volume of production was in the hands of non-Saami oligopolies. Thus, the Saami reindeer herders lost revenues from the production of meat, which had traditionally been the basis of both their culture and their economic livelihoods. This situation shows the devastating impact on aboriginal culture as a result of Norway's exclusive use of modern agricultural science to manage such systems.

5. Fishing in the Arctic is transforming under the influence of climate change. Thanks to the constantly growing demand for fish on the world market and the fact that most of the catch regions in other places are depleted, in recent years, the production of industrial fishing in the Arctic has remained relatively stable at about 5 million tons per year. For comparison, the global annual volume is unchanged on average, at about 80 million tons since 2003. These figures apply, of course, to industrial fishing, which accounts for more than 90 percent of all fishing. In Arctic fishing, small-scale fishing accounts for 5–7%. Subsistence fishing is between 1 and 2 percent, and recreational fishing is still an extremely small part of the total. Industrial fishing is an activity worth considering from the point of view of understanding the overall development of the green economy in the Arctic, as the ice recedes, and warmer waters are displacing commercial species to the north. By 2014, in the Barents Sea, for example, Norwegian fishermen were catching more than 11 percent of their annual quota in waters previously inaccessible to them. Greenpeace, after analyzing 18 million satellite location signals, reported that more than 100 fishing vessels had entered waters previously covered with ice [27].

3 Conclusions

The novelty of studying and covering the problems of Arctic entrepreneurship is primarily the recognition of it as a complex and multidimensional phenomenon, within which we can distinguish elements of oil and gas, ethnic, social, rural, innovative, and Arctic types of entrepreneurship, which have essential specificity.

Entrepreneurship in the Arctic zone of the Russian Federation is a unique type of business, which is distinguished by the "belt" criterion, in contrast to the "equatorial" or "tropical" entrepreneurship, which do not exist. Only for Arctic conditions, it has been recognized as necessary to single out a special concept of "Arctic entrepreneurship". And this is quite understandable, since it is in the Artic that e classic unified standards of state policy in the field of business support demonstrate their inoperability. Therefore, it becomes logical to separately study such phenomenon as Arctic entrepreneurship, as carried out in specific social and climatic conditions, which include the following:

1. The high cost of transporting goods to the Arctic territories (therefore, for example, the cost of vegetables and fruit in the stores of the Yamalo-Nenets Autonomous Region is 2.5–3 times higher than in stores of non-Arctic regions)
2. High risks of environmental accidents that are difficult to eliminate
3. Need to conduct business activities taking into account the northern deliveries in a limited period of Arctic sea and river navigation
4. Harsh natural and climatic conditions that leave their mark on all types of activities
5. Accelerated depreciation of fixed assets
6. Poorly developed roads, engineering, municipal infrastructure for doing business
7. Shortage of qualified personnel.

However, new business opportunities are emerging, and parties interested in the Arctic are playing a key role in facilitating and creating favorable environment for increased economic activity in the Arctic [29]. Currently, there is a wide range of business activities, of which the oil and gas industry, cargo transportation, and shipping are well developed. Recently, tourism and animal husbandry have been developing well. Nevertheless, the Arctic region has great economic potential for implementation in other areas, such as the green economy, tourism, innovation, and indigenous cultural business, through entrepreneurial activities and policies that address the specific opportunities and needs of the Arctic zone of the Russian Federation. Business development in the Arctic must rely on human capital and strive for innovative development in which the unique resources of the Arctic become competitive on a global scale. Urbanization particularly highlights the need for business development in the Arctic, as the region's younger generation wants services, opportunities and education just like those offered in major metropolitan areas. The development of business in the Arctic is also an existential issue, since the creation of new modern economic centers could halt the outflow of young people from the Arctic regions.

Entrepreneurial activities in the Arctic, unlike similar activities in other regions, have significantly higher costs due to the specifics of local climate and geography.

At first glance, there is a lot of conditions in the Arctic making entrepreneurship unprofitable and sometimes simply impossible. However, the purpose of this article was to demonstrate the opposite: despite the harsh climate and the seemingly high costs and high risks compared to the "mainland" business, the Arctic today offers the greatest opportunities for the introduction and development of innovations, both in the oil and gas industries, and in tourism and animal husbandry. Following the principles of sustainable development and adhering to the principles of the green economy, entrepreneurship in the Arctic should be conducted taking into account the following provisions:

- The use of innovations to introduce energy-efficient clean technologies and renewable energy sources (among which wind energy (areas along the northern sea borders of the Russian Federation are characterized by an average wind speeds of over 6–7 m/s, which is quite attractive for the implementation of wind generation), solar energy, and in some areas, plant biomass energy and energy of sea waves and tides are key [28]). In its turn, support for enterprises engaged in innovative activities should become a priority in entrepreneurship development programs. In this context, it is necessary to continue the digitalization of public services; the efforts of the authorities and residents should be directed to retraining and re-education, taking into account the development of digital technology and training in entrepreneurship, reducing the digital divide.
- Restoring, protecting and maintaining the diversity and basic functions of marine ecosystems, introducing the principles of ecosystem policy, providing assistance to indigenous peoples in business development, due to the fact that despite the development of entrepreneurship in the Arctic territories, their natural, cultural and recreational authenticity should be preserved.

References

1. Federal Law "On State Support of Entrepreneurial Activity in the Arctic Zone of the Russian Federation" dated 13.07.2020 No. 193-FZ. Available online: http://www.consultant.ru/document/cons_doc_LAW_357078/ (2020)
2. D.V. Kolpakidi, Features of the development of small and medium-sized businesses in the Arctic zone, *St. Petersburg Journal of Economics*, **3**, pp. 76-89 Available online: https://www.elibrary.ru/item.asp?id=44039693 (2020)
3. A.P. Shikhverdiyev (Ed.), Models for the development of small and medium-sized businesses in the Arctic: All-Russian (with international participation), *Youth Scientific and Practical Conference (October 23–25, 2019): collection of articles*, p. 316 (2019)
4. T. Taivalantti and S.N. Polbitsyn, Entrepreneurship in the arctic regions: reviewing arctic policies of Finland, Russia and Sweden, *Materials of the V International Scientific and Practical Conference*, pp. 273-277 Available online: https://elibrary.ru/item.asp?id=40076424 (2019)
5. The Norwegian Government 2021 The Norwegian Government's Arctic Policy. People, opportunities and Norwegian interests in the Arctic – Abstract. Report Date: 2021 Available online: https://www.regjeringen.no/en/dokumenter/arctic_policy/id2830120/#tocNode_43 (2021)
6. A. Pilyasov, Arctic entrepreneurship development factors, *Geography, environment, sustainability*, **13**, pp. 46-56 doi:10.24057/2071-9388-2019-91 (2020)
7. S.E. Zhura and V.V. Markin, Features of the development of entrepreneurship in the Russian, *Arctic Sociopolitical Studies*, **3(8)**, pp. 97-109 (2020)
8. World Shipping Council 2017 About the industry. Available online: www.worldshipping.org/about-the-industry/global-trade/top-50-world-container-ports (2017)
9. M. Liu and J. Kronbak, The potential economic viability of using the Northern sea route (NSR) as an alternative route between Asia and Europe, *Journal of Transport Geography*, **18(3)**, pp. 434-444 (2010)
10. H. Lindstad, et al., Economic savings linked to future Arctic shipping trade are at odds with climate change mitigation, *Transport Policy*, **45**, pp. 24-30 (2016)
11. N. Hong, The melting Arctic and its impact on China's maritime transport, *Research in Transportation Economics*, **35(1)**, pp. 50-57 (2012)
12. A Russian tanker under the Cyprus flag for the first time crossed the Arctic without an icebreaker 2017 Available online: https://evropakipr.com/novosti/rossiyskiy-tanker-pod-kiprskim-flagom-vpervye-preodolel-arktiku-bez-ledokola (2017)
13. P. Tseng and N. Pilcher, Assessing the shipping in the Northern Sea Route: a qualitative approach, *Maritime Business Review*, 2(4), pp. 389-409 Available online: https://doi-

org.ezproxy.unecon.ru/10.1108/MABR-06-2017-0013 (2017)

14. B.S. Sergi and A. Berezin, *Oil and Gas Industry's Technological and Sustainable Development: Where Does Russia Stand? Exploring the Future of Russia's Economy and Markets* (Emerald Publishing Limited, Bingley, 2018), pp. 161-182 doi: 10.1108/978-1-78769-397-520181009
15. O. Demiroglu, et al., The last resort? Ski tourism and climate change in Arctic Sweden, *Journal of Tourism Futures*, **6(1)**, pp. 91-101, doi: 10.1108/JTF-05-2019-0046 (2020)
16. E. Reinert, The economics of reindeer herding: Saami entrepreneurship between cyclical sustainability and the powers of state and oligopolies, *British Food Journal*, **108(7)**, pp. 522-540, doi: 10.1108/00070700610676352 (2006)
17. DOE Helps Entrepreneurs Deploy Arctic Solutions 2021 Available online: https://www.energy.gov/eere/articles/doe-helps-entrepreneurs-deploy-arctic-solutions (2021)
18. D. A. Ksenofontova and A.V. Sokolov, The most important expeditions in the period of development of the Arctic from ancient times to the XX century. Available online: https://morproekt.ru/articles/science-artiles/obzornye-stati/181-vazhnejshie-ekspeditsii-v-period-osvoeniya-arktiki-s-drevnejshikh-vremen-do-xx-veka (2021)
19. M.Y. Zelenkov, Assessment of prospective logistics costs in the operation of the Northern Sea Route, *River transport (XXI century)*, **2(90)**, pp. 40-43 (2019)
20. L.E. Ametistova and A.Yu. Knizhnikov, Environmental Aspects of Arctic LN G Projects Development, *WWF Moscow*, Available online: https://wwf.ru/en/resources/publications/booklets/environmental-aspects-of-arctic-lng-projects-development/ (2016)
21. Energy Information Administration, US Dept. of Energy. Available online: https://www.eia.gov/analysis/studies/archive/2009/arctic/ (2009)
22. A. Taylor, et al., Arctic expedition cruise tourism and citizen science: A vision for the future of polar tourism, *Journal of Tourism Futures*, **6(1)**, pp. 102-111 doi: 10.1108/JTF-06-2019-0051 (2020)
23. Þ. Barðadóttir, The receiving of cruise ships in Northern Iceland: findings from interview research, Available online: www.rmf.is/static/research/files/mottaka-skemmtiferdaskipa-rmf-2017-03.pdf (2017)
24. Cruise Industry News 2019 expedition market report Available online: www.cruiseindustrynews.com/store/product/digital-reports/2019-expedition-market-report/ (2019)
25. J. Saarinen and A. Varnajot, The Arctic in tourism: complementing and contesting perspectives on tourism in the Arctic, *Polar Geography*, **42(2)**, pp. 109-124 (2019)
26. A. Varnajot, Digital Rovaniemi: contemporary and future arctic tourist experiences, *Journal of Tourism Futures*, **6(1)**, pp. 6-23, doi:10.1108/JTF-01-2019-0009 (2020)
27. International Legal Management of Fisheries in the Arctic Available online: https://pro-arctic.ru/07/07/2015/resources/17289 (2015)
28. A.V. Mitko and V.K. Sidorov, Prospects of Renewable Energy Sources in the Arctic. Available online: https://magazine.neftegaz.ru/articles/arktika/624988-perspektivy-vie-v-arktike/ (2020)
29. A. Kuchumov, et al., Essential, legal and environmental aspects of sustainable development of arctic tourism, IOP Conference Series: Earth and Environmental Science, **678(1)**, 012042 (2021)
30. I.G. Gerasimova, I.S. Oblova, E.I. Golovina, The Demographic Factor Impact on the Economics of the Arctic Region, Resources, 10, 117. doi: 10.3390/ resources10110117 (2021)
31. Yakovleva J.A., Sharok V.V., Vakhnina E.G. Sport and health aspects of adaptation in northern regions of Russia. *Teoriya i praktika fizicheskoy kultury=Theory and Practice of Physical Culture*, **4**, pp. 20-21 (2020)

Problems and prospects of economic growth in the context of the environmental component of sustainable development in the Arctic zone of the Russian Federation

Artur Kuchumov[1,*], *Elena* Pecheritsa[2,*], *Natalia* Blazhenkova[3,*] and *Alexandra* Chaikovskaya[4,*]

[1]Candidate of Sciences in Economics, Associate Professor, Department of Economics and Management in the Service Sector, St. Petersburg State University of Economics, 191023, St. Petersburg, Russian Federation
[2]Candidate of Sciences in Sociology, Associate Professor, Department of Economic Security, St. Petersburg State University of Economics, 191023, St. Petersburg, Russian Federation
[3]Doctor of Economics, Professor, Department of Finances and Credit, Ufa State Petroleum Technical University, 450000, Ufa, Russian Federation
[4]Candidate of Sciences in Economics, Associate Professor, Department of Regional Cooperation, St. Petersburg State University of Industrial Technologies and Design, 191186, St. Petersburg, Russian Federation

Abstract. The article examines the perceived drivers of economic growth and the corresponding implications for proactive, evidence-based policies, discusses national issues and shapes a shared vision for the future among key interested parties. The authors highlight the main drivers of change and economic growth in the Arctic, including energy security, climate change, and transportation development, which have a huge direct impact on the Arctic region across all three dimensions of sustainable development: economic, social, and environmental. The authors consider the following as the problems facing the creation of conditions for economic growth in the Arctic region of the Russian Federation, in terms of sustainable development: environment, problems of the Arctic population, management problems in the changing Arctic, and the adaptation of technologies and production processes to the Arctic climate. In conclusion, the authors deduce that the creation of a joint group consisting of relevant private, public, governmental and non-governmental organizations would be the best strategy for considering effective mechanisms for multinational and multilateral governance of sustainable development in the Russian Arctic zone. The authors also believe that in order to create conditions for economic growth in the Arctic zone of the Russian Federation and to facilitate further transition to sustainable development, the authorities at the first stage should provide economic support measures for doing business, especially in two key areas for the Arctic: the development of animal husbandry and the development of various types of tourism.

1 Introduction

Over the past 2 decades, Russia has identified two key national interests in the Arctic: development of hydrocarbons and control over sea routes, and consistently works on their promotion [1]. In 2021, with the beginning of the UN decade of activities on achieving Sustainable Development Goals by 2030, the problems of sustainable development of the Arctic territories of the Russian Federation become particularly relevant. Although some progress in achieving the indicators contained in the previous Arctic Development Strategy for the period up to 2020, Russia is still facing global challenges to develop the already achieved goals, namely, the formation of energy and transport and information and telecommunication infrastructure, addressing the issues of environmental security and social development, advancing sciences and technologies for the development of the Arctic, and improving the investment climate [2, 3].

We support the opinion of the authors of [4,5] that the natural ecosystems characteristic of the Arctic, the presence of indigenous communities, and commercial interest in the region require an innovative model of development based on a high level of responsible exploitation, diplomacy, regulation, and the development of scientific methods, based on the general principles of the economy-nature-society system, linking environmental, social, and economic processes.

According to the authors of [6], the Arctic currently has great potential for economic growth, despite the fact that at present it is a region with high levels of poverty, social instability, untrained workforce, and environmental vulnerability.

It is known that 20 percent of the world's energy resources in the form of fossil fuels, such as oil, gas, and coal, are located in the Arctic. In addition to the

* Corresponding author: arturspb1@yandex.ru, helene8@yandex.ru, bnm61129@yandrx.ru, sasha_chaikovska@list.ru

traditional range of hydrocarbon resources, the Arctic region has enormous potential and wealth for alternative and renewable energy projects, including wind, hydropower, tidal, and geothermal. Due to the current high demand for energy around the world, several countries are turning their attention to circumpolar international relations and resource exploration, but no discussion of Arctic development would be complete without a thorough understanding of the concept, features and application of the principles of sustainable development.

The authors of [7] and [8] believe that creation of conditions for economic growth of the Arctic zone of the Russian Federation is the basis for its transition to sustainable development.

According to A.I. Bedritsky [9], the existing management system certainly needs to be updated and adjusted in accordance with binding rules for all participants in order to counter a number of threats to the integrity of the Arctic ecosystems and the serious consequences of climate change. The author of [10] agrees with this statement and adds that production of hydrocarbons in the Arctic region is a challenge for the implementation of the concept and features of sustainable development.

The Arctic region contains huge amounts of extractable energy resources, most of which are offshore, in areas that have long been inaccessible but are now becoming open due to climate change. Some of the most important regions for energy production include the Northern Slope in Alaska (USA), Eastern and Western Siberia, the Timan-Pechora District, South/ North Barents (Russia), the territory east of Norway, and East Greenland. Relevant regions in the energy sector also include Russian territories: The Vilkitsky Islands, the Laptev Sea, the Vilyui sector, the Khatanga district and the North Kara Sea; the western part of the Barents Sea north of Russia, belonging to Denmark and Norway; West Greenland region; the Sverdrup Islands north of Canada; and the Beaufort-Mackenzie Basin in the Canadian Arctic.

Some authors single out energy security, climate change, and transport development as the main drivers of change and economic growth in the Arctic; they have a huge direct impact on the region from all three components of sustainable development: social, economic, and environmental [11], while T.O. Ediseeva [12] highlights three main factors affecting the current and future economic growth of the Arctic region: the availability of important Arctic sea routes, the growing importance of hydrocarbons and mineral resources in the Arctic, and, as a result, the management of the Arctic. Another controversial aspect is the legal framework that must be developed for the implementation of local and international environmental governance, understood as the means by which society will define and act in accordance with the objectives and priorities related to the management of natural resources and economic growth in the Arctic. This statement includes rules, both formal and informal, that regulate the people's behavior in decision-making, as well as their implementation. An appropriate political and legal framework at the global, national, regional, and local levels is a prerequisite for an adequate level of governance.

The purpose of this study is to examine the opportunities for economic growth in the context of the environmental component of sustainable development of Russia's Arctic regions. The objectives of the study are to study and analyze the problems that hinder the creation of conditions for economic growth of the Russian Arctic regions in terms of environmental goals of sustainable development.

2 Results

The regions of the Arctic zone of the Russian Federation (AZ RF) have an invaluable opportunity to introduce genuine and effective changes in the perception of the Arctic and to create new models of economic growth for the future of the Arctic region. We will consider the challenges to the creating conditions for economic growth in the Arctic zone of the Russian Federation from a sustainable development perspective.

1. Environment.

The Arctic region is prone to persistent organic pollution. Mostly, mid-latitude pollutants migrate to the Arctic by wind and ocean currents. This is a local problem of transboundary pollution that requires urgent regional regulation and attention from global governance.

solution of the problem of the USSR environmental legacy. During the annual environmental rehabilitation of the Arctic (suffering from domestic and military landfills of 70 years ago), through the efforts of the environmental platoon of the Northern Fleet, hundreds of tons of technogenic waste, including on island territories, are collected and sent for processing [13].

Energy production is both a risk and an opportunity. Models for the development of oil and gas and other fossil energy resources in the Arctic will be determined by the industry's ability to develop technological strategies to minimize damage, pollution, and natural disasters, acting mainly preventively and actively responding to pollution phenomena. To restore contaminated areas, it is necessary to develop innovative methods of environmental restoration, as well as corporate social responsibility (CSR) schemes that take into account the needs of the local population.

Since 2019, RAEX-Europe ranking agency has started to compile ESG ratings of Russian regions. This ranking evaluates not only the managerial and social components, but also changes in environmental risks and measures taken to eliminate them. When calculating the rankings, risks are assessed using per capita pollution indicators of each region, while effectiveness in eliminating environmental risks is assessed according to environmental protection indicators. The higher the environmental risks, the higher the environmental protection indicators should be. In this case, the system achieves a balance. In the case of imbalance, we can conclude that due attention to risks is not paid [14]. The ESG ranking of the AZ RF regions is presented in Table 1.

Table 1. ESG ranking of the Arctic zone regions. Source: compiled according to [15]

Region / Position	E	S	G	ESG
Yamalo-Nenets Autonomous District	76	2	12	19
Nenets Autonomous District	81	1	13	21
Krasnoyarsk Territory	17	43	34	28
Republic of Sakha (Yakutia)	10	15	74	35
Chukotka Autonomous District	58	4	65	41
Murmansk Region	78	14	20	50
Arkhangelsk Region	73	41	66	71
Komi Republic	82	42	48	78

The maximum volumes of gross pollutant emissions from stationary sources are metallurgical (Krasnoyarsk Territory) and oil and gas (Yamalo-Nenets Autonomous District, Komi Republic) regions. The "antileader" in terms of indicators is the Krasnoyarsk Territory, in which more than 13% of all emissions of enterprises into the atmosphere of the Russian Federation occur [15].

2. Indigenous peoples, environmental pollution and climate change consequences. In addition to the economic consequences, which are generally positive for the region and depend on equitable governance through the return of income from exploitation to local communities, we can mention the enormous social and environmental consequences, which seem more difficult to manage. Indigenous communities in resource-rich arctic areas are increasingly exposed to the serious impacts of climate change, as well as to external development pressures advocated by the government and its industrial partners. With the discovery of vast energy resources in previously inaccessible areas of the Arctic, new measures are being taken at the federal level to develop these regions, including bringing modern infrastructure to remote settlements. Thus, climate change and modernization have become two inextricably linked forces that are greatly altering the context in which the indigenous populations of the region exist. Recent ice modelling results from polar-orbiting satellites have revealed actual evidence of a significant decrease in the arctic sea ice since 2010. The loss of sea ice opens new prospects for offshore oil production and new sea routes. Modernization is perhaps the main driver of all changes that synthesizes the causes and consequences of physical and social transformations. Increased industrial activity in the Arctic, aimed at facilitating access to oil and gas reserves, is further aggravating already severe climate impacts, which include changes in weather patterns, wildlife and vegetation, as well as in the availability and quality of food and water. As a result, the changes affect not only the physical space, but also the sociocultural landscape of the Arctic. Over the past 60 years, Arctic indigenous peoples have undergone tremendous changes that have affected their culture and social organization. As terrestrial and marine habitats are disturbed and fragmented by climate change and industrial development, the traditional livelihoods of indigenous communities have become increasingly limited [16].

3. the impact of the coronavirus pandemic, which that has spread over the past two years, also cannot be overlooked. Due to their isolation and low population density, the Arctic regions of the Russian Federation should be in a much safer position compared to the more densely populated territories. However, this pattern has not worked for the AZ RFThe Arctic is home to some of Russia's hardest-hit regions. For example, the highest per capita incidence of coronavirus among all 85 subjects of the Russian Federation was registered in the Yamalo-Nenets Autonomous District, and four of the ten federal subjects with the highest number of deaths per capita are also in the AZ RF. Another seemingly illogical fact is that the incidence of coronavirus in the Murmansk region at times was greater than in neighboring Finland or Norway, which have 7 times the population of Murmansk [17]. This fact can be explained by the fact that COVID-19 propagated to remote Arctic regions of the Russian Federation as a result of the migration of workers coming from all regions of Russia and the neighboring countries to work at industrial mining enterprises in the Arctic. An example is the second largest outbreak of COVID-19 in Russia, in Belokamenka, Murmansk Oblast, registered in mid-April 2020, at Russia's second-largest liquefied natural gas plant, owned by Novatek, where 20% of 1,100 employees were infected.

Furthermore, the pandemic undoubtedly has a strong impact on small and medium businesses in the AZ RF. The introduction of lockdowns and QR codes, the shift of some workers to a remote format, all these reasons certainly do not have the best affect entrepreneurship in the best way: people use catering services to a much lesser extent, buy fewer clothes, cosmetics, etc., the tourism sector has also suffered seriously, even though domestic tourism in Russia began to develop rapidly due to the closure of a number of borders, nevertheless the income from inbound tourism has dropped significantly.

As for the energy and oil and gas sectors, which are the most relevant in the AZ RF, the pandemic makes its adjustments there as well, as given the decline in production, the sector's demand for services is seriously decreasing. To solve this problem, the government of the Russian Federation decided to support the energy sector by creating a fund of unfinished wells. This is a separate mechanism with state support measures to ensure the operation of the sector when production is restored in full. It is about additional possibilities of crediting oil companies, which allocated these resources under an additional drilling order. The total volume of crediting of this sector for two years may amount to about 400 billion rubles [18]. In the context of the environmental component of sustainable development, the restrictions on business activity associated with the pandemic, in our opinion, rather have a beneficial effect on the environment. And here we can talk about the imbalance of interests between the environmental and economic components of the goals of sustainable development.

4. Management aspects. Governance challenges in the changing Arctic are specific, characterized by a constant search for efficiency in a completely fragmented, multidimensional and changeable reality. Russia, despite having the largest geographical share of the Arctic territory, has the least clear-cut type of regulation, which exposes the environment and local communities at a considerable level of vulnerability, and the existing legal framework represents regional and local legal mechanisms that need to be harmonized.

4.1. Management of human potential and the labor market. The number of unemployed in the territories of the AZ RF on average is at the level of all-Russian indicators, however, this applies, firstly, not to all regions, and, secondly, the average Russian indicator of 7–9% is almost 2 times higher than the threshold value of 4%. For example, in the Nenets Autonomous District the unemployment rate is 9.2%, in the Murmansk region and the Republic of Sakha (Yakutia) it is 8.8% [19,20]. The peculiarity of this region is its fairly high level of wages, which is sometimes 2 times higher than, for example, wages in the central part of Russia, while the number of jobs, concentrated mainly in the industrial sector, is much lower, and working conditions are not the easiest, with the prevalence of shift work and physical labor in factories, enterprises for the extraction of gas, oil, and other minerals. Not to mention the climate of the regions of AZ RF, which is not suitable for everyone. On the plus side, we can note the regional social support for young specialists in the regions of the AZ RF, such as payments to young specialists and young families.

4.2. Interrelations and interactions between both the economic entities of the regions of the AZ RF and regions as a whole. The creation of a single effective system of links for the interaction of the Arctic regions requires new methods of management based on "netization" and digitalization using information technology, which will ultimately lead to the maintenance of a balance between all three components of sustainable development: environmental, economic, and social.

5. Adaptation of technologies and production processes to the Arctic climate. Arctic cities and towns, usually located in rural areas with harsh climates, sparsely populated, over long distances and with scarce infrastructure, struggle to find innovative solutions to attract existing and potential residents, tourists, entrepreneurs, and investors. Their industrial base is constantly changing, with the traditional industries such as mining, forestry, and hydropower becoming more technologically advanced, while service industries such as tourism and culture are expanding rapidly [21, 27-29].

3 Discussion

According to various estimates, about five hundred million people live in the Arctic, including 2.5 in the AZ RF. In some Arctic regions, the majority of the population is indigenous in origin and has lived there for several thousand years [22]. Approximately 7 percent of the world's population lives in the Arctic, an area that makes up 10 percent of the earth's surface, and is believed to contain more than 20 percent of undiscovered oil resources. This region is also rich in fish resources, minerals, and freshwater. The Arctic is also home to vast areas untouched by modern human activities and unique terrestrial and marine ecosystems of global importance that are attractive to tourists. The Arctic plays an important role in the global climate system, so aligning hydrocarbon development, climate change policy, and the specifics of the region is challenging, as the resources of the Arctic ecosystem can only support the region's population with moderate resource consumption in a sustainable development process. The challenge is to try to simultaneously protect the natural and social values of the Arctic by developing opportunities for local populations, using the available surpluses wisely, while ensuring economic growth. The economic development of arctic resources can serve growing global needs, but it can also serve the global process of exploring a real model for sustainable development.

Respect, shared values, traditional knowledge and competency development programmes can ensure that northerners have a strong capacity to implement policies and at the same time benefit from development outcomes. Partnership in energy education and research will strengthen knowledge and capacity to meet the needs of the Arctic. A sustainable future of the AZ RF and its residents depends on competent local people and partnerships with the non-Arctic world. In terms of education in the oil and gas sector, the biggest challenge in training personnel in the Arctic is to develop a world-class training and development system that meets the needs of business and industry, but provides the most prepared workforce that meets local environmental and social needs. Leading educational institutions in the Russian Federation have a crucial role to play in building the capacity in Arctic communities by sharing knowledge and experience and ensuring broad access to relevant oil and gas education for indigenous peoples at all levels. In fact, such training should be based on the study of business strategies, exploring policies between partners, and managing with sustainable business models. a sustainable future for the Arctic in terms of training also depends on an educational plan focused on environmental management and a better understanding of sustainable development models. Models for the development of oil, gas, and other fossil energy resources in the Arctic must be driven by the industry's ability to develop technology strategies to minimize damage, pollution, and disasters, as well as to respond to events and restore contaminated areas [23-26].

4 Conclusions

Achieving a balance between preservation and development is a major theme of the current Arctic debate. In this balance, the ongoing oil and gas exploitation seems to be the most contentious interest, as environmental considerations collide with economic interests. Thus, the decision of whether to develop new offshore areas for exploitation includes not only energy issues, but also how the development of economic interests will affect indigenous populations, the marine environment, regional development, climate change and energy security. The most common explanations for the increased interest in energy development in the Arctic in recent years are rising commodity prices, new technologies to reduce the cost of Arctic operations, and increased availability due to the melting Arctic sea ice. Additional impetus for oil and gas development in the Arctic has come from the depletion of more southern oil and gas fields in Alaska, Norway and Russia, as well as a reference to promising Arctic resource potential in a 2008 report by the U.S. Geological Survey. While these factors create a backdrop for growing commercial interest in the region, they do not provide a full picture. The decision on how and when to open new Arctic sea areas for prospecting and exploratory drilling is perhaps as much a result of the relationship between the local and federal governments as it is result of commercial interests. Since the final decision to authorize exploratory drilling is made by the respective federal and state governments, understanding the inner workings of new site discoveries is critical to understanding arctic oil and gas development.

In formulating the general problems of sustainable development in the AZ RF, they are as follows:

• environmental, associated with both transboundary pollution and permanent emissions from industrial (mainly energy) enterprises

• social, primarily associated with the problems of the indigenous population, which is increasingly exposed to the serious impacts of climate change, as well as to external pressure from the development of the territories of the AZ RF, and, secondly, the high level of unemployment in a number of Arctic regions

• economic, currently associated with the impact of the Covid-19 pandemic.

In addition, it can be noted that there is an imbalance between development and conservation. For example, economic development seriously impedes the development of the environmental factor (the impact of the oil and gas industry on the environment), and the social factor may hinder economic growth (it is quite difficult to simultaneously take into account the interests of both indigenous peoples living in the RF AZ and mining companies in the exploration and damage to the vulnerable nature of the region).

Current trends suggest that a joint group consisting of relevant private, public, governmental and non-governmental organizations, rather than one single entity, would be the best strategy for considering effective mechanisms for multinational and multilateral management of economic growth and sustainable development of the Arctic zone of the Russian Federation. This strategic form of governance could be a good way to protect the common interests, priorities and programmes of stakeholders. to create conditions for economic growth in the Arctic region of the Russian Federation and further the transition to sustainable development, the authorities at the first stage need to ensure measures of economic support for business, especially in the two main areas for the Arctic: the development of animal husbandry (for example, reindeer herding) and the development of various types of tourism (especially environmental tourism), which have become especially relevant after the closure of borders due to the spread of coronavirus infection, while respecting the principles of minimizing anthropogenic impact on the surrounding environment

References

1. Oxford Analytica 2019 Sea route and energy are key to Russia's Arctic vision, Expert Briefings. Available online: https://doi-org.ezproxy.unecon.ru/10.1108/OXAN-DB246950 (2019)
2. Sustainable Development Goals. Available from: Available online: https://www.un.org/sustainabledevelopment/ru/sustainable-development-goals/ (2021)
3. National set of SDG indicators. Available online: https://rosstat.gov.ru/sdg/national (2021)
4. S.S. Gutman and A.A. Basova, Indicators of sustainable development of the Arctic zone of the Russian Federation: problems of choice and measurements, Arctic: Ecology and Economy, **4(28)**, pp 32-48 (2017)
5. S.M. Nikonorov, On conceptual approaches to the development: Polar Index, Barents Index and Index of Sustainable Development of the Arctic Council countries. Available online: https://www.econ.msu.ru/sys/raw.php?o=73806&p=attachment (2021)
6. M. Koman, et al., *How Far is the European Union from Sustainable Development Goals? Challenges on the Path Toward Sustainability in Europe* (Emerald Publishing Limited, Bingley, 2020) pp. 31-53
7. R.S. Czarny, et al., *The North American Arctic Games: Between Cooperation and Rivalry, Politics and Development in the North American Arctic* (Emerald Publishing Limited: Bingley, 2021) pp. 73-107
8. R. V. Knaub, Development of the Arctic territories of the Russian Federation in the 21st century in the context of security from emergency situations of various genesis, *Geopolitics and Ecogeodynamics of Regions*, **7(2)**, pp. 58-72 (2021)
9. A.I. Bedritsky, Sustainable development of the Arctic zone of the Russian Federation and climatic aspects of environmental and hydrometeorological safety, *Energy Policy*, **4**, pp. 3-10 (2018)
10. J.S. Higginbotham, The North American Arctic. Energizing Regional Collaboration and Governance. Special Report. Available online:

https://doi.org/10.1051/e3sconf/202337806003

https://www.cigionline.org/sites/default/files/documents/NAA%20Special%20ReportWEB.pdf (2018)

11. S.A. Karaganov, et al. Arctic Policy of Russia: International Aspects, *Documents of XXII April International Scientific Conference on the Problems of Economic and Social Development, Moscow, 2021 National Research University "Higher School of Economics* (Publishing House of the Higher School of Economics, Moscow, 2021) p. 56
12. T.O. Ediseeva, Features of sustainable development of the Arctic regions, *Economics and Business: Theory and Practice*, **5-1(63)**, pp. 173-176 (2020)
13. Development of the Arctic. Available online: https://fussing-Arctic.rf/razvitie-arktiki/ (2021)
14. RBC Available online: https://trends.rbc.ru/trends/green/5f172e609a79476e61071503 (2021)
15. ESG ranking of Russian regions 2020. Available online: https://raex-a.ru/rankings/regions/ESG_raiting (2020)
16. G.M. Arruda and S. Krutkowski, Social impacts of climate change and resource development in the Arctic: Implications for Arctic governance, *Journal of Enterprising Communities: People and Places in the Global Economy*, **11(2)**, pp. 277-288 (2017)
17. P. Devyatkin, The impact of the coronavirus pandemic on the life in the Arctic region. Indigenous Peoples. Available online: https://russian-arctic.info/info/articles/redaktsionnye-stati/vliyanie-pandemii-koronavirusa-na-zhizn-v-arkticheskom-regione-korennye-narody/ (2021)
18. How Russian oil industry survived the pandemic. Available online: https://www.eprussia.ru/epr/393-394/4021861.htm (2021)
19. Calendar. Publication of official statistical information on the socioeconomic development of the Arctic zone of the Russian Federation in 2020. Available online:https://gks.ru/free_doc/new_site/region_stat/calendar1-2020.htm (2020)
20. Rosstat: The report "Socioeconomic situation of Russia" for January-July 2021, the final version (online, pdf). Available online: https://nangs.org/analytics/rosstat-doklad-sotsialno-ekonomicheskoe-polozhenie-rossii-itogovyj-pdf (2021)
21. M. Lindberg, et al., Co-creative place innovation in an arctic town, *Journal of Place Management and Development*, 13(4), pp .447-463 (2020)
22. S.F. Lukin, Arctic Encyclopedia: The Population of the Arctic. Available from: https://docplayer.com/428382-Arkticheskaya-enciklopediya-naselenie-arktiki-arctic-encyclopedia-arctic-population.html (2014)
23. G.M. Arruda, Arctic governance regime: the last frontier for hydrocarbons exploitation, *International Journal of Law and Management*, 57(5), pp. 498-521 (2015)
24. B.S. Sergi and A. Berezin, Oil and Gas Industry's Technological and Sustainable Development: Where Does Russia Stand?, Sergi, B.S. (Ed.) *Exploring the Future of Russia's Economy and Markets* (Emerald Publishing Limited, Bingley, 2018) pp 161-182
25. A. Østhagen, Developing North American Arctic Offshore Oil and Gas: A Comparative Study – Part 1 By Article, Canada, Denmark and Greenland, Natural Resources and Energy, United States. Available online: https://www. thearcticinstitute.org/developing-north-american-arctic_part1/ (2012)
26. A. Kuchumov, et al., Essential, legal and environmental aspects of sustainable development of arctic tourism, *IOP Conf. Ser.: Earth Environ. Sci.*, **678**, 012042 (2021)
27. K.V. Matrokhina., V.Y. Trofimets, E.B. Mazakov , A.B. Makhovikov and M.M. Khaykin, Development of methodology for scenario analysis of investment projects of enterprises of the mineral resource complex, Journal of Mining Institute, 259, pp. 112-124, doi: 10.31897/PMI.2023.3 (2023)
28. I.G. Gerasimova, I.S. Oblova, E.I. Golovina, The Demographic Factor Impact on the Economics of the Arctic Region, Resources, 10, 117, doi: 10.3390/resources10110117 (2021)
29. V.V. Sharok, E.G. Vakhnina, Yu.A. Yakovleva, Health Resource of National Physical Education and Sport System: Northern Dimension, *Teoriya i praktika fizicheskoy kultury=Theory and Practice of Physical Culture*, **3**, pp. 45-46 (2019)

A paradigm shift in the agricultural system in the Arctic zone of the Russian Federation

Andrey Ingiri[1,*], *Polina* Ageeva[2], *Natalia* Ikko[3], *Anna* Yashkina[4]

[1]Junior researcher, Laboratory of the medical and Biological Technology, FRC KSC RAS, 184209, Apatity, Russian Federation
[2]Head of the scientific direction of lupin narrow-leaved breeding, All-Russian Lupine Research Institute, a branch of the FWRC FPA, 241524, Bryansk, Russian Federation
[3]Ph.D. of Biological Sciences, docent, Department of Natural Sciences, MASU, 183038, Murmansk, Russian Federation
[4]Senior lecturer, Department of Technosphere Safety, MSTU, 183010, Murmansk, Russian Federation

Abstract. The article presents an overview of the environmental and economic consequences of chemicalization in crop production of the Murmansk region. Proposed is a way to overcome the current situation in the industry through alternative biological sources of increasing soil fertility. Three original varieties of lupines inoculated with a highly effective strain of nodule bacteria Rhizobium lupini were studied in a production experiment. The fodder value of green mass was calculated. Microbiological and agrochemical analyzes of soil in areas with alternative and traditional agricultural technologies were carried out.

1 Introduction

The first steps in the direction of agricultural development of Russia's Polar territories date back to the 1920s. And, in a fairly short time on a historical scale, the ambitious goal set by the young Soviet state was achieved. Already in 1930, the first state farm was organized on the Kola Peninsula, which made it possible to test experimental work in production conditions [1, 2]. Later on, especially in the post-war years, the agriculture of the Arctic only developed. This became possible with the full support and interest of the state in the development of the Northern territories of the country, as well as the mobilization of modern achievements of science and technology [3].

The unpreparedness of the agro-industrial complex of the Murmansk region to the change of political and economic paradigm in the 1990s in the country led to bankruptcy and liquidation of a large number of agricultural enterprises, which caused a number of negative social and economic consequences and threatened the food security of the region. The reduction in the production of local products inevitably led to an increase in the share of food imports from other regions and abroad.

To date, a comprehensive analysis of the economic and socio-economic integral indicators of the industry characterizes it as unstable. The economic efficiency of agricultural enterprises is at an unacceptably low level; most farms are unprofitable even taking into account subsidies. The provision of the local population with locally produced food is also insufficient. All this suggests the need to revise the agrarian policy of the Murmansk region [4].

Pioneers in the direction of agricultural development of the northern territories had to overcome the difficulties associated with the agro-environmental and climatic features of the region. Insufficient heat supply of short vegetation period, poor soil microflora and other factors significantly slow down the processes of mineralization of plant nutrition elements occurring in the soil. In this regard, application of increased doses of mineral fertilizers has become a determinant condition for increasing productivity of arable land.

E.E. Kislykh points out that the average NPK fertilizer system in the region should be at least 260-320 kg/ha of active substance (a.s.) or N120-180P60-80K60-80 in mineral fertilizers and 50-100 t/ha of organic [5].

According to the data of the Murmansk agrochemical service station, the highest yield of forage grasses in the region was achieved in 1986-1990 when NPK was applied with mineral fertilizers in doses of 323 kg/ha a.s., and 400 kg/ha a.s. with organics. With this amount of ameliorants applied, the yield of perennial grasses averaged 126 h/ha in the region, and annual grasses reached 171 h/ha. Harvested areas amounted to more than 17 thousand hectares. Since 1990, there has been a sharp decline in the use of organic and mineral fertilizers. This led to an imbalance of basic nutrients in soils, which negatively affected the yield of annual and perennial forage crops. In 2016, application doses were already 117 kg/ha a.s. of organic and 52 kg/ha a.s. of NPK mineral fertilizers, and the yield of annual grasses decreased to 90 h/ha and perennial to 51 h/ha [6]. For 2020, the average yield of annual grasses is 65 h/ha, perennial grasses - 45 h/ha, with the application of fertilizers of 111 kg/ha of a.s. NPK organic and 4 kg/ha in a.s. NPK mineral.

* Corresponding author: dr.varuly@inbox.ru

Thus, we can conclude that the regulation of the productivity of artificial agrophytocenoses in the Murmansk region is ineffective. The inability to replenish soil reserves of nutrients used by plants for crop formation leads to a violation of the environmental balance and, ultimately soil degradation.

The solutions proposed today have three main directions: a return to the intensive model by increasing funding, restoration and modernization of technologies serving the industry; the use of foreign experience in the development of agriculture in similar agro-climatic conditions; the extensive model implying emphasis on traditional trades such as reindeer husbandry, fishing and gathering wild plants [7, 8]. Such strategies are not able to solve the complex of accumulated socio-economic problems, to ensure environmental and food security of the Arctic territories.

The preservation and restoration of soil fertility by traditional means for Polar agriculture, with the help of chemicals and organic fertilizers, is associated with significant material investments and environmental risks. The use of mineral and organic fertilizers, plant protection chemical products in crop production puts agriculture on a par with industry, transport and energy in terms of the level of negative impact on the environment. The delayed course of biochemical reactions in the Arctic ecosystems, associated with the predominance of low temperatures and other factors, significantly affects their ability to self-purification and recovery rate, therefore, the search for environmentally safe technologies in all spheres of economic activity is most relevant for the region in conditions of high industrial and anthropogenic load.

Crop production in the Murmansk region is mainly focused on the production of green and preserved non-transportable feed for dairy farming. The species composition of crops used in fodder production is not very diverse. Annual fodder grasses, such as oats, in pure form or grass mixtures with vetch and rapeseed are uses as green fodder and for silage preparation. Perennial grasses, such as meadow fescue, red fescue, sown 20 or more years ago, are used for green fodder, hay, haylage and preparation of silage.

The most effective and universal solution may be the use of symbiotic nitrogen fixation by legume crops. Narrow-leaved lupine (Lupinus angustifolius) is considered as the most promising crop for introduction due to its biological and environmental characteristics (deeply penetrating root system, ability to grow in a wide range of soil conditions, high productivity of green mass). In addition to its excellent fodder qualities, which are ensured by high content of high-grade protein, vegetable fat, small amount of anti-nutrients, etc., this plant is known for its environmental and health-improving properties [9].

Earlier studies of the effectiveness of the introduction of annual and perennial legume components into the structure of crop production have shown that inoculation of seeds is a mandatory technique, because there are no species-specific microsymbionts in the soils of the Murmansk region. Pre-sowing treatment of seeds with a biological preparation containing bacteria of the genus Rhizobium provided a significant increase in yield and, more than doubled the accumulations of crude protein, compared with non-inoculated samples [10].

The purpose of this work was to study the agrobiological and feed potential of three varieties of narrow-leaved lupines with inoculation of the seed material by Rhizobium lupini in the Murmansk region without applying of mineral and organic fertilizers. The tasks of the study included: calculation of the energy nutritional value of the aboveground mass of L. angustifolius varieties; determination of the number of diazotrophs in soils, as well as microorganisms assimilating mineral forms of nitrogen; determination of agrochemical indicators in the areas under L. angustifolius and under Avena sativa after harvesting.

2 Materials and methods

The objects of research were three original varieties of seeds of the second reproduction of Lupinus angustifolius obtained at the All-Russian Research Institute of Lupine - a branch of the V.R. Williams Federal Research Center, Bryansk: Vityaz, Belorozovy 144 and Bryanskiy kormovoy, in the amount of 72 kg (24 kg of each variety). To inoculate the seed material with species-specific symbiotic nitrogen-fixing bacteria, the biological preparation "Rhizotorphin brand A" produced by the ECOS Biopreparations Company, St. Petersburg was used.

To test the varieties of narrow-leaved lupines under production conditions, an area of 0.4 hectares on the land of a farm in the Kola district of the Murmansk region was determined. According to archival materials of field soil survey conducted by Murmansk prospecting expedition of the Sevzapgiprozem institute in 1981, the soil at the site is sandy sod-weakly podzolic illuvial-ferruginous.

According to the comprehensive agrochemical survey conducted in 2020, the soil at the site has the following agrochemical characteristics (Table 1).

Table 1. Agrochemical characteristics of the soil before laying the production experiment.

pH_{KCl}	P_2O_5, mg/100g	K_2O, mg/100g	Org. matter, %
6.3	362.5	19.6	5.21

Soil samplings were carried out by a soil drill according to GOST R 58595-2019.

Laboratory tests of soils were carried out according to standard methods: determination of pH in salt extract (KCl) by the CSRIAS method, GOST 26483-85; determination of mobile compounds of phosphorus and potassium by the Kirsanov method in modification of the CSRIAS, GOST R 54650-2011; determination of soil organic matter by the Tyurin method in the modification of the CSRIAS, GOST 26213-91.

The soil at the site has a high content of mobile phosphorus, very low content of exchangeable potassium, high content of organic matter and neutral acidity of the soil solution (gradations according to the M.g. for conducting ...) [11].

Laboratory tests of the aboveground mass of lupines were carried out in the Lupine Research Institute, Bryansk. The following indicators were studied: crude protein (GOST 13496.4-2019); crude fiber (GOST 31675-2012); crude fat (GOST 13496.15-2016); crude ash (GOST 32933-2014).

To study the microbiological activity of soils, we used elective nutrient media: Ashby, Bean agar, and Starch-Ammonia agar.

After harvesting, the following agrochemical parameters were determined: alkaline hydrolyzable nitrogen by the Cornfield method: M.g., 1985, and exchange phosphorus by the Kirsanov method, GOST 26207-91.

Seeding was carried out on 06.29.20. seeding rate was determined according to the standard methodology based on the seeding qualities of the seeds.

Before sowing, the soil was mechanically tillaged twice with disc harrows to remove weeds and improve the air regime. seeds immediately before sowing were treated with the preparation "Rhizotorphin brand A". Mineral and organic fertilizers were not applied.

3 Results

Harvesting was carried out on 08.09.20. The vegetation period was 70 days (Table 2).

Table 2. Characteristics of the yield of L. angustifolius.

Variety	Yield of green mass, t/ha	Height of herbage, cm	Vegetation phase
Vityaz	18	84	End of flowering, beginning of beans formation
Belorozovy 144	31	108	End of flowering, beginning of beans formation
Bryanskiy kormovoy	24	84	End of flowering, beginning of beans formation

At the time of harvesting, the plants reached the bean tying phase (the length of the beans is 3-5 cm), but flowering was still present in the upper tier. The formation of large nodules of light pink color was located on the roots.

To compose a balanced ration for feeding livestock, it is necessary to know the biochemical composition of the harvested feed (Table 3).

Table 3. Results of laboratory studies of varieties L. angustifolius

Variety	Mass fraction, % on a.d.m.*			
	Protein	Fat	Cellulose	Ash
Vityaz	19.3	1.50	21.1	11.75
Belorozovy 144	16.5	1.75	20.4	9.90
Bryanskiy kormovoy	21.2	1.91	19.0	10.48

* on absolutely dry matter.

Based on the data obtained, calculations of the energy value of feed for cattle were made (Table 4).

The regression equation developed at the L. K. Ernst Institute of Animal Husbandry was used to calculate the exchange energy in coarse fodder from raw nutrients for cattle:

$$EEC = 0.012 * P + 0.020486 * F + 0.00159 * C + 0.0105 * NFES, \quad (1)$$

where EEC – exchange energy for cattle in MJ per 1 kg of feed (hay, hay cutting, haylage, straw, silage up to 50% moisture and other coarse feed); P, F, C, NFES – protein, fat, cellulose, nitrogen-free extractives in grams per 1 kg of feed.

The content of nitrogen - free extractives is calculated by the formula:

$$NFES = 1000 - (P + F + C + A), \quad (2)$$

Energy Feed Unit (EFU = 10 MJ): EFU = EE (MJ)/10

Table 4. Assessment of energy nutritional value of coarse L. angustifolius fodder for cattle

Variety	g/kg, on a.d.m.					EEC, MJ	EFU
	P	F	C	A	NFES		
Vityaz	193	150	211	117	329	9.17	0.92
Belorozovy 144	165	175	204	99	357	9.62	0.96
Bryanskiy kormovoy	212	191	190	105	302	9.92	0.99

The "Methodical guidance for assessment of quality and nutritional value of feed raw materials", adopted by the Ministry of Agriculture of the Russian Federation on 20.06.2002, proposed formulas for calculation of EE and FU (feed unit) in forages based on mass fraction of crude cellulose content in dry matter and constant coefficients [12] (Table 5).

For grass artificially dried forages (grass flour, cutting, briquettes, pellets) prepared from annual and perennial legumes and cereals rich in protein, the formula for calculating energy nutrition is as follows:

$$EE = 13.71 - 16 * C; FU = EE2 * 0.0081, \quad (3)$$

https://doi.org/10.1051/e3sconf/202337806004

where EE is the amount of exchange energy (for cattle) in 1 kg d.m. (dry matter), MJ; FU is the amount of forage units in 1 kg d.m., kg; C is the mass fraction of crude cellulose in 1 kg of d.m., kg.

Table 5. Assessment of the energy nutritional value of lupinus angustifolius varieties, according to the "Methodical guidance for assessment of quality and nutritional value of feed raw materials".

Variety	EE, MJ/kg	FU, kg
Vityaz	13.37	1.45
Belorozovy 144	13.38	1.45
Bryanskiy kormovoy	13.41	1.46

The data of laboratory tests and calculated indicators of energy nutritionness make it possible to classify all studied samples under study to the first quality class according to the regulatory requirements for artificially dried forages (M.g. for assessing…).

For microbiological sowing, mixed soil samples were selected from a plot under lupine, as well as, for comparison, from a plot under oats, traditionally cultivated in the Murmansk region for fodder. The plots are located on the same array, in close proximity, and have similar agrochemical characteristics and area. Sampling was carried out after harvesting. Elective media (bean agar and Ashby medium) and starch-ammonia agar (SAA) for cultivating microorganisms using mineral forms of nitrogen were used for cultivation of aerobic diazotrophs. Microbial counts were performed using the Koch's cup method. The results of the study are presented in Table 6.

Table 6. The number of the main physiological groups of microorganisms under crops L. angustifilius and Avena sativa

Plot	Number CFU per 1 g of absolutely dry soil		
	Ashby	Bean agar	SAA
under lupines	3465±308	2502±316	7728±2818
under oats	575±201	3367±30	3137±557
t	7.8	4.3	1.6
p	< 0.05	> 0.05	> 0.05

We can note a marked increase in the number of nitrogen fixers in the plot under the lupine. The increase in the number of this group of microorganisms in relation to the control (the plot under oats) was about 4 thousand CFU/1 g of soil. At the same time, the number of other physiological groups of microorganisms did not change significantly. From the results obtained, we can conclude that the use of symbiotic strain of microorganisms for pre-sowing inoculation of L. angustifilius seeds affected the direction of microbiological processes in the soil. Narrow-leaved lupine, covering its own needs in nitrogen nutrition through symbiosis, returns to the soil from 30 kg/ha of nitrogen with crop residues, thereby providing a nutrient substrate for soil microbiota for further transformation of biogenic elements. It should also be noted that the reaction of the soil solution in the plots is neutral, close to alkaline, pH 7.2, 7.5, under oats and lupine, respectively. This is the most favorable environment for the life activity of soil microorganisms.

Mixed soil samples were taken after harvest to determine alkaline hydrolyzable nitrogen and exchangeable phosphorus (Figure 1).

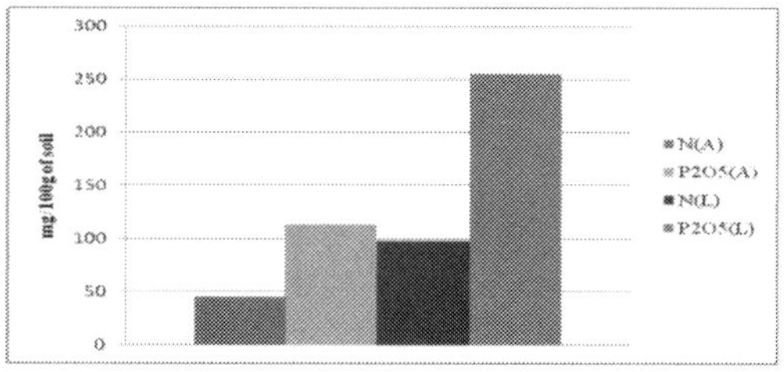

Figure. 1. Nitrogen and phosphorus content in experimental plots after harvesting.

Legend: N, P2O5(A) – the content of nitrogen, phosphorus in the area under Avena sativa; N, P2O5(L) – under L. angustifolius.

The content of nitrogen and phosphorus compounds potentially available for plant nutrition is at a very low level, but in the area under the narrow-leaved lupine, the indicators are more than twice as high (N(A) – 45.5; N(L) – 98; P2O5(A) – 113.6; P2O5(L) – 254.5 mg /100g). This may indicate lower costs of soil resources for the formation of a crop of narrow-leaved lupine in comparison with oats.

4 Discussion

Narrow-leaved lupine is a promising fodder crop for cultivation in the agro-climatic conditions of the Arctic. The inclusion of narrow-leaved lupine in crop rotations will reduce fodder consumption in the farms of the region, improve the quality and quantity of livestock products, significantly reduce the need for the use of chemical meliorants. Introduction of legume components, with mandatory inoculation of seed material with species-specific highly effective strains of microsymbionts, into the crop production of the Arctic, will allow without significant financial costs to approach the problem of conservation and reproduction of soil fertility, stop soil degradation, increase the stability of artificial agrocenoses at the northern limit of agriculture.

The problem of the development of the agro-industrial complex of the Murmansk region is complex and requires a similar approach to its solution. It is necessary, first of all, to revise the system of agriculture, to restructure it in accordance with modern achievements of fundamental and applied science, increase the knowledge intensity of the industry through the integration of precision farming technologies, remote monitoring, etc. It is necessary to work on the development of local varieties of annual and perennial legumes, sidereal and fodder directions, which are able to ensure their growth at the expense of biologically fixed atmospheric nitrogen.

This approach is consistent with the "State program for the development of agriculture and regulation of agricultural products, raw materials and food markets" (approved by the Decree of the Government of the Russian Federation from July 14, 2012 No. 717), one of the goals of which is to reproduce and increase the efficiency of agricultural use of land and other natural resources of the Russian Federation. Realization of the natural potential of legume-rhizobial symbiosis in the agro climatic conditions of the Arctic territories will contribute to the environmentalization of production and will solve a number of problems related to ensuring food security of the Murmansk region.

The introduction of environmentally oriented natural technologies in production cycles should contribute to the speedy progressive development of a "green economy" in the Russian Federation.

References

1. S.A. Dyuzhilov, Polar agriculture: problem statement and its solution in the 1920s Kola North, *Proc. of the Kola Scientific Center of the Russian Academy of Sciences*, **3**, pp. 71-88 (2016)
2. S.A. Dyuzhilov and Y.E. Bogdan, The Polar Department of the VIR at the turn of the 1920s-1930s: new vectors of development in scientific activity, *Proc. of the Kola Scientific Center of the Russian Academy of Sciences*, **4-11(8)**, pp. 35-55 (2017)
3. A.Y. Maretskaya and V.N. Maretskaya, Development of agriculture in the Murmansk region - historical and economic aspect, *Modern problems of regional development: Abstracts of the 6th International Scientific Conference ed. E.J. Frisman* (Birobidzhan, Institute of Complex Analysis of Regional Problems of the Far Eastern Branch of the Russian Academy of Sciences, 2016) pp. 440-43
4. A.Y. Maretskaya and V.N. Maretskaya, *Agricultural Sector of the Murmansk Region: History of Development, Current State* (Apatity, FRSC KSC RAS, 2018) p.108
5. E.E. Kislykh, *Environmental Bases of Soil Fertility Regulation in Conditions of Intensive Anthropogenic Impact* (Petrozavodsk, Publishing House of PGU, 2002) p. 38
6. M.G. Nesterkin and N.N. Khludneva, The state of soil fertility in the Murmansk region, *Achievement of science and technology of the agro-industrial complex,* **6**, pp. 10-14 (2018)
7. V.A. Ivanov, Conditions and opportunities for realizing the potential of agriculture in the zone of the North, *Arctic and the North*, **35**, pp. 25-45 (2019)
8. Y.A. Maretsky and V.N. Maretskaya, Problems of development of agriculture and crop-growing industries of the Murmansk, *The North and the market: the formation of an economic order*, **1(27)**, pp. 116-20 (2011)
9. N.S. Kuptsov and I.P. Takunov, *Lupine: Genetics, Breeding, Heterogeneous Crops* (Bryansk, Klintsy city printing house, 2006) p. 576
10. P.V. Laskin, et al., Inoculation of annual lupines by nodule bacteria in the conditions of the Far North, *Fertility*, **2**, pp. 53-54 (2010)
11. "Methodological guidelines for conducting comprehensive monitoring of soil fertility of agricultural lands" (approved by Ministry of Agriculture of Russia on September 24, 2003), p. 304 (2003)
12. Methodological guidelines for assessing the quality and nutritional value of feed (Moscow, CIAS, 2002) p.76

Prospective industrial complexes in the Russian Arctic: focus on rare-earth metals

Victoria Solovyova[1,*] and *Alina* Cherepovitsyna[2]

[1]PhD in Economics, Assistant, St. Petersburg Mining University, 199106, St. Petersburg, Russia
[2]PhD in Economics, Senior Researcher, Luzin Institute for Economic Studies – Subdivision of the Federal Research Centre, Kola Science Centre of the Russian Academy of Science, 184209, Apatity, Russia

Abstract. At present, issues related to the development of rare-earth metal industrial complexes are becoming particularly relevant for Russia. Rare-earth metals (REMs) play a vital role in high technology. Therefore, success in technological progress largely depends on the availability of these metals. However, the country's position remains uncertain – on the one hand, Russia possesses huge REM reserves, on the other hand, they are not exploited, and the existing level of import dependence of metals on supplies from foreign countries exceeds 80%. The main national REM reserves are in the Arctic region. Nevertheless, even rich deposits are still not exploited. This study aims to investigate the prospects for the development of REM industrial complexes in the Russian Arctic, taking into account both opportunities and constraints. The resource potential of REMs in the Russian Arctic is investigated, a comparative analysis of the characteristics of foreign and national REM objects is provided. The factors influencing the prospects of development of REM industrial raw material complexes are determined. Data on the main REM projects are systematized. The study makes it possible to draw some important conclusions about the current prospects for the development of rare-earth metal industrial raw materials complexes in the Russian Arctic.

1 Introduction

In recent years, the importance of rare-earth metals for the global economy has increased. This is primarily due to the intensive development of new technologies [1, 2]. Currently, these metals contribute to the climate agenda issues and to the advance of low-carbon path, as they are used in electric cars, wind turbines, and other progressive "green" technologies.

Russia, on a par with other developed countries, is trying to expand science-intensive industries, which include rocket and space, radioelectronic, aviation industries, as well as a number of industries related to information and communication technologies and medicine [3]. Currently, rare-earth metals are required in production of modern engines PD-14, PD-35, GTU (aviation industry), electronic devices and equipment, microcircuits (radio-electronic industry), etc. [1, 2].

The rare-earth crisis (2011-2012) proved the unreliability of the global supply system of REMs from China. Recently, rare-earths have become a "geopolitical tool". In this way, China as the main monopolist of the world REM market, is trying to influence the economics of other countries. Therefore, under current conditions, the need to develop national mining and processing industries to produce the required rare-earth metal products seems obvious [4]. In this context, the Arctic region is of particular interest. It is here that the largest sources of rare-earth raw materials are located – the deposits of the Khibiny massif, as well as Tomtor, Lovozero and Africanda deposits [5].

At the same time, the development of the Arctic territories is one of the most difficult directions in a view of geological features, specific natural and climatic conditions, infrastructure constraints, high level of environmental risks, etc. [6-10]. It is assumed that the Northern eco-systems are fragile and unstable. This requires strict compliance with the environmental standards and demands. The situation is aggravated by the specificity of rare-earth metal sources, imperfection of the existing system of subsoil use management, lack of experience in implementing mega projects, insufficient level of technological availability.

The "Strategy for the Development of the Arctic Zone of the Russian Federation and National Security for the period up to 2035" states the necessity to develop rare-earth metal sources (in particular, the Lovozero deposit) as one of the key priority areas of development the Arctic [11]. Meanwhile, specific measures to implement this direction have not been established. It is not clear what are the prospects for the exploitation of REM deposits in the Russian Arctic, when the planned projects will be implemented, whether the Arctic region will become the basis for the development of REM industry in the country, what is needed to create REM raw material complexes with a full technological cycle, how to ensure environmental safety. All these unresolved questions

* Corresponding author: vikasolovyova9@gmail.com

determine the relevance and the significance of the research offered.

2 Literature review and research design

There is a large volume of published research describing the role of rare-earth metal industry development. Rare-earths are among the most important and strategic minerals and have been called "vitamins of modern society". It is predicted that in the near future competition for access to these metals will be comparable to the competition in the world oil and gas market [12-15].

Russian researchers investigate key trends of the national and global REM market. Much of the current literature focuses on the identification of existing problems and barriers associated with the exploitation of the rare-earth mineral resource base. These include the high level of environmental risks, the low level of economic efficiency of REM projects, the declarative nature of legal regulation, and low indicators of domestic demand for rare-earth products [16-19]. Meanwhile, despite of all the difficulties, national researchers agree that the development of rare-earth metal industry is essential to ensure the sustainable development of high-tech industries [20]. Given the complicity of the organization of rare-earth metal industrial complexes, questions related to reliable prospects for their creation and further development remain open.

One of Russia's key advantages in the development of its rare-earth metal mineral resource base is its rich resource potential, much of which is concentrated in the Arctic region [7,9,21-22]. Domestic scientists have repeatedly emphasized the strategic importance of the Arctic territories in capacity as the main source of critically and strategically important minerals, including rare-earth metals [23-25]. According to the theses of the report by Yu. Ampilov at the 6th International Conference "Arctic 2021", in order to efficiently explore the Arctic region, it is advisable to focus not on the intensification of the exploitation of hydrocarbon resources, but on the implementation of REM projects [26]. This is explained by the changing market conditions – global trends of the energy transition, the spread of "green" technologies, etc.

The methodology of this study includes case studies, system-oriented analysis, decomposition method and comparative analysis. A content-analysis of academic literature and analytical reports on the topic was used. The paper is organized as follows: (1) the role of rare-earth metals for the Russian economy is considered and the necessity of creation of national REM production facilities (mining and refining) is shown, (2) the resource potential REMs in the Arctic region (key deposits, their characteristics) is analysed, (3) the existing environmental restrictions are discussed, (4) reliable prospects for the development of the rare-earth metal industry in the Russian Arctic according to the plans established at the state level are studied and the main factors influencing the creation and further development of the competitive industrial complexes of REMs in the Arctic region are identified.

3 Main part

3.1 Rare-earth industrial complexes: is there a need for development?

In today's world, the significance of rare-earth metals is becoming more and more prominent. These metals are successfully used in the production of catalysts, glass, metallurgical alloys, permanent magnets, etc. [1]. Recent trends related to the global energy transition have proven an overriding importance of this group of metals in terms of the production of "green" technologies [27]. According to estimates by Grand View Research, the annual growth rate of the global rare-earth market reaches 10.4%, which can be explained by the constant expansion of demand for these valuable components [28]. For many countries, rare-earth metals have become "critical" materials. This indicates the limited access to REMs, or rather to the raw sources of their mining. For example, Japan, European countries, and South Korea have no direct access to these resources. China, the main monopolist of the global REM market, cannot be considered as a reliable supplier: constant changes in the quota system, threats of sanctions, etc. This is why other countries are forced to develop long-term strategies on how to ensure their own demand for the necessary metals within the framework of growing needs for REMs [1]. China's share has fallen from 92% in 2010 to 58% in 2020, but the role of this country remains significant [28-29].

Russia, along with other world developed countries is trying to take high positions in the high-tech market [31]. However, the level of import dependence on foreign REM supplies is more than 80% [32]. This can affect the sustainability of such national industries as aircraft and engine construction, shipbuilding industry, nuclear power complex, rocket and space industry, information and communication technologies. Domestic production capacities do not allow to meet the current demand in high value-added products. Russia's share in the global market is only 1.2% [28].

The need to develop the Russian rare-earth industry is reported by the state program "Industry Development and Improving its Competitiveness", the Order of the RF Government "On Approval of the Consolidated Development Strategy of the Manufacturing Industry of the Russian Federation for the Period to 2024 and for the period until 2035" and other regulatory documents [31]. Nevertheless, the REM industry in Russia has not been properly developed.

3.2 Is the Russian Arctic a "growth point" for REM industrial complexes in Russia?

The Arctic territories are known by their abundance of mineral resources [9, 10]. There is an opinion that the Arctic will become a framework for the development of the REM industry in Russia, and it is not accidental [23]. More than 60% of national reserves rare-earth metals (categories A+B+C1) and about 90% of C2 reserves are located there. The total reserves exceed 18 000 thousand tons of ΣTR2O3. This fact allows us to consider the Arctic region as one of the most prospective sources of strategically important minerals [5]. Figure 1 shows the key rare-earth deposits and production capacity of REM products in Russia.

Industrial exploration of rare-earth metals in the Arctic region began during the existence of the USSR. However, despite the presence of a number of promising objects for development, the only source for these metals was and remains the Lovozero deposit (located in the Murmansk region) [33]. Exploitation of this deposit is provided by the Lovozersky mining and processing plant (GOK). Total reserves are estimated at 2600 thousand tons of ΣTR2O3. In 2019, Lovozersky GOK mined 2.7 thousand tons of ΣTR2O3. The concentrate produced here is sent for further processing to Solikamsk Magnesium Plant, which produces light rare-earth metal carbonates. Then, one part of the products is sent to LIT, where REMs are separated into individual components, while another part is exported (due to the limited processing capacity in the country) [5, 33].

In addition to the Lovozero deposit, the apatite-nepheline ores of the Khibiny group of deposits are also known for their reserves of rare-earths. These deposits are developed by Apatit (PhosAgro) and North-West Phosphorus Company (Akron). The commercial products are apatite concentrates and phosphate fertilizers. Rare-earth metals are not extracted into concentrate. REMs are partially transferred to fertilizers, partially to large-tonnage phosphogypsum waste. Nevertheless, a significant portion of rare-earths remains in the tailings [5].

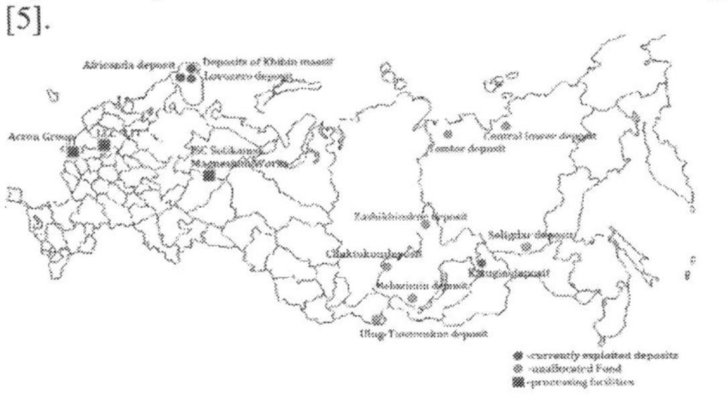

Fig. 1. Rare-earth metal deposits and processing facilities in Russia [5]

Large inferred resources of rare-earths are also located in Africanda perovskite-titanomagnetite deposit – about 860 thousand tons of ΣTR2O3 [5]. This deposit has a long history. In the 1950-s, it was put into operation. An ore processing plant was built to produce perovskite and titanomagnetite concentrates. It was planned to expand capacities for the production of various types of ferroalloys [34]. Nevertheless, later, an increased content of radionuclides in the slag was found (they did not meet the existing sanitary standards). That was the reason why the production was stopped.

The Tomtor deposit is represented by monazite-pyrochlore ores in the weathering crusts of carbonatites. In 2017, reserves of categories A+B+C1 of the Buranny site accounted only for 0.4% of the total national reserves. But the complex ores of the Tomtor deposit are metal-rich. The average content of REMs in the ores of the Tomtor deposit is higher than in the ores of the most known deposits in the world, and they contain the most valuable components in terms of the end-use opportunities – REMs of the yttrium group [5]. Figure 2 presents an overview of the comparative characteristics of Russian Arctic and foreign promising objects of rare-earth metals.

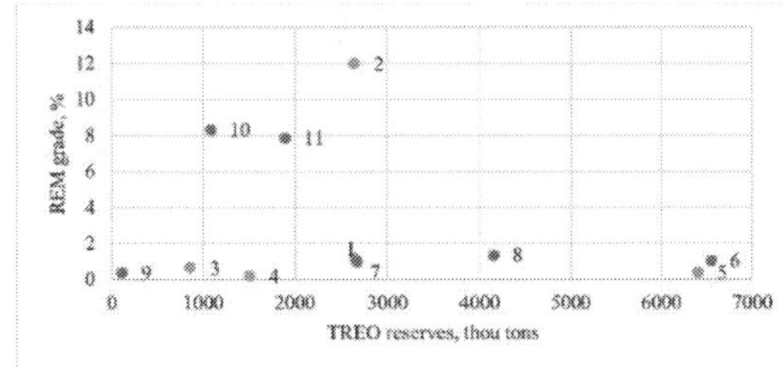

1 – Lovozero deposit (Russia), 2 – Tomtor deposit (Russia), 3 - Afrikanda deposit (Russia), 4 – Partomchorr deposit (Russia), 5 - Khibin group of deposits (Russia), 6 - Kvanefield (Greenland), 7 - Sørensen deposit (Greenland), 8 - Thor Lake (Canada), 9 - Kipawa (Canada), 10 - Mountain Pass (USA), 11 - Mt. Weld (Australia)

Fig. 2 Comparative characteristics of the Russian and foreign promising objects of rare-earth metals (foreign objects are taken selectively) [5, 35,36]

It is important to note that in terms of reserves, a number of Arctic rare-earth fields are ahead of foreign sites (considered promising). This fact creates certain prospects for their exploration. As an example, in the USA, the Mountain Pass deposit is successfully developed. However, the volume of reserves and the REM content of this deposit are lower than similar indicators of the Tomtor field and Khibiny group of deposits [5, 35, 36].

Therefore, in terms of available reserves, the Arctic territories can become a base for the development of the national rare-earth metal industry. At the same time, as practice shows, solely the availability of the resource potential in this case is not the only criterion for determining the reliable prospects for creating competitive industrial complexes for the production of REMs [37]. It is necessary to find out whether efforts are being made to develop the Arctic REM deposits, whether the implementation of rare-earth projects is planned in the near future, and what factors may influence this process.

3.3 REM production in the Arctic: environmental constraints

The development of arctic deposits is associated with certain features: severe weather conditions, infrastructure constraints (transport, telecommunications), lack of logistics systems and others [7-10, 38-40]. But one of the most important issues remains environmental one.

The increase in the extraction and processing of metals can surely affect the fragile Arctic eco-systems. The paradox is the following: on the one hand, rare-earths contribute to the "green" economy, but, on the other hand, their mining can be considered unsustainable in terms of environmental constraints [41]. This leads to a discussion about whether an energy transition based on technologies containing "dirty" metals can reduce environmental damage, emissions of CO2 and other harmful substances [41-42].

The indicator of radioactivity of rare-earth raw materials - the content of such components as thorium and uranium - is mandatory for accounting. Their neutralization is possible only with the use of special technologies for raw material purification [43]. The higher the content of these components, the higher the risk of environmental threat.

It should be noted that the apatite present in the Russian Arctic has relatively low level of radioactivity, which can be attributed to the positive characteristics of the national resource base. Nevertheless, environmental risks remain high [43]. In 2018, Russia approved the Information and Technical Handbook BAT (ITS BAT) "Mining of Rare and Rare-Earth Metals", which contains a description of production and technological processes, machinery and equipment, methods and technics of mining rare-earth elements, allowing to minimize the potential risks of negative impact on the environment and reduce the material and energy consumption of production facilities [44].

We argue that in case of the environmental factor should be taken into account as a key factor in the development of Arctic REM reserves. Current projects do not raise the question of whether or not the technologies planned for use comply with modern environmental standards, whether or not they are among the best available technologies, etc. However, without compliance with environmental restraints, the prospects of REM projects may remain uncertain.

3.4 Prospects for the development of rare-earth industrial complexes in the Russian Arctic

Extraction of rare-earth raw materials is only the first step on the way to the creation of finished REM products. This is followed by the stages of obtaining concentrates, producing oxides, isolating individual components, etc. Although the metallurgy of rare and rare-earth metals is mainly based on general principles, it differs in the use of individual chemical and metallurgical technologies [44]. Therefore, the of formation of such industrial complexes without system-oriented solutions is simply impossible.

From China's experience, to be competitive in the global market, it is necessary to develop not only the upstream segment (extraction and production of primary concentrates), but also the downstream segment (production of oxides, individual metals, and magnets, which are widely demanded in the market). Of course, the formation of such productions is impossible without an appropriate technological base – research and development (R&D), search for new sources and possibilities for the extraction of rare-earth metals, approbation of processing technologies.

The complexity, high - science and capital-intensity nature of such multi-stage production facilities explain the reluctance of subsoil users to develop REM deposits and the lack of intentions on the part of investors to finance such projects [40]. Therefore, the state policy (mechanisms, measures and support tools) is essential in this context. In 2020, the Russian Parliament provided incentives to the development of rare-earth industry by reducing the mineral extraction tax from 8% to 4.8% [45].

Special measures to support Arctic REM projects are currently being developed. For example, in February 2020, the Order of the Ministry of Natural Resources of the Russian Federation for the Murmansk region reduced, the infrastructure coefficient for the development of solid minerals by 25% - to 1.0 [45]. This measure was considered as one of the main intensives for auctioning the right to use the subsoil plots in the central part of the Afrikanda deposit. In fact, the state support determined new prospects for the exploitation of this field. For a long period of time, the implementation of the Africanda project was considered unprofitable from an economic point of view [46].

On the example of the Africanda project, we can identify a number of other effective measures that can be applied to the development rare-earth deposits. These include tax incentives for profits, reduction of the mineral extraction tax (when creating infrastructure), and economic support for mineral processing to increase the added value of finished products [46-47]. The orientation of the proposed support measures is focused not only on the development of upstream, but also on downstream segments. It forms preconditions for advancement along the technological chain. Exceptional performance of the mining phase without implementing processing steps is economically inefficient in the case of rare-earth metals.

Based on the key features of the creation of rare-earth industrial complexes, a number of main factors that may determine the prospects for the development of such complexes in the Arctic have been identified:

(1) infrastructure (production, transport)

(2) availability of technologies (mining, processing, extraction technologies)

(3) investments (capital investments)

(4) compliance with environmental standards (best available technologies, compliance with environmental constraints)

(5) availability of plans for creation of facilities for production of rare-earth products with high added value

(6) state policy (measures implemented to support the REM industry - both general and at the level of individual regions).

Based on the analysis of the provisions of the draft "Strategy for the Development of Rare and Rare-Earth Metals Industry", the state program "Industry Development and Increasing its Competitiveness", the roadmap for the development of rare and rare-earth metals mining until 2024 (prepared by the Ministry of Industry and Trade of the Russian Federation jointly with Rosatom), it was found that in the medium and long term, two major projects that can affect the development of the national REM industry – the Afrikanda and Tomtor fields [31, 34]. The State Report "On the state and use of the Mineral Resources of the Russian Federation in 2019" states that the Partomchorr deposit is also at the stage of preparation for exploration. However, despite of the

https://doi.org/10.1051/e3sconf/202337806005

existing rare-earth potential, it has been established that there are no plans to extract rare-earth metals into commercial products. This may indicate a lack of appropriate technology [5].

Figure 3 presents characteristics of the prospective REM objects in the Russian Arctic.

Criteria	Objects		
Name of the deposit	Tomtor deposit	Africanda deposit	Partomchorr field (Khibin massif)
Main characteristics			
Location	Yakutia (Republic of Sakha)	Murmansk region	Murmansk region
Geological-industrial type	weathering crust of carbonites	perovskite-titanomagnetite ores	apatite-nepheline
REM reserves (categories A+B+C1)	2 640.4 thou. tons	860 thou. tons	1505 thou. tons
Rare-earth metal content	12%	0.67%	0.2%
Stage of implementation	Engineering	Appraisal	Appraisal
Possible production volumes **-REM**	2.4 thou. tons REM (concentrate)	8.5 thou. tons REM (concentrate)	extraction of rare-earth metals into commercial products is not planned
-other products	10 thou. tons ferroniobium; 3.5 thou. tons didim	75 thou. tons titanium dioxide 1.3 thou. tons niobium; 71 tons tantalum	phosphate raw materials
Time frames (launch of the project)	2030 year (previously planned to 2025)	2024 year	2027 year
Project operator	Triark Mining	Arkmineral	Fosagro
Key factors			
(1) Infrastructure	-	+	+
(2) The availability of technologies	+	+	-
(3) Required investments	53 bln rbl	19.4 bln rbl	NA
(4) Compliance with environmental standards	Minimizing environmental consequences (special mode of operation, abandonment of large-scale exploitation)	Effective environmental methods have been developed to organize the extraction and enrichment of ore ecological threats (Kola Scientific Centre of the Russian Academy of Sciences)	NA
(5) Plans for REM high added value products' manufacture	+	+ (production of REM oxides and individual metals)	-
(6) Support measures (the state policy)	It is planned to use the measure of state support to subsidize the interest rate of the loan.	-Reduction of the starting payment -reduction of the infrastructure coefficient for the development of solid minerals, - income tax benefits, - reduction of mineral extraction tax	State support measures are not defined

Fig. 3 Characteristics of prospective REM objects in the Russian Arctic [5, 32, 46-47]

As can be seen from the figure 3, the most "prepared" REM project in the Russian Artic is the Africanda project–availability of technology, developed infrastructure of the region. According to the proposed plans, its implementation will provide employment for about 450 residents of the region, and tax revenues to the budget after reaching production volumes are planned at more than 1.5 billion rubles per year. Thus, not only economic but also social effect is expected.

The prospects for growth of REM production are also associated with the exploitation of the Buranny site of the Tomtor field. Given the availability of appropriate technologies, the main "barrier" for the development of this deposit remains the undeveloped infrastructure of the region, which determines the high level of capital investments (53 bln rbl). Complicated logistical solutions, as well as the extremely unfavorable location of the planned facility sharply increase the cost of the finished product. Even with favorable market conditions, it is difficult to ensure the cost-effective development of these valuable components. It is no coincidence that the timing of the project was postponed - initially it was announced for 2025, however, later the terms were shifted for five years, which was explained by the impossibility of applying state support measures in the medium term.

Overall, according to the plans reviewed, the main increase in REM production in the Russian Arctic will occur only in 2030. Figure 4 shows the projected production volumes of rare-earth concentrates (on the assumption that Africanda and Lovozero projects are implemented). The production volumes of the Lovozersky GOK are taken based on the indicators for 2019.

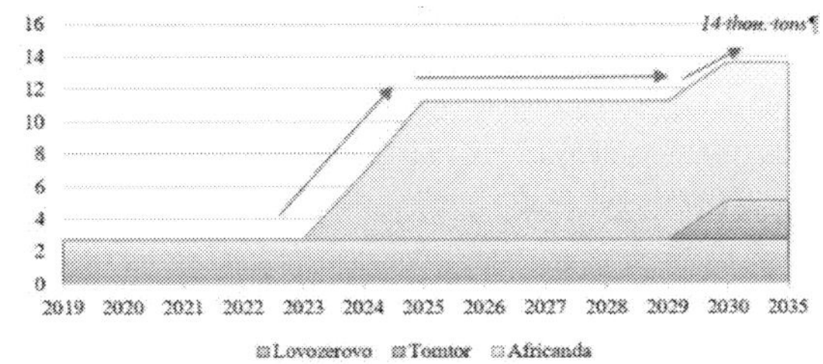

Fig. 4 Forecasting production volumes of rare-earth concentrates (on the assumption that projects planned will be implemented), thou. tons of ΣTR_2O_3

At the same time, this is only the first stage in the production of rare-earth goods mentioned earlier. Reliable prospects for creation and development of industrial REM complexes will be formed only if the plans for launching the production and technological chain - the creation of high added value products that are in demand on the domestic and world market - are implemented. Otherwise, the situation with the exploitation of the Lovozero field may repeat itself, when most of the products have to be exported due to the lack of opportunities (a) to use concentrates obtained, (b) to process them into finished products with high added value [5]. Therefore, the focus should be on technology. In the conditions of the Arctic region, it is technological advance that can help to reduce production costs, minimize environmental risks, and ensure the competitiveness of the finished products of the national REM.

Taking into account the high level of capital investment required to activate the creation of rare-earth industrial complexes in the Russian Arctic, it is necessary to ensure consistency in the applied measures of state support. In this case, the solution can be the use of such instruments as financial and non-financial development institutions, as well as mechanisms aimed at creating stable conditions for business functioning - signing of special project investment contracts (SPICs), inclusion in regional investment projects (RIPs), implementation of projects with special economic zones – territories of advanced development [31]. It is also necessary to revise the provisions of the current system of subsoil use in order to expand the possibilities for involvement in the production process of technogenic objects and already developed fields (reduction of one-time payments, incentives for subsoil users to increase the complex use of raw materials, etc.).

4 Conclusion

To summarize the research, we would like to underline the following points:

1. The development of the Russian rare-earth metal industry is one of the essential priorities for the national economy on the way to overcome the high level of import dependence. The success in this case can affect the possibility of the technological advance of the country.
2. The Arctic region has a large mineral base of rare-earth metals. It is established, that the main sources of REM in the country are concentrated here - Tomtor, Lovozero and Africanda and other deposits. However, having such potential, the rate of development of domestic rare-earth industry remains low: new facilities are not being developed, the available production facilities in the country are not able to provide the domestic industries with required components (both in quality and volume). This is explained by both the complexity of creation and development of REM industrial complexes and by the special features of the exploration of Arctic territories.
3. It has been established that environmental issues in REM exploitation remain particularly important. New projects must follow environmental standards and constraints to ensure sustainable mining and processing of rare-earth metals, taking into consideration the specific characteristics of raw materials.
4. Based on the study of the specifics of creation and development of REM industrial complexes, the factors that can determine their prospects in the Arctic region were determined (1) the availability of infrastructure (production, transport), (2) the availability of technology (mining, processing, extraction technologies), (3) the required amount of investment (capital investments), (4) compliance with environmental standards, (5) the availability of plans to create production facilities to produce rare-earth products with high added value, (6) state policy (implementable measures to support the REM industry).
5. Analysis of the most promising rare-earth projects in the Russian Arctic showed that the most "well-prepared" are the projects for the development of the Afrikanda and Tomtor deposits (Buranny site). They are planned to be launched in the medium term (until 2030) and will provide a significant increase in rare-earth products for the national industry. Despite the difficulties in the development of the Arctic territories, the availability of technology and the possibility to form a complete chain of technological cycle for the production of products with high added value will allow such projects to maintain an acceptable level of economic efficiency.
6. To intensify the process of creating and developing rare-earth industrial complexes in the Arctic, it is necessary to ensure the consistency of applied measures of state support, as well as to revise the provisions of the current system of subsoil management to expand the opportunities for involving in the production process of technogenic objects and already developed deposits.

References

1. D.S. Abraham, *The Elements of Power: Gadgets, Guns, and the Struggle for a Sustainable Future in the Rare Metal Age* (Paperback, Yale University Press) 336 p. (2015)
2. J. Kynický, et al., Editorial for Special Issue "Rare Earth Deposits and Challenges of World REE Demand for High-Tech and Green-Tech at the Beginning of the 3rd Millennium", *Minerals*, **11(4)**, p. 378 (2021)
3. National report: high-tech business in the regions of Russia 2019, ed. S.P. Zemtsova (Moscow, RANEPA, 2019) 109 p (in Russian)
4. Y. Chen and B. Zheng, What Happens after the Rare Earth Crisis: A Systematic Literature Review, *Sustainability*, **11(5)**, p. 1288 (2019)
5. State report on the state and use of mineral resources Russian Federation in 2019 Ministry of Natural Resources and Environment of the Russian Federation, 494 p. (2019)
6. M. Shabalov and D. Dmitrieva, Implementation of cluster scenario approach for economic development of the arctic zone of the Russian Federation International, *Journal of Applied Business and Economic Research*, **15(2)**, pp. 281-289 (2017)
7. E. Rutenko and A. Cherepovitsyn, Strategic Planning of Oil and Gas Companies The Decarbonization Transition, *Energies*, **15(17)**, pp. 6163 (2022)
8. K.V. Matrokhina, et al., Development of methodology for scenario analysis of investment projects of enterprises of the mineral resource complex, *Journal of Mining Institute,* **259**, pp. 112-124 (2023)
9. A.E. Cherepovitsyn, et al., Innovative approach to the development of the mineral resource potential of the Arctic zone of the Russian Federation, *Journal of Mining Institute*, **232**, pp. 438-444 (2018)
10. A. Tsvetkova and E. Katysheva, Present problems of mineral and raw materials resources replenishment in Russia, *International Multidisciplinary Scientific GeoConference Surveying Geology and Mining Ecology Management, SGEM 2019*, **(5.3)**, pp. 573-578 (2019)
11. Strategy for Developing the Russian Arctic Zone and Ensuring National Security until 2035 Available online: http://en.kremlin.ru/acts/news/64274 (2021) (in Russian).
12. Rare earth elements – The vitamins of modern industry 2019 Chemicals Research & Analysis. HIS

Markit Available online: https://ihsmarkit.com/research-analysis/rare-earth-elements--the-vitamins-of-modern-industry.html (2019)

13. Rare earths in the energy transition: what threats are there for the "vitamins of modern society"? 2020 Energies Nouvelles Available online: https://www.ifpenergiesnouvelles.com/article/les-terres-rares-transition-energetique-quelles-menaces-les-vitamines-lere-moderne (2020)
14. C. Church and A. Crawford, Minerals and the Metals for the Energy Transition: Exploring the Conflict Implications for Mineral-Rich, Fragile States *The Geopolitics of the Global Energy Transition, Lecture Notes in Energy*, **73**, pp. 279-304 (2020)
15. E. Brennan, The Next Oil?: Rare Earth Metals The diplomat Available online: https://thediplomat.com/2013/01/the-new-prize-china-and-indias-rare-earth-scramble/ (2013)
16. Y.V. Kalashnikova, Innovative and strategic problems of the russian industry and rare earth metals, *Omsk Scientific Bulletin*, **4(121)**, pp. 61-64 (2015) (in Russian)
17. A.L. Bilin, The problem of revival of the rare earth industry in Russia Innovative geotechnologies in the development of ore and non-metal deposits, *Collection of reports of the VII International Scientific and Technical Conference in the framework of the Ural Mining Decade*, pp. 373-379 (2018)
18. V.A. Kryukov and S.A. Zubkova, Reindustrialization without REMs?, *ECO*, **46(8)**, p. 5-24 (2018) (In Russian)
19. Kryukov V A et al. Rare Earth Industry – how to take advantage of opportunities. *Russian Mining Industry*, **5**, p. 68-84 (2020) (in Russian)
20. A.A. Gasanov, et al., Some trends of the world ree market and prospects for Russia, *Izvestiya Vuzov Tsvetnaya Metallurgiya (Izvestiya. Non-Ferrous Metallurgy)*, **4**, pp. 31–44 (2018) (In Russian)
21. A.S. Buynovskiy, et al., Current state of the rare earth industry in Russia and Siberia, *1st International Symposium on Inorganic Fluorides: Chemistry and Technology, ISIF*, **11**, pp. 126-132 (2014)
22. A.V. Lalomov and A.A. Bochneva, Rare-metal potential of placer deposits and weathering crusts of the Russian Arctic Arctic: ecology and economics, **4(32)**, pp. 111-122 (2018) (in Russian)
23. V.S. Vasiltsov, On the matter of the economic development of the Russian Arctic Management of innovative development of the Arctic zone of the Russian Federation, *Collection of selected works based on materials of the All-Russian scientific and practical conference with international participation*, pp. 81-84 (2017) (in Russian)
24. In the Arctic, the mining of rare earth metals and the development of alternative energy are promising – opinion 2021 PortNews Available online: https://www.nationalgeographic.com/news/2010/10/101001-energy-rare-earth-metals/ (2021) (in Russian)
25. A.V. Volkov, et al., Deposits of strategic metals in the Arctic region, *Proceedings of the Fersmanov Scientific Session of the State Institute of the KSC RAS*, 16, pp. 80-84 (2019) (in Russian)
26. Outcomes of the 5th International Arctic Forum "Arctic – territory of dialogue" (2019) 17 p.
27. M. Paschke et al. The supply of rare earths for "green" energy and sustainable development, *Bulletin of St. Petersburg State University*, **3(5)**, pp. 56-73 (2016)
28. U.S. Geological Survey Mineral Commodity Summaries 2020: U.S. Geological Survey (2020) 200 p.
29. Rare Earth Elements Market Size, Share & Trends Analysis Report By Product (Cerium, Dysprosium, Erbium), By Application (Magnets, Catalyst), By Region, And Segment Forecasts, 2019 – 2025 Grand view research Available online: https://www.grandviewresearch.com/industry-analysis/rare-earth-elements-market (2021)
30. J. Mitchell, Can the West break China's stranglehold on rare earth supply chains? Available online: https://www.investmentmonitor.ai/business-activities/extractive-industries/china-rare-earths-supply-chain-west (2021)
31. Ministry of Industry and Trade of the Russian Federation Available online: http://government.ru/en/department/54/events/ (2021) (in Russian)
32. S.Yu. Scripachev, Current state and prospects for the development of rare earth industry, *Federal State unitary Enterprise All-Russian scientific research Institute of aviation materials Presentation*, 15 p. (2018) (in Russian)
33. G. Meleniev Russian rare earth priorities Available online: http://rareearth.ru/ru/pub/20150420/01578.html (2015) (in Russian)
34. Afrikanda: the beginning of history Available online: http://discoverkola.com/afrikanda-nachalo-istorii (2021) (in Russian)
35. J. Dostal, Rare Earth Element Deposits of Alkaline Igneous Rocks, Resources, **6(3)**, p. 34 (2016)
36. J. Snow, et al., The Value of REEs Presentation IAMGOLD Corporation, 24 p. (2012)
37. With rhenium never seen before. How Russia plans to become a leader in the production of rare metals 2020 Available online: https://www.kommersant.ru/doc/4596106 (2020) (in Russian)
38. M.A. Nevskaya and O.A. Marinina, Challenges and opportunities of state regulation of the innovation process in the russian mineral resource sector, *Mining waste products in the mineral resources sector of the Russian Federation*, **16(1)**, pp. 149-159 (2017)

39. T. Ponomarenko and I. Sergeev, Problems and opportunities of mining company's strategic development, *Management Theory and Studies for Rural Business and Infrastructure*, **39(4)**, pp. 453-464 (2017)
40. A.O. Nedosekin, et al., Strategic Approach to Assessing Economic Sustainability Objects of Mineral Resources Sector of Russia, *Journal of Mining Institute*, **237**, pp. 354-360 (2019)
41. G. Pitron, The Rare Metals War: The Dark Side of Clean Energy and Digital Technologies Scribe, 284 p. (2020)
42. C. Blumenthal, The dirty side of clean energy technologies: a call for action Available online: https://www.cleanenergyleaders.org/celi/the-dirty-side-of-clean-energy-technologies-a-call-to-action (2016)
43. Rare earth minerals and the geochemistry: the implications of their respective uranium and thorium contents. Thesis for: BSc Applied GeologyAdvisor: Frances Wall Jonathan O'Callaghan (2012)
44. Best Available Technology Information Technology Handbook 24-2020 Production of rare and rare earth metals Moscow: Russian BAT Bureau, p. 338 (in Russian)
45. Tax Code of the Russian Federation (part two) - CIS Legislation Available online: https://cis-legislation.com/document.fwx?rgn=1717 (2022)
46. Heading to the North: why the Murmansk region is interesting for investors Available online: https://plus.rbc.ru/news/5f887c6a7a8aa9d7d5ef6673 (2020) (in Russian)
47. A.A. Zlobin, New history of development of the Africanda perovskite ores deposit of the Murmansk region Modern science: current issues, achievements and innovations, *XV International Scientific and Practical Conference*, pp. 102-107 (2020) (in Russian)

Associated Petroleum Gas Utilization: New Opportunities for the Oil and Gas Complex in a Circular Economy

Arina Riadinskaia[1,*] and *Alina* Cherepovitsyna[2]

[1]Ph.D, student, St. Petersburg Mining University, 199106, St. Petersburg, Russian Federation
[2]PhD in Economics, Senior Researcher, Luzin Institute for Economic Studies – Subdivision of the Federal Research Centre, Kola Science Centre of the Russian Academy of Science, 184209, Apatity, Russia

Abstract. The topic of the beneficial use of associated petroleum gas (APG) remains relevant due to the limited resources and environmental problems. This is the engine for the active development of management theory and practice with a focus on such concepts as resource saving, energy efficiency and circular economy, which determine the vector of environmentally oriented development of industrial enterprises. In this paper, the existing methods of APG utilization are considered, a qualitative comparison of technologies in terms of such parameters as CAPEX, marketable output, prospects for the sale of commercial products, loss of profits, environmental damage, extraction of petrochemical raw materials from APG, technology implementation period is carried out. Based on the example of the Messoyakha fields, the economic efficiency of APG utilization technologies, which correspond to the characteristics of the field (injection into the reservoir, shallow and deep processing) was calculated. The calculations showed that deep conversion is the most effective method of utilization of APG stored in underground gas storage (UGS) of the Messoyakha fields. This method of APG utilization is integrated into the circular economy system and meets its principles.

1 Introduction

Currently, there is a significant growth of interest in the circular economy in industry, including oil and gas [1]. This trend is due to the fact that natural resources are limited. In addition, the problem of environmental pollution is quite acute.

APG flaring annually accounts for more than 35% of all emissions into the atmosphere of the oil and gas industry as a whole [2]. emissions contain hundreds of thousands of tons of NO (nitrogen oxide), CO (carbon monoxide), about 350 million tons of CO_2 (carbon dioxide), sulfur dioxide and other products of incomplete combustion of hydrocarbons. Another important environmental problem is the melting of Arctic ice due to the burning of APG, which leads to the deposition of a large amount of soot on the snow cover, which increases the absorption of solar energy by Arctic ice [3, 4].

At the same time, APG flaring causes enormous economic damage, since APG is a potential raw material that can be processed to produce marketable products, electricity, or used to maintain reservoir pressure in order to increase oil recovery. In Russia, more than 20 billion m^3 APG are flared annually [2].

The limited resources and environmental problems of industry, including in the oil and gas complex, determine active development of the theory and practice with a focus on such concepts as resource conservation, energy efficiency and circular economy [5]. These approaches determine the vector of environmentally-oriented development of industrial enterprises. The APG utilization methods studied in this paper are an example of circular production chains that correspond to such a vector and ensure minimal impact on the environment.

2 Literature review

Based on the literature review, we can conclude that the number of scientific studies on the topic of circular economy (or closed-loop economy) has increased significantly in recent years. Thus, for the query "circular economy" (keywords), SienceDirect displays 17,148 results for the period since 2018. This topic is explored by Jonas Grafström, Siri Aasma (2021), Roberta De Angelis (2021), Ricardo Weigend (2020) and many others.

Circular economy is also associated with sustainable development of organizations. Murillo Vetroni Barros (2021) discusses this topic in his research.

A significant number of scientific publications are devoted to the topic of the closed cycle of water consumption. For example, C. E. Nika (2020), Massimiliano Sgroi (2018), Nikolaos Voulvoulis (2018).

Circular water consumption also refers to the organization of production processes in oil and gas fields. Most large oil and gas complexes now pay sufficient attention to water purification and reuse (for instance, Gazprom Neft [6], Rosneft [7], Lukoil [8]).

* Corresponding author: arina_german@mail.ru

https://doi.org/10.1051/e3sconf/202337806006

Quite a lot of scientific research is devoted to the topic of APG utilization. Both domestic [9,10] and foreign [11,13] scientists write about the dangers of APG flaring. For example, [2] consider the effectiveness of APG utilization methods in small oil fields; [13] investigate how the role of CO_2 is changing in the transition to a circular economy.

Also, quantitative data and innovative technologies for APG utilization are highlighted in the annual reports of oil and gas companies (for example, [14,15]), Annual Statistical Compilations of the Fuel and Energy Complex [4], research by SIBUR Holding [16].

the literature review revealed that some scientists associate the circular economy and APG utilization in scientific studies [17,18], but, nevertheless, the topic remains understudied. In this regard, the authors made an attempt to investigate methods of APG utilization in the context of circular economy and to present a practical case of the choice of APG utilization method at the Messoyakha fields. The Messoyakha fields are owned by Rosneft and Gazprom Neft (the project operator) and are located in the Tazovsky district.

3 Circular economy in the oil and gas complex

The circular economy is based on the 3R – Reduce, Reuse, and Recycle - principle [19].

In modern economics, the circular economy is understood as a model in which used materials are recycled or released into the biosphere without harmful effects. The tendency to perceive waste not as garbage but as a useful resource is a key feature of the circular economy. The relevance of this approach is also due to the fact that there is a threat of exhaustion of many natural resources.

The introduction of circular economy into production entails changes in economic, environmental and social aspects [20].

Examples of circular production chains in oil and gas production can be water consumption, waste use and beneficial use of APG.

The closed cycle of water consumption eliminates the discharge of contaminated water. construction of large water treatment complexes may allow taking wastewater from various sources for its further use in industrial steam generation, necessary for production of high-viscosity oil [21].

Another option for implementing a circular economy in oil and gas production is the use or sale of company waste. One part of the company's waste is recycled for safety and environmental reasons (e.g., used batteries, fluorescent lamps). Another part (waste of polyethylene, polypropylene, spent catalysts) can be sold as secondary raw materials.

Another option for introducing the circular economy into production is the use of APG. Previously, APG was flared as a by-product of oil production, but there are ways to use APG to reduce emissions and increase the economic efficiency of oil and gas projects (through the sale of the obtained marketable products).

4 Comparative analysis of associated petroleum gas utilization methods

4.1 APG in Russia: measures to reduce the level of APG flaring and statistics

In 2009, a decree "On measures to stimulate the reduction of atmospheric air pollution by products of associated petroleum gas combustion at flare installations" was signed in Russia [23, 24].

This decree establishes the rate of APG flaring at no more than 5% of the total APG production for 2012 and subsequent years. Also, this decree regulates the increase in payments for over-limit APG flaring (an additional factor of 4.5 is introduced) [25].

Figure 1 shows statistics of APG production, APG flaring volumes in Russia, and the APG utilization rate for the period 2000-2019.

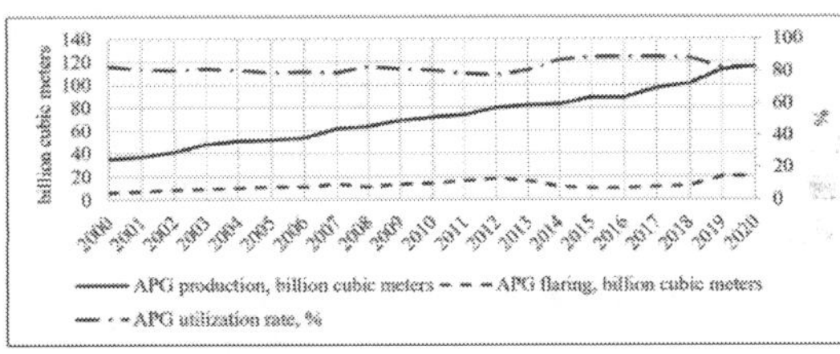

Fig. 1 APG production and flaring in Russia for 2000-2020, billion m^3. The level of APG utilization in Russia in 2000-2020, %. Compiled by the authors on the basis of [4]

During the analyzed period from 2000 to 2020, APG production in Russia is increasing, which is associated with the growth of oil production. From 2012 to 2015, the national average APG utilization rate increased by 13%. The change in this indicator is primarily due to the entry into force of the decree "On measures to stimulate reduction of atmospheric air pollution by combustion products of associated petroleum gas in flares ".

In 2019, the APG utilization rate in Russia was 82.9%, which is 5% lower than in 2018. Thus, the level of APG utilization in Russia had been declining for 2 years. This may be due to the growth of oil production at new fields, which are characterized by an insufficient level of development of the infrastructure necessary for APG utilization.

At the same time, at some large oil and gas fields in Russia the utilization rate reaches 99.6% (Surgetneftegaz) [14]; 97.6% (LUKOIL) [26]; 95.6% (Tatneft) [27].

4.2 APG utilization methods

In world practice today, the main methods of APG utilization are the following [2]: re-injection of APG into the reservoir, injection into the Unified Gas Supply System (UGSS), power generation, simple and deep conversion.

Figure 2 shows APG utilization methods, their description, reflecting environmental and economic aspects of technologies.

Thus, the main methods of APG utilization in the world practice and measures to reduce the level of APG utilization are identified.

APG utilization methods	Description	Economic aspect	Environmental aspect
Re-injection into oil reservoir	This process involves the injection of APG into the gas cap. The goal is to increase oil recovery by increasing the intra-reservoir pressure.	No large capital investments are required. The economic effect depends on the growth of oil recovery in the field.	Environmental impact is considered to be zero.
Injection into the UGSS	Sending part of the APG to the main gas pipeline for sale to consumers. The implementation of such a process is associated with technological limitations (for instance, the distance from the gas pipeline; a certain ratio of APG and natural gas).	Monetizing APG as a conventional fuel gas.	Environmental impact is considered to be zero.
Power generation	APG is sent to gas turbine plants. It can be used in the heat supply systems of the field facilities. When a significant amount of APG is released at the field, the organization of power plants with the delivery of electricity to the regional power supply networks is considered effective.	Sale of electricity or its use for the field's own needs	Gas turbine emissions: NO_2, NO, CO, etc.
Chemical processing (for example, Cyclar and APG in BTK processes)	The Cyclar process is the production of a mixture of aromatic hydrocarbons from the propane-pentane fraction of APG. APG in BTK process: APG is processed into a mixture of aromatic hydrocarbons (for instance, it can be mixed with the main oil stream).	Aromatic hydrocarbon concentrate is a valuable raw material for the petrochemical industry, its independent implementation is possible.	Greenhouse gas emissions CH_4, CO_2 from gas processing plants and petrochemical plants
GTL technology	Technology is based on the Fischer-Tropsch synthesis and the conversion of hydrocarbons into synthesis gas. Syngas is a mixture of carbon dioxide and hydrogen.	Converting gas to liquid to produce liquid hydrocarbons such as naphtha, kerosene, diesel, jet fuel, gasoline, etc.	
Fractional (non-chemical) processing	As a result of fractional processing, the following commercial products are obtained at the output: stable condensate, gas gasoline, propane-butane fraction, etc.	Sales of the received commercial products.	

Fig. 2 Environmental and economic aspects of APG utilization methods. Compiled by the authors based on [16, 28-32]

5 The case of the Messoyakha fields

5.1 Characteristics of the Messoyakha fields

Messoyakhaneftegaz is a joint venture between Rosneft and Gazprom Neft (the project operator). The nearest settlement is the village of Tazovsky, located 140 km away from the Messoyakha fields. The distance to the nearest town (Novy Urengoy) is 340 km.

Production at the Vostochno-Messoyakhskoye field started in 2016, the geological structure of the reservoir enables production at a depth of 800-1000 meters [33].

The gas factor (gas content in the products of oil wells) in the Messoyakha fields averages 360-380 m^3/t. Also, APG is similar in composition to natural gas, since the methane content in APG is 94%.

In 2021, the company commissioned several gas cycle infrastructure facilities at once. At the Vostochno-Messoyakhskoye field there is a compressor station with a preliminary gas discharge unit with annual capacity of 1.5 billion m^3 per year. At the Zapadno-Messoyakhskoye field, there is an underground gas storage (UGS), where APG will be injected until a decision is made on its further useful use. A 47-kilometer gas pipeline will connect these two key facilities.

5.2 Choice of APG utilization methods for the Messoyakha group of fields

Since no decision has been made on the utilization of APG from UGSs at the Messoyakha fields, the authors made an attempt to compare the following methods of APG utilization for the group of Messoyakha fields: re-injection of APG into the oil reservoir, simple conversion and deep conversion.

These 3 methods were chosen for the following reasons:

1. Injection into the Gazprom's UGSS is associated with a number of technological limitations: remoteness from the main gas pipeline (more than 500 km) [35], a certain ratio with natural gas production.

2. Re-injection of APG into the reservoir for enhanced oil recovery is the most widespread method of APG utilization in Russia, therefore it was chosen for comparison.

3. Commercial products obtained after simple and deep APG conversion are considered to be products with high added value.

5.2.1 Simple and deep APG conversion

The process of conversion of APG into gas chemistry products is carried out by separating the gas from water and oil, then it is pre-processed and sent to a gas processing plant (GPP). APG is re-processed and light fractions of gases are extracted from it, such as: hydrocarbons, dry gasoline-free gas, a mixture of methane and propane.

As a result of gas fractionation, gas mixtures are emitted from NGLs, which are subsequently used as fuel or as raw materials for gas chemistry (BGS and LPG). This concludes the simple conversion process (Figure 3).

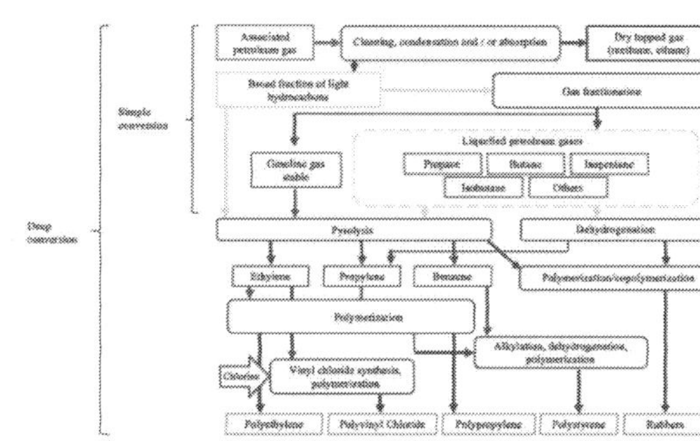

Fig. 3 Associated petroleum gas processing scheme. Compiled by the authors on the basis of [36]

Then, at the GPP, the gas passes through another stage of conversion, which is part of the deep APG conversion. This stage includes pyrolysis, dehydrogenation and polymerization.

5.2.2 Re-injection of APG into the oil reservoir

Due to the pressure difference in the well and in the formation, oil moves to the well. To maintain reservoir pressure, water is injected into the reservoir. But APG can also play the role of the working agent (when injected into the gas cap of the field).

Before re-injection, the APG is separated from the crude oil, it is dried, compressed and is injected into the gas cap in this form (Figure 4).

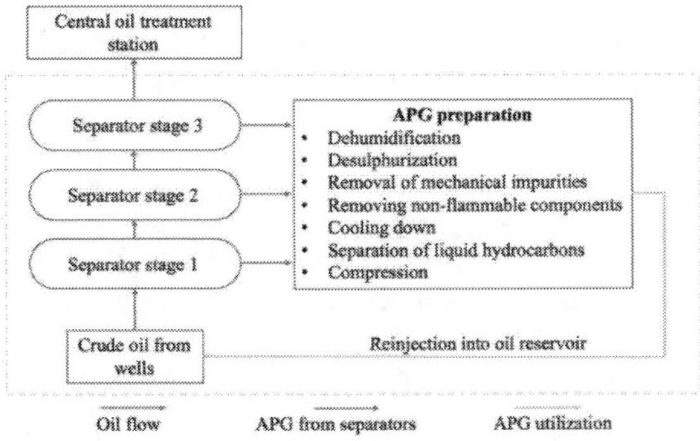

Fig. 4 Technological scheme of APG injection into the reservoir. Compiled by the authors on the basis of [32]

Thus, the characteristics of the Messoyakha fields and possible methods of APG utilization have been identified – APG re-injection into oil reservoir, simple and deep conversion.

5.2.3 Selection of the most effective method of APG utilization for the Messoyakha group of fields

To select the most effective method of APG utilization use at the Messoyakha fields, a qualitative comparative analysis was conducted on such parameters as CAPEX, marketability products, sales prospects, lost profits, environmental damage, extraction of petrochemical raw materials from APG and the technology implementation period (Figure 5). it should be noted that the key parameters are applicability of the technology to the geological conditions of the field, CAPEX, economic benefit, the level of impact on the environment, and the period of implementation of the technology.

Indicators	Re-injection into oil reservoir	Simple conversion	Deep conversion
CAPEX	Construction of a compressor and pipeline facilities	Construction of compressor and pipeline facilities and mini gas fractionating plant	Construction of compressor and pipeline facilities. GPP and gas chemical production.
Market products	Additional volume of hydrocarbon production	Dry gasoline-free gas, LPG and gasoline gas stable	Base polymers (polyethylene and polypropylene), elastomers (polybutadiene), etc.
Sales prospects	Processing at an oil refinery into petroleum products	Sales to petrochemical companies	Automotive industry, construction companies, chemical industry etc.
Lost profits	It is defined as the difference between the values of the economic effects of the most effective method and the calculated one		
Environmental damage	0	Greenhouse gas emissions CO_2, CH_4 from gas processing plants and gas chemical plants	
Extraction of petrochemical raw materials from APG, %	0	Up to 60	Up to 100
Period of technology implementation	Under a year	2-3 years	3-5 years

Fig. 5 Qualitative comparison of APG utilization method. Compiled by the authors on the basis of [16, 32, 37, 38, 39]

Calculations were made based on the following initial data:

1. Discount rate. Since Messoyakhaneftegaz is a joint venture of Gazprom Neft and Rosneft in equal shares, the discount rate is calculated as the weighted average WACC of Gazprom Neft and Rosneft - 10.7%.

2. CAPEX. For reservoir re-injection, CAPEX was calculated by summing the costs of individual units, works and services. When calculating CAPEX for the introduction of simple and deep conversion at the Messoyakha fields, gas processing plants similar in conversion volumes were found, and an amendment was made to the construction of capacities due to the harsh climatic conditions [40].

3. The estimated period is from 2022 to 2043, since according to the project plan, the completion of hydrocarbon production is 2043.

a comparison of disposal methods by quantitative indicators is presented in Table 1.

Table 1. Assessment of the economic efficiency of APG utilization methods. Compiled by the authors.

Indicators	Re-injection into oil reservoir	Simple conversion	Deep conversion
CAPEX, million rub.	458.9	6 000	30 000
NPV, million rub.	88.3	316.7	2920,2
IRR, %	13.9	11.6	12,2
PI	1.19	1.06	1.18
PP, years	14.9	18.9	17.11
Lost profits, million rub.	2 831.9	2 603.5	0

The calculations have shown that for utilization of APG stored in underground gas storages of the Messoyakha deposits, the most effective method is deep processing, which, in its turn, is integrated into the system of the circular economy.

6 Discussion and Conclusion

Thus, methods of APG utilization in the context of circular economy were investigated and the environmental and economic aspects of the main methods were highlighted. The characteristics of the Messoyakha fields were considered, the three most suitable methods of APG utilization were identified, and their economic efficiency was calculated. According to the results of the calculations, the project of deep processing of APG n has the highest NPV indicator (2 920,2 million rubles).

It should be noted that the authors allow an error in the economic calculations performed. This is due to the fact that the initial data were collected from open sources of information, which, in its turn, may be approximate. The authors also considered an ideal situation that takes into account only minimal risks when implementing such technologies at the Messoyakha fields.

References

1. P. Dobson and E. Favaro, The rise of the circular economy in corporate waste disclosures Deloitte Available online: https://www2.deloitte.com/global/en/blog/responsible-business-blog/2020/rise-of-circular-economy-in-corporate-waste-disclosures.html (2020)
2. A.V. Ivanov, et al., environmental and economic justification of the utilization of associated petroleum gas at oil fields of Russian Federation, *Geology and Geophysics of Russian South*, **10(1)**, pp. 114-126, doi: 10.23671/VNC.2020.1.59069 (2020) (in Russian)
3. D.M. Dmitrieva and N.V. Romasheva, Sustainable Development of Oil and Gas Potential of the Arctic and its Shelf Zone: The Role of Innovations, *Journal of Marine Science and Engineering*, **8(12)**, pp. 1 – 18 (2020)
4. Annual Statistical Collection «Fuel and Energy Complex of Russia-2019» 2020 Available online: https://ac.gov.ru/uploads/2-Publications/TEK_annual/TEK.2019.pdf (2020)
5. A.E. Cherepovitsyn, et al., Prospects for the exploration of hydrocarbon deposits in the Arctic based on socio-economic evaluation, *International Journal of Civil Engineering and Technology*, **9(13)**, pp. 938-948 (2018)
6. Gazprom Neft Annual Report 2017 Available online: https://csr2017.gazprom-neft.ru/safe-development/environment-and-resources/water-management (2017)
7. Official Website of Rosneft 2021 Available online: https://www.rosneft.ru/press/news/item/205587/ (2021)
8. Official Website of Lukoil Available online: https://csr2019.lukoil.ru/environmental/water-resource/water-use (2019)
9. A. Vorobyev and E. Shchesnyak, Associated Petroleum Gas Flaring: The Problem and Possible Solution, *14th International Congress for Applied Mineralogy (ICAM2019)*, pp. 227-230, doi: 10.1007/978-3-030-22974-0_55 (2019)
10. I.V. Burenina and G.Z. Mukhametyanova, Russian Gas Utilization: Problems and Prospects, *OGBUS*, **3**, pp. 524-542, doi: 10.17122/ogbus-2015-3-524-542 (2015) (in Russian)
11. E.E. Okoro, et al., Gas flaring, ineffective utilization of energy resource and associated economic impact in Nigeria: Evidence from ARDL and Bayer-Hanck cointegration techniques, *Energy Policy*, 153, 112260, doi: 10.1016/j.enpol.2021.112260 (2021)
12. S. Beigiparast, et al., Flare gas reduction in an olefin plant under different start-up procedures, *Energy*, **214**, 118927, doi: 10.1016/j.energy.2020.118927 (2021)
13. P. Tcvetkov, et al., The Changing Role of CO2 in the Transition to a Circular Economy: Review of Carbon Sequestration Projects, *Sustainability*, **11**, 5834, doi: 10.3390/su11205834 (2019)
14. Annual Report of Gazprom neft 2019 Available online: https://ar2019.gazprom-neft.ru (2019)
15. Annual Report of Surgutneftegaz 2019 Available online: https://www.surgutneftegas.ru/investors/reporting/ (2019)
16. Associated Petroleum Gas Utilisation in Russia 2017 WWF Available online: https://www.sibur.ru/upload/iblock/a70/a70036cc7e90e0b2be004a04efb7bf3a.pdf (2017)
17. D. de Selliers and C. Spataru, The Role of the Oil and Gas Industry in the Transition Towards a Circular Economy, *785th International Conference on Economics, Management and Social Study* (ICEMSS) (2020)
18. Z. Jian and H. Kun, Research on Early-Warning Method and its Application of Complex System of Circular Economy for Oil and Gas Exploitation, *Energy Procedia*, **5**, pp. 2040–2047 doi: 10.1016/j.egypro.2011.03.352 (2011)
19. I.S. Jawahir and R. Bradley, Technological Elements of Circular Economy and the Principles of 6R-Based Closed-loop Material Flow in Sustainable Manufacturing, *Procedia CIRP*, **40**, pp. 103–108 doi: 10.1016/j.procir.2016.01.067 (2016)
20. I. Taranic, et al., Understanding the Circular Economy in Europe, from Resource Efficiency to Sharing Platforms: The CEPS Framework Centre for European Policy Studies, pp. 3-19 doi: 10.13140/RG.2.2.14272.94728 (2016)
21. C.E. Nika, et al., Water Cycle and Circular Economy: Developing a Circularity Assessment Framework for Complex Water Systems, Water *Research*, **187**, 116423, doi: 10.1016/j.watres.2020.116423 (2020)
22. Official Website of Lukoil Available online: https://lukoil.ru (2021)
23. Resolution "On measures to stimulate the reduction of atmospheric air pollution by products of associated petroleum gas combustion at flare installations" Available online: https://docs.cntd.ru/document/902137658 (2021)
24. V.P. Skobelina and I.S. Tremasova, Measures Of Stimulation Of Use Of Associatedpetroleum Gas Use, *Journal of Mining Insitute*, **205**, p. 158 (2013)
25. Associated Profit 2018 Available online: https://www.gazprom-neft.ru/press-center/sibneft-online/archive/2018-june/1715822/ (2018)
26. Annual Report of Lukoil 2019 Available online: https://lukoil.ru/FileSystem/9/469460.pdf (2019) (in Russian)
27. Annual Report of Tatneft 2019 Available online: https://www.tatneft.ru/aktsioneram-i-investoram/raskritie-informatsii/godovie-otcheti/?lang=ru (2019) (in Russian)
28. Solving the problem of APG flaring 2017 Neftegaz.ru Available online: https://neftegaz.ru/science/ecology/331519-

reshenie-problemy-szhiganiya-poputnogo-neftyanogo-gaza/ (2017) (in Russian)

29. V.B. Borugadda, et al., Techno-economic and life-cycle assessment of integrated Fischer-Tropsch process in ethanol industry for bio-diesel and bio-gasoline production, *Energy,* **195**, 116985, doi: 0.1016/j.energy.2020.116985 (2020)
30. V. Rajović et al. 2016 Environmental flows and life cycle assessment of associated petroleum gas utilization via combined heat and power plants and heat boilers at oil fields, *Energy Conversion and Management*, **118**, pp. 96-104, doi: 10.1016/j.enconman.2016.03.084
31. A.P. German, Environmental and economic aspects of APG utilization projects in the circular economy system Ecology and society: balance of interests Available online: http://eco2020.volnc.ru (2020) (in Russian)
32. Technology of re-injection of gas into the oil reservoir 2018 CH Available online: https://www.gazprom-neft.ru/press-center/sibneft-online/archive/2018-june/1715824/ (2018) (in Russian)
33. A.E. Cherepovitsyn, et al., Innovative Approach to the Development of Mineral Raw Materials of the Arctic Zone of the Russian Federation, *Journal of Mining Insitute*, **232**, p. 438, doi: 10.31897/pmi.2018.4.438 (2018)
34. Gas To The Reservoir 2020 Available online: https://messoyaha.neftegaz.ru/?utm_source=article_util&utm_content=ng (2020) (in Russian)
35. Unified Gas Supply System of Russia Available online: https://www.gazprom.ru/about/production/transportation/ (2021) (in Russian)
36. Associated gas of the last stages of separation 2018 Neftegaz.ru Available online: http://energas.ru/pressroom/articles/poputnyy-gaz-poslednikh-stupeney-separatsii-komprimirovanie-nizkonapornogo-png/ (2018) (in Russian)
37. Effective Process 2018 CH Available online: https://www.gazprom-neft.ru/press-center/sibneft-online/archive/2018-april/1533008/ (2018) (in Russian)
38. Sibur Annual Report 2017 Available online: https://www.sibur.ru/upload/pdf/SIBUR-AR-2017-RU.pdf (2017) (in Russian)
39. K.V. Matrokhina, V.Y. Trofimetss, E.B. Mazakov, A.B. Makhovikov, M.M Khaykin, Development of methodology for scenario analysis of investment projects of enterprises of the mineral resource complex, *Journal of Mining Institute*, **259**, pp. 112-124, doi: 10.31897/PMI.2023.3 (2023)
40. A.A. Ilinova and V.M. Solovyova, Technological problems of the development of hydrocarbon shelf resources of the Russian Arctic St. Petersburg, *Oil and gas of Western Siberia. Materials of the International Scientific and Technical Conference*, pp. 104-106 (2017) (in Russian)

Global challenges and opportunities for Arctic oil and gas projects

Alexey Cherepovitsyn[1], *Evgeniya* Rutenko[2,*] and *Sergey* Yudin[3]

[1] Doctor of Economic Sciences, Professor, St. Petersburg Mining University, 199106, St. Petersburg, Russian Federation
[2] Postgraduate, St. Petersburg Mining University, 199106, St. Petersburg, Russian Federation
[3]First Deputy, VIS Group, 121096, Moscow, Russian Federation
Candidate, St. Petersburg Mining University, 199106, St. Petersburg, Russian Federation

Abstract. Integrated development of hydrocarbon potential is one of the high-priority tasks of the government of Russia policy in the Arctic region. When determining the prospects of oil and gas projects, it may be necessary to envisage ways of responding to large-scale changes in the global energy system, which significantly affect the timing, commercial and budgetary efficiency of project activities in the Arctic. The high level of turbulence creates new challenges for the Russian and global oil and gas complex associated with the strengthening of competition in traditional markets of hydrocarbon resources, the progressive development of technological innovation, and changes in the price environment. The main direction of development of the Arctic oil and gas sector should be quick adaptation to new conditions, reassessment of strategic planning systems of key market players, as well as a targeted state policy in relation to the formation of a favorable investment climate and flexible taxation system in the region. For the purpose of the study, an analysis of the factors of instability of the global energy system and their influence on the efficiency of operation of arctic hydrocarbon fields is conducted. The characteristics of the main directions of government policy in the Arctic zone, stimulating the production of hydrocarbons, are given. The potential limitations and opportunities for the implementation of oil and gas projects resulting from the transformation of the global raw materials market are identified and systematized.

1 Introduction

In the long term, the geostrategic interests of the Russian Federation are associated with the unique mineral resource base and economic potential of the Arctic zone, the development of which is important for the effective growth of the country's economy, as well as for ensuring national security and state sovereignty. production of hydrocarbons in the Arctic zone will play an important role in the energy balance of Russia, being the basis for replacing the declining production at existing fields over the time horizon of 2035. In particular, the Arctic shelf can become the main source of oil and gas resources in the future both for Russia and for the world market as a whole [1].

The Russian sector contains about 41% of the world's Arctic oil reserves and 70% of gas reserves [2]. At the same time, the undiscovered potential is 91% on the shelf and 53% on land. While more than 350 oil and gas fields have been discovered in the inland zone of the Russian Arctic, there are about 30 on the Arctic shelf [3], [4].

The main onshore oil and gas reserves are confined to the Yamalo-Nenets Autonomous District. Good geological knowledge, close location to the existing transport infrastructure makes it possible to produce almost 6% of Russian oil, 77% of gas condensate and 81% of natural gas in the region [5]. In addition, the oil producing complex of the Nenets Autonomous District, gas fields of the Krasnoyarsk Krai and the Republic of Sakha (Yakutia) have high production potential [4, 6].

The development of offshore oil and gas fields is at an early stage, however, by 2050, the Arctic shelf, whose reserves are estimated at more than $20 trillion, will provide about 20%-30% of Russian production [6]. At the moment, more than 20 large oil and gas-bearing basins have been identified here. At the same time, hydrocarbon resources are distributed unevenly over the sea territories, and most of them fall on the western Arctic shelf (Kara, Pechora and Barents Seas), which can rightfully be considered a strategic reserve of the oil and gas complex of Russia [2], [7].

Today, the implementation of Arctic oil and gas projects is carried out in the conditions of global instability in the world energy market and structural reorganization of the oil and gas complex. Significant changes are taking place in the energy policy of countries with an emphasis on solving environmental problems and intensifying the use of energy-saving and resource-saving technologies. New key players emerge to begin to compete for traditional markets. Renewable energy sources significantly increase their share in the

* Corresponding author: evgewe@mail.ru

global energy balance [8]. The liquefied natural gas (LNG) market is expanding [9]. There are significant fluctuations in the supply and demand for energy resources, as well as in the price environment. The COVID-19 pandemic is changing the course of the global economy and the prospects of the oil and gas industry. In addition, the efficiency of the oil and gas fields exploitation is currently directly related to the transition to a new technological structure and the introduction of advanced technological solutions, including digital [10].

When developing Arctic oil and gas resources, special attention should be paid to environmental safety. Large-scale development of hydrocarbons, an increase in trans-Arctic traffic have a significant negative impact on the unique natural systems of the Far North. Consistent reduction of the impact on the environment, organization of environmentally friendly production and introduction of the best world practices in the field of environmental safety, compliance with environmental requirements in the extraction and transportation of resources are the targets of oil and gas companies. Climate change is also one of the major global challenges. The climate agenda requires players in the oil and gas market to reduce greenhouse gas emissions and expand the use of renewable energy in production activities. If no decisive action is taken today, subsequent adaptation to climate change will require significantly more effort and costs.

With such turbulence in the global oil and gas industry, the consolidation of efforts of the Russian government and Russian oil and gas companies indicates a disinclination to give up the implementation of complex projects in the Arctic region. It is important to assess and use the opportunities that large-scale market changes provide, as well as to develop measures to minimize the negative consequences of the transformation of the global energy system.

2 Trends in the global oil and gas sector

2.1 Strengthening the US position in the global oil and gas market

Over the past few years, the United States has demonstrated the largest steady growth in hydrocarbon production in the history of the market. to a significant expansion of the mineral resource base allowed for development. Not only has this changed the proportions of the market and the mechanism for setting prices for resources, but also intensified competition for traditional and promising sales markets for Russia. Thus, in 2012, due to an increase in production and a decrease in import of natural gas to the United States due to the "shale revolution", the adoption of an investment decision on the largest gas condensate field in the Russian Arctic – Shtokmanovskoye field - was postponed. global shocks in the oil and gas sector has negatively impacted the US shale segment in 2020. The annual long-term forecast of the World Energy Outlook (WEO) International Energy Agency (IEA) predicts a sharp decline in investments and production volumes for this industry [11]. Nevertheless, the United States remains a key player in the global hydrocarbon production market.

2.2 Expansion of the LNG market

The leading trend in recent years has been the large-scale expansion of the global market for liquefied natural gas. In 2019, Russia became one of the five largest market players, and the Russian Arctic zone has become an important center for LNG production [12]. According to the plans of the Ministry of Energy, the share of Russia in the world market of liquefied gas will increase from the current 4% to 15%-20% by 2035 [13]. At the same time, other large players enter the global LNG market, acting as objective competitors of Russian Arctic projects in the sales markets. For example, Australia and Canada have big plans to increase their LNG production capacities. There are forecasts that by 2035 the US will rank first in LNG production, overtaking Russia, Qatar and Australia [11, 14].

2.3 Updating environmental protection issues

Currently, the structure of global oil and gas consumption is changing significantly under the influence of trends of decarbonization and the struggle for the use of environmentally friendly fuels from renewable energy sources [15, 16]. Thus, the share of renewable energy in electricity generation in Europe in 2020 exceeded the share of fossil fuels for the first time in history; Asia-Pacific Region (APR) plans to increase renewable energy capacity by 50% by 2025; renewable energy by capacity will overtake gas-fired power plants in the United States by 2040. Due to a wide range of government support measures, total investment in global renewable energy industry over the past 15 years has exceeded $4 trillion [11, 14]. The ambitions of many countries to achieve climate neutrality increase the likelihood of introducing a carbon tax on fossil fuel imports in the coming years [17, 18, 19]. Although hydrocarbon fuels are not expected to be completely abandoned in the near future, the development of renewable energy sources can have a significant impact on the implementation of Arctic oil and gas projects focused primarily on export.

2.4 Decrease in global demand for hydrocarbons

According to the results of 2020, the global oil sector suffered more than the global economy as a whole. The world consumption of liquid hydrocarbons decreased by 8-12%, and world GDP – by 5-6% [11, 20]. The decline in oil consumption may be sustainable in the long term, for example, in May 2020, the heads of Royal Dutch Shell and BP declared that global demand may never recover to pre-crisis levels, meaning that peak oil demand has already been reached in 2019 [20]. In this regard, the top scenario of the Energy Strategy of the

Russian Federation until 2035, adopted in June 2020, implies not an increase in oil production, but rather maintaining values at the level of 2018 (555 million tons). The lower scenario implies a decline in production by 11.8% to 490 million tons [13]. The global natural gas market is undergoing negative changes as well. After a significant increase of 5.3% in 2018, global gas demand grew by 1.8% in 2019 and decreased by about 3% in 2020 [20]. This is a revolutionary turnaround in an industry oriented to dynamic growth. According to IEA's forecasts, demand for natural gas will continue to grow at 1.2% per year during the period of 2030-2040 [11], which is slightly lower than pre-crisis forecasts, since gas faces increasing competition from renewable energy sources, which are becoming more competitive in terms of the cost.

2.5 Changes in the price environment for hydrocarbon raw materials

The largest collapse in global oil prices in 2020 was triggered both by outbreak of the COVID-19 affair earlier that year and price wars of the key players. In April 2020, the price fell to an all-time low of $ 8.48 per barrel in 1998. The IEA predicts an increase in oil prices to $ 0 per barrel by 2025 and to $75 by 2030 [11]. A drastic decline in resource prices will adversely affect the launch and implementation of Russian Arctic oil and gas projects, which have high economic and environmental risks [21]. For most fields, the low price combined with high costs will make production unprofitable, even though the presence of fiscal preferences. significant price increases are not expected; this is due to the fact that major producers of conventional oil will act in such a way as to prevent prices from rising to a level that would provoke either a rapid increase in shale production in the United States, or a faster shift of consumers towards other energy sources.

2.6 Decrease in investment to the oil and gas sector

The decline in global demand for hydrocarbons and price volatility increase uncertainty and risks for investors and have an adverse impact on the level of investment attractiveness of the oil and gas industry [22]. According to IEA estimates, spending on hydrocarbons decreased by USD 1 trillion in 2020, which is equivalent to a decrease in market size by about 28%-30%. As a result, by the end of 2020, global investments in the oil and gas sector decreased by 30%-35%. By 2030, against the background of an increase in oil prices to $75 per barrel, investments will start to recover [11]. At the same time, it is expected that the bulk of capital investments will be directed to compensate for the decline in production at the existing fields.

2.7 Development of technologies, digitalization of the mineral and raw materials sector

Hydrocarbon production in the Russian Federation is an area that today is difficult to provide exclusively with domestic technologies today. Moreover, the Arctic region and its extreme climatic conditions impose increased requirements for oil and gas equipment [23]. Currently, Russian subsoil users have a shortage of environmental and commercially viable technological solutions and engineering surveys for geological exploration and development of hydrocarbon deposits, and the US and EU sanctions introduced in 2014 have significantly limited access to foreign technology and investment [24]. In addition, often the matter is in replacing foreign equipment with domestic equipment, but about creating new technologies, since they have no analogues in the world. This requires significant financial resources and time.

2.8 The emergence of new sales markets

The decline in demand for hydrocarbons occurs simultaneously with it shift from developed economies to the countries of the Asia-Pacific region [25]. China, South Korea, Japan and India may become promising export destinations for Russian Arctic resources, which will compensate for the reduction in supplies to the countries of the European region. However, even entering new markets does not guarantee stable exports, so the feasibility of implementing Arctic projects should be considered in terms of future demand for energy resources [26].

3 State policy of the Russian Federation in the Arctic zone

Global challenges and constraints require efficient government policy to support complex oil and gas projects. The complexity of e projects is due to the harsh climatic conditions of the Arctic region, the need to comply with stringent environmental requirements and the use of expensive technology and equipment [27]. This results in high capital intensity of projects and a decrease in investors' interest [28]. underdeveloped or, in some places, completely absent transport and logistics infrastructure also leads to higher costs [29].

The last decade is a time of key decisions for the Arctic zone of the Russian Federation, the implementation of large-scale infrastructure and mining projects, attraction of investments, and the growth of the role of the Northern Sea Route. Consolidation of efforts of state authorities and major oil and gas companies is aimed at developing national programs and strategic documents for efficient development of the resource and economic potential of the region.

Today, Russia is creating the most favorable conditions for doing business and implementing investment projects for oil and gas production in the Arctic region [30]. The country has developed a basic system of tax preferences for Arctic oil and gas projects. In the conditions of low hydrocarbon prices, these measures allow developing previously unprofitable reserves, including the gas condensate part and Achimov deposits of gas fields and oil rims. In addition, the government encourages new approaches to financing Arctic projects. One of these approaches is the introduction of a public-private partnership mechanism. This scheme has already proved its efficiency and effectiveness, especially in terms of constructing infrastructure facilities in the Arctic region. In 2020, a bill was drafted that establishes a new model for the activities of investors on the Arctic shelf. This solution guesses the Norway's experience in state regulation of activities in the Arctic region, allowing foreign and private companies to operate on their sites in consortium with a state operator.

Since 2017, on the basis of the principle of regional and sectoral development, there has been a mechanism to develop support zones in the Arctic region. This mechanism is focused on creating favorable conditions for the implementation of large-scale oil and gas projects along with the development of transport, energy and social infrastructure. The number of investment projects in the field of exploration, exploitation and processing of mineral resources is more than 60% of priority projects within the framework of developing support zones [31].

Also within the framework of the state policy for the development of the Arctic zone, infrastructural restrictions on the growth of mineral resources production will be removed. Today, there is an active development of marine, aviation, and coastal transport in the Arctic, modernization of the infrastructure of Arctic seaports, new port-industrial complexes are being created, and the capacity of sections of the railway network is expanding [32].

In addition, the issue of the producing domestic equipment for production of hydrocarbons is being actively addressed at the state level in Russia. The result is active design and testing of technological systems for exploration, production and processing of minerals in the Arctic. A significant achievement was the creation of prototypes of the first Russian subsea production equipment. The Russian technological base for the LNG liquefaction, storage and transportation, as well as the supply of spare parts, is actively developing. The departure of foreign service companies from the Russian market contributed to the start of intensive activities to build up our own service competencies, search for import substitution opportunities and new partners, primarily in the Asian region. In the last decade, the quick improvement of oil and gas technologies has become a key factor in the development of the industry. Due to the intensified introduction of Russian innovative technologies, including digital technologies, it becomes possible to develop complex Arctic reserves and significantly reduce the costs of hydrocarbon production. In the future, advanced knowledge-intensive developments are able to significantly increase the volume of profitable reserves, reduce the cost of their exploitation and transportation, thereby increasing the benefits of subsoil users and the budgetary efficiency of oil and gas projects.

4 The influence of factors of global instability on the implementation of oil and gas projects in the Arctic region

The analysis of the factors of global instability of the world energy market makes it possible to identify the most adverse factors for the implementation of production in the Arctic, as well as factors that can have a favorable impact (Figure 1).

Global Trends	Challenges and Opportunities for Arctic Oil and Gas Projects
Strengthening competition in hydrocarbon markets	Challenges: - loss of traditional sales markets for Arctic resources; - postponement of the launch of new oil and gas projects.
Expansion of the LNG market	Opportunities: - increasing LNG capacities; - entering new sales markets; - diversification of the activities of oil and gas companies.
Transition to a low-carbon economy, development of RES	Challenges: - decrease in the export of Russian resources; Opportunities: - development of natural gas production projects.
Decrease in global demand for hydrocarbons.	Challenges: - decrease in the export of Russian resources; - postponement of the launch of new oil and gas projects.
Changes in the price environment for hydrocarbon raw materials	Challenges: - decrease in the commercial and budgetary efficiency of existing projects, postponement of the launch of new projects; - price wars between competitors.
Decrease in investment to the oil and gas sector	Challenges: - postponement of the launch of new oil and gas projects; Opportunities: - development of public-private partnerships, attraction of foreign investment.
Development of technological innovations, including digital technologies	Challenges: - changes in the timing of projects due to the lack of Russian technologies and equipment for activity in arctic conditions; Opportunities: - reducing dependence on imported equipment, machinery, technologies, service; - formation of our own scientific and technological groundwork.

Fig. 1. Global trends in the world energy complex and their impact on the implementation of Russian Arctic projects (supplemented by the authors).

5 Conclusions

Thus, today, there is a twofold situation in the implementation of oil and gas projects in the Russian Arctic zone. On the one hand, in the past few years, an intensification of project activity for the development of the hydrocarbon potential of the Arctic zone, supported by state support in the framework of defining the macro-region as the main tool for implementing geopolitical interests and ensuring national security of the Russian Federation has been observed. For these purposes, the Arctic infrastructure facilities are being modernized and updated, a set of incentives for subsoil users is being developed, conditions are being created for investment and business activities in the region, and the Russian Northern Sea Route is being transformed into an international commercial transit route.

On the other hand, serious changes in the organization of the world economy and energy markets, trends of decarbonization, technological transformation and digitalization of the oil and gas complex, strengthening the competitive position of renewable energy in the market can create threats to the stability and sustainable development of capital-intensive, high-risk, requiring technological renewal Arctic projects.

Tough price competition provoked by the imbalance of supply and demand and, as a result, a sharp drop in hydrocarbon prices, and a significant change in the structure of energy consumption, call into question the effectiveness of new project activities in the Arctic and increase the risks of changing the timing and reducing the profitability of current projects. Long-term forecasts do not guarantee a quick stabilization of the market, reducing the investment attractiveness of Arctic projects.

But the low competitiveness of Arctic projects, in particular offshore projects, does not mean that further exploration and development of this extremely promising oil and gas region should be abandoned. In addition, some changes in the global oil and gas market can be used for more efficient development of the resource potential of the Arctic.

For example, Russian Arctic LNG projects have significant prospects. This will be possible due to their favorable geographic location and the use of the shortest delivery route – the Northern Sea Route. In addition, LNG projects have low operating costs and also enjoy tax incentives provided by the state. huge reserves of natural gas and the experience of the Yamal LNG project, which has demonstrated the technological feasibility of building and operating LNG plants in the Arctic, provide increased attention to the region and arouse interest in the implementation of new projects.

The development of renewable energy reduces the demand for hydrocarbons. Of traditional raw materials, natural gas can compete with renewable energy, which is the only energy source that simultaneously ensures energy security and sustainable development on a global scale. This is due to the economic and environmental advantages of natural gas against the background of the growing importance of climate policy in most countries of the world. In this regard, there are significant prospects for the development of Arctic gas fields.

It is also important to take advantage of the growing market opportunities in the Asia-Pacific Region. Expanding the geography of supplies and increasing cooperation with customers from APR countries, whose economies are actively recovering from the consequences of the pandemic, is an important strategic benchmark for of Arctic oil and gas projects, as well as an opportunity to ensure the throughput capacity of the Northern Sea Route.

The main factors of global instability identified in the study, as well as the resulting restrictions on project activities in the Arctic and the emerging prospects, should be taken into account both in the strategic planning of Russian oil and gas companies and in the framework of the state policy of subsurface use in the Arctic zone of the Russian Federation.

The implementation of Arctic projects will be a response to the challenges facing the Russian production sector in the long term. Integrated development of the northern hydrocarbon potential will create an opportunity to form a new production base for the oil and gas industry. In combination with the geographical proximity to the largest sales markets, the EU and. APR, and the developed transport and logistics system, the operation of Arctic fields will increase the competitiveness of the Russian oil and gas industry and ensure the leading position of Russia in the unstable global energy sector.

References

1. D.L. Gautier, et al., Assessment of Undiscovered Oil and Gas in the Arctic, *Science*, **324**, p. 1175-9 (2009)
2. E.S. Balashova and E.A. Gromova, Arctic shelf development as a driver of the progress of the Russian energy system, *MATEC Web of Conf.*, **106**, 06008 (2017)
3. A.A. Ilinova, et al., Arctic oil and gas offshore projects: how to forecast their future, *IOP Conf. Ser.: Earth Environ. Sci.*, **539**, 012153 (2020)
4. A.E. Kontorovich, et al., *Vestn. Ross. Akad. Nauk*, **89**, pp. 1095-1104 (2019) (in Russian)
5. Project of Information Agency Neftegaz.RU: Yamal, the Heart of the Oil and Gas Industry of Russia Available online: https://yanao.neftegaz.ru/en/history/ (2021)
6. WMT Consult Research 2019 The Arctic: how to wake up a "sleeping" region Available online: http://www.ngv.ru/magazines/article/arktika-kak-razbudit-spyashchiy-region/ (2019)
7. B.A. Nikitin and A.D. Dzublo, Prospects for the development of gas resources of the shelf of the Arctic seas of Russia, *News of the Gas Science*, **4(32)**, pp. 15-24 (2017) (in Russian)
8. B.W. Brook, et al., Silver Buckshot or Bullet: Is a Future "Energy Mix" Necessary?, *Sustainability*, **10(2)**, 302, doi: 10.3390/su10020302 (2018)
9. R.M. Zhdanovskih, Expansion of international trade in Russian gas due to its diversification and use of innovative directions, *Int. Trade and Trade Policy*, **2**, pp. 133-148 (2018)
10. D. Dmitrieva and N. Romasheva, Sustainable Development of Oil and Gas Potential of the Arctic and Its Shelf Zone: The Role of Innovations, *J. Mar. Sci. Eng.*, **8**, p. 1003, doi: 10.3390/jmse8121003 (2020)
11. World Energy Outlook International Energy Agency IEA Available online: https://www.iea.org/topics/world-energy-outlook (2021)
12. J. Henderson and V. Yermakov, *Russian LNG: Becoming a global force* (Oxford Institute for Energy Studies, Oxford, 2019)
13. Energy Strategy of the Russian Federation until 2035 Available online: http://static.government.ru/media/files/w4sigFOiDjGVDYT4IgsApssm6mZRb7wx.pdf (2021)
14. International Energy Outlook U.S. Energy Information Administration's EIA Available online: https://www.eia.gov/outlooks/ieo/ (2021)
15. D. Gielen, et al., The role of renewable energy in the global energy transformation, *Ener. Strategy Rev.*, **24**, pp. 38-50, doi: 10.1016/j.esr.2019.01.006 (2019)
16. M.J. Pickl, The renewable energy strategies of oil majors – From oil to energy?, *Ener. Strategy Rev.*,

26, 100370, ttps://doi.org/10.1016/j.esr.2019.100370 (2019)

17. M. Grayson, On a global scale, reducing fossil-fuel dependence is more than just a technological challenge, *Nature*, **551,** S133 doi: 10.1038/d41586-017-07507-y (2017)
18. B.P. Heard, et al., Burden of proof: A comprehensive review of the feasibility of 100% renewable-electricity systems, *Renew. Sustain. Energy Rev.*, **76**, pp. 1122–1133 doi: 10.1016/j.rser.2017.03.114 (2017)
19. A. Bourgain and S. Zanaj, Tax Competition Approach to Resource Taxation in Developing Countries, *Res. Policy*, **65**, 101519, doi: 10.1016/j.resourpol.2019.101519 (2020)
20. BP Energy Outlook 2020 Available online: https://www.bp.com/en/global/corporate/energy-economics/energy-outlook.html (2020)
21. A. Cherepovitsyn, et al., Innovative approach to the development of the mineral resource potential of the Arctic zone of the Russian Federation, *Jour. of Min. Inst.*, **232**, pp. 438-444 (2018)
22. A.M. Fadeev, et al., Features of strategic management of the oil and gas complex and transportation of hydrocarbon products in the development of offshore oil and gas fields in the Arctic, *Vest. MGTU*, **20(4)**, pp. 742-754 (2017) (in Russian)
23. A.A. Ilinova and V.M. Solovyova, Technological aspects of development of Russian Arctic shelf resources, *The North and the Market: Form. the Econ. Order*, **4(60)**, pp. 32-42 (2018)(in Russian)
24. A.Y. Nikulina and M.N. Kruk, Impact of Sanctions of European Union and Unites States of America on the Development of Russian Oil and Gaz Complex, *Int. J. Econ. Fin. Issues*, **6(4)**, pp. 1379-1382 (2016)
25. R.G. Bertelsen and V. Gallucci, The return of China, post-Cold War Russia, and the Arctic: Changes on land and at sea, *Mar. Policy*, **72**, pp. 240-245, doi:10.1016/j.marpol.2016.04.034 (2016)
26. E.G. Katysheva, Developing Gas Fields in the Yamal Peninsula as a Factor of Economic Development of the Arctic Zone of Russia, *IOP Conf. Ser.: Earth Environ. Sci.*, **302**, 012127 (2019)
27. M.B. Alekseeva, et al., Systemic Diagnostics of the Arctic Industry Development Strategy, *Jour. of Min. Inst.*, **238**, pp. 450-458 (2019)
28. K.V. Matrokhina., V.Y. Trofimets, E.B. Mazakov, A.B. Makhovikov and M.M. Khaykin, Development of methodology for scenario analysis of investment projects of enterprises of the mineral resource complex, *Journal of Mining Institute*, **259**, pp. 112-124, doi: 10.31897/PMI.2023.3 (2023)
29. A.M. Fadeev, The shelf has a bright future Available online: https://pro-arctic.ru/10/03/2016/expert/20514 (2016) (in Russian)
30. S.A. Agarkov, et al., Major trends in efficiency upgrading of the economic activity in the Arctic zone of Russian Federation, *Jour. of Min. Inst.*, **230**, pp. 209-216 (2018)
31. State Program Socio-economic Development of the Arctic Zone of the Russian Federation Available online: http://government.ru/rugovclassifier/830/events/ (2021)
32. E.G. Katysheva, The role of the Northern Sea Route in Russian LNG Projects Development, *IOP Conf. Ser.: Earth Environ. Sci.*, **180**, 012008 (2018)

Arctic climate policy development: taking public beliefs into account

Elena Klyuchnikova[1,*] and *Gleb* Klyuchnikov[2,*]

[1] PhD (economic), head of department of international relation, Institute of Industrial Ecology of the North, Kola Research Center at the Russian Academy of Sciences, 184209, Apatity, Russian Federation
[2] Journalist, State-financed organization Gorod+, 191015, Sankt Petersburg, Russian Federation

Abstract. For the effective implementation of climate policy in the Arctic, public beliefs must be taken into account in its development. The research presented in this report shows that the majority are inclined to believe that climate change is occurring, where the younger generation associates these changes with anthropogenic impact, while the older respondents with natural causes. The majority are confident that climate change in the Russian Arctic will bring positive change, increase the standard of living in the Arctic, improve conditions for farming, and increase the intensity of traffic along the Northern Sea Route. Thus, the population is not ready to bear the possible costs when the climate policy is tightened; people's expectations are aimed at increasing the comfort of living in the Arctic.

1 Introduction

Public discourse about climate change is based on the fact that climate change is global in nature and brings economic, social, and environmental risks. Therefore, one of the global goals of sustainable development [1] is taking urgent action to combat climate change and its impacts. In the Arctic, the climate is changing faster than in other global regions. Over the past 20-25 years, the area and thickness of the Arctic sea ice has been decreasing [2]. Indicators describing extreme heat in winter continue to grow, and it is predicted that in 10-15 years the average winter temperature will be 4-5 °C higher than in the late 20th century [3]. Based on the projected climate changes in the Arctic, such risks are predicted as more frequent catastrophic weather events, degradation of permafrost, long-term biodiversity stress, and threats to public health [4]. It is well known that concepts adopted by the public have an influence on the practice through the policy choices made by governments, companies, research organizations, NGOs. This was stressed by John Maynard Keynes in his "General Theory of Employment, Interest and Money": "The ideas of economists and political thinkers — both when they are right and when they are wrong — matter much more than is commonly thought. In reality, they alone rule the world" [5]. Therefore, it is relevant to understand public beliefs about the ongoing climate change. This knowledge will help predict the success of public policy implementation. Moreover, adaptation measures are already being developed, such as the creation of technologies to reduce the risks from global climate change and ensure the stability of remote Arctic facilities located in the zone of permafrost spreading [6]. Many authors predict greater use of carbon-free energy sources in the Arctic [7] and increasing economic attractiveness of measures to mitigate the impact on climate [8].

In frame of the study 40 semi-structured interviews were conducted with 40 experts (scientists, journalists, workers of mining enterprises, students of technical specialties) aged 23 to 73 years. The experts were selected based on the intention to reveal the beliefs of specialists and decision-makers in the mining industry, which forms the backbone of the Russian Arctic economy. The survey involved residents of Russia's Murmansk Region, Krasnoyarsk Territory, and Yamalo-Nenets Autonomous District. The responses were analyzed according to two criteria — age group and occupation. Age group was determined based on the definition of climate: "climate is long-term (several decades) weather" [9]. It was assumed that "several decades" means thirty years. Therefore, three age groups were identified: aged 23-30 years, 30-60 years, and over 60 years old.

2 Climate change

* Corresponding author: e.klyuchnikova@ksc.ru, gleb.klyuchnikov97@yandex.ru

It was found that 62% are convinced that the climate is currently changing, 10% that it is not changing, and 28% are not sure or do not have their own point of view.

A higher percentage of those having doubts was found among students aged 25 to 30 years (55%). The majority of surveyed researchers are convinced that the climate is changing. Almost all respondents aged 23-30 are confident that the climate is warming and that the process is linear. In the second age group (30-60 years old), the prevailing opinion is that the climate changes cyclically. In this age group, three opinions were found that the climate is too unstable to highlight specific repetitive patterns or a single trend. In the older age group, the respondents believe that it is impossible to trace the cyclical nature of climate change over a human life. Although, in general, they tend to believe that the change is cyclical: "I believe that it changes cyclically, but this cyclical pattern is long-term. In a short-term perspective, climate change tends to unfold for the worse for humanity. But the global cycles are very long-term — here I mean natural cycles."

The opinions of the interviewed researchers about the causes of climate change also differed. In the first age group, regardless of profession and occupation, 90% of respondents believe that the climate is changing due to anthropogenic causes. In the second age group, there is a diversity of opinions: eight respondents (six researchers, one student, and one journalist) put the anthropogenic factor first place in the list of causes of climate change. Twelve survey participants are inclined to believe that the anthropogenic factor is either overestimated or does not lead to climate change at all (eight scientists and four journalists). Another two respondents consider both factors to be responsible for climate change (one student and one journalist).

Thus, there is no consensus among the respondents about the existence of climate change, the direction of climate change, and the cause thereof. Despite the fact that the majority of respondents believe that the climate changes cyclically, they found it difficult to indicate the duration of one climatic cycle. Only one stable tendency was revealed — the younger the respondents, the more of them believe in anthropogenic causes of climate change, while the older generation is inclined to believe that if the climate is changing, it is due to natural reasons.

3 Economic implications

The second group of questions was devoted to economics. The respondents were asked how climate change is expected to affect the Arctic economy and what positive or negative consequences that can have. It should be noted that almost all respondents agreed that climate warming is currently taking place, regardless of the cyclical or linear nature of the process.

Among 39 respondents, 27 admitted the impact of climate change on the economy of the Arctic. In the first age group, half of the respondents did that, in the second age group the impact of climate change on the economy was admitted by 78%, and in the third age group, 66% believe that climate change will affect the economy of the region.

Respondents who do not consider climate change to be a factor affecting the economy explained this by their lack of knowledge on this issue, or by the fact that political forces influence the economy much more strongly: "I think the economy will not be particularly affected, unless the people who set the political agenda decide that climate change should somehow affect the economy," said one respondent.

Among the 27 respondents who answered affirmatively to the question about the capability of climate change to affect the Arctic economy, 15 respondents (56%) positively assess the hypothetical warming in the region. The main benefits for the economy, according to the respondents, will be the improvement of the situation in the agricultural industry. This opinion is shared by six representatives of the second age group, two representatives of the first age group, and one of the oldest age group. "Of course it will. Even during the harsh climate period, in the 1970s and 1980s, agriculture, which used unpaid labor, flourished. Murmansk Region achieved 80% self-sufficiency in agricultural products. Obviously, agriculture in this region will be more profitable, even without unpaid labor," said a mining industry employee belonging to the third age group.

Some of the respondents pin great hopes on the growth of traffic along the Northern Sea Route — this expected advantage was noted by six respondents, among them two journalists (from age groups 2 and 3) and four researchers (three from age group 2 and one from age group 3): "So far, we see that year-round navigation along the Northern Sea Route has not yet opened, but if there is climate warming, the route will be easier to navigate, which will lead to an increase in ship traffic," replied a journalist from age group 3.

The most important challenge for the economy caused by climate change, according to the respondents, is the melting of permafrost. This problem is especially acute for the residents of Norilsk — all respondents from Krasnoyarsk Territory named this problem. "We see facades collapsing due to soil thawing — this is a problem. A lot of money will have to be invested in stabilizing the foundations. Since the city was designed assuming permafrost, looks like it is suffering. Many buildings are falling apart. If this continues, the city may cease to exist," noted a journalist from Norilsk.

Thus, respondents, regardless of age or profession, for the most part, positively assess the possible economic consequences of climate change in the Arctic. It is assumed that warming will cause melting of glaciers, which, in turn, will increase cargo traffic along the Northern Sea Route. Positive changes for agriculture are also stressed. In the opinion of the majority of respondents, climate change will expand the arable land range beyond the Arctic Circle, and will make it possible to grow a number of crops outside of greenhouses.

4 Sources of information

Since the public perception of climate change is studied, it is important to find out from which sources respondents receive information about climate change, who they consider to be experts in this issue, and which sources they trust most.

Our study shows that the younger age group is more inclined to consider scientists as subject-matter experts — this opinion is shared by four students. The only researcher in this age group believes that Green Peace activists are experts in this field, but he does not follow their work. A mining industry employee considers Discovery Channel and National Geographic TV show hosts as subject-matter experts. Another student thinks that relevant international organizations are experts on climate change. Three survey participants from this age group did give any answer.

The second age group, perhaps due to the prevailing number of journalists and researchers, believes that climatologists and other related scientists are experts in the field of climate. Another six respondents were unable to give an answer due to either lack of information or mistrust of any unfamiliar sources of information. Only two respondents from the youngest age group named activists or opinion leaders (e.g. Greta Thunberg) as experts on climate change.

In the older age group, almost all respondents named researchers or relevant scientific organizations as climate experts.

Concerning trust (or mistrust) in information sources, the respondents were more unanimous — 69% (27 respondents) of them believe that only scientific publications can be trusted. At the same time, the percentage of those trusting scientific sources noticeably prevails in each group over the rest — 60% in age group 1, 70% in age group 2, and 90% in age group 3.

It is important to note that when asked specific questions about scientific sources of information, many respondents could not name specific articles, publications, journals, etc. It can be concluded that the very fact of the presence of the word "scientific" significantly increases the level tolerance in the perception of information about the climate.

Distrust in any sources of information about the climate was expressed by four people - one researcher and one student from age group 1, and one researcher and one journalist from the age group 2. They refer to the bias of scientific research in the field of climate: "All scientific articles on climate change, in my opinion, are highly politicized. Therefore, many of them cannot be considered independent. I would be very wary of any article related to climate change," noted one of the respondents.

Among trustworthy sources of information on climate change, mass media, such as groups on the social media platform VKontakte, Telegram channels, or even the Google search engine, were singled out — this is the opinion of one young student, one student from age group 2, and a young mining industry employee. Two respondents — one mining employee from the older age group and a researcher from age group 2 — responded that one can trust colleagues or friends in this matter. At the same time, the mining industry employee believes that the scientific literature on this topic is biased. Information from official sources (regional, municipal, or corporation leadership) was considered trustworthy by only one respondent — a researcher from age group 2. Mass media (meaning conventional media such as print, radio, television, licensed Internet publications) were described as a trustworthy source of information about the climate by only one respondent — a researcher from age group 2.

Thus, we can conclude that people, regardless of profession and age, are willing to trust scientific sources and scientists in matters of climate change. At the same time, it should be noted that the majority of respondents are not directly interested in scientific literature on climate. They trust the information they receive from scientists. Also, regardless of age, respondents do not trust conventional mass media.

5 Conclusions

Thus, the study showed that young people mostly believe in climate warming, believe that the climate is changing under the influence of anthropogenic factors. Opinions about the direction of climate change divided almost equally — 60% believe that it moves linearly towards a global warming, while 40% believe that it has a cyclical nature. Exactly half of the respondents believe that climate change will affect the economy of the Arctic. The majority were unable to name any positive aspects of this process. Among the young respondents, there is no conviction that the climate discussion is in any way biased — this is supported by the fact that some respondents are willing to trust activists, international organizations, and always scientists. Only two out of ten respondents did not trust anyone in this matter. They get updates on the issue from the Internet and, very rarely, from conventional mass media.

The age group 30-60 predominantly believe that the climate changes cyclically. This is the opinion shared by scientists. Among those who believe that the climate is changing in a single direction, the majority are journalists. Most of the respondents are inclined to believe that the anthropogenic factor does not affect the climate as much as natural processes, however, almost 40% of the anthropogenic factor supporters are students and researchers. The impact of climate change on the economy of the Arctic, in the opinion of the majority of representatives of the second group, is positive, mainly in agriculture. Since in this age group researchers represent the greatest share, they can only trust their colleagues. Same opinion is shared by journalists and students.

The older age group believes that the human impact on the climate is overestimated, but the climate itself is changing towards warming. Among the benefits for the economy of the Arctic, they also highlight agriculture and an increase in the ship traffic along the Northern Sea Route. Older respondents trusted only researchers. The older generation prefers to get updates on climate change from scientific articles, ignoring mass media, social media, and other sources.

Thus, the population is not ready to bear the possible costs when the climate policy is tightened; people's expectations are aimed at increasing the comfort of living in the Arctic.

This research was funded by Russian government grant no. 0226-2019-0047.

References

1. UN. Sustainable Development Goals Available online: https://www.un.org/sustainabledevelopment/ru/climate-change (2021)
2. Second assessment report Climate Change in the Russian Federation Available online: http://voeikovmgo.ru/download/2014/od/resume_teh.pdfLink (2014)
3. Annual reports on the state of the climate in the Russian Federation, Available online: http://climatechange.igce.ru/index.php?option=com_docman&Itemid=73&gid=27&lang=ru (2021)
4. V.M. Katsov and B.N. Porfiriev, Climatic change in the Arctic: consequences for the environment and economy, *Arctic: Environment and Economics*, **2(6)**, pp. 66 Available from: http://www.ibrae.ac.ru/docs/2(6)/66-79.pdf (2012)
5. J.M. Keynes, General theory of employment, interest and money (Moscow, Progress, 1978) p.494
6. G. Buslaev, et al., Ensuring the sustainability of arctic industrial facilities under conditions of global climate change, *Resources*, **10(12)**, 128, doi:10.3390/resources10120128 (2021)
7. Y. Zhukovskiy, et al., Scenario modeling of sustainable development of energy supply in the Arctic, Resources, **10(12)**, 124, doi: 10.3390/resources10120124 (2021)
8. P.S. Tcvetkov, et al., The Changing Role of CO2 in the Transition to a Circular Economy: Review of Carbon Sequestration Projects, *Sustainability*, **1(20)**, 5834; doi: 10.3390/su11205834 (2019)
9. O.A. Drozdov, et al, Climatology (Leningrad, Gidrometizdat, 1989) p. 568

The Karelians: securing the right to preserve traditional habitats (experience of the republic of Karelia)

Svetlana Yalovitsyna[1,*]

[1] PhD in History, Vice-Director for Science, Institute of Linguistics, Literature and History of the Karelian Scientific Center of the Russian Academy of Sciences, 185035, Petrozavodsk, Russian Federation

Abstract. The article is devoted to the problem of preserving the places of traditional residence of Karelians in modern conditions of significant reduction in the number of Karelians and dispersed residence in the Republic of Karelia (RK). As one of the tools to solve this problem the method of creating the List of Settlements of Traditional Karelian Residence (STKR) in the Republic of Karelia (RK) is considered. The article presents the current regulatory framework of the Russian Federation on the issues of indigenous minorities, including legislative acts and other regulatory documents concerning the development of the Arctic zone of the Russian Federation (AZRF), which includes six municipalities of the Republic of Karelia, inhabited by, among others, Karelians. Historical documents that illustrate the connection of specific territories with the Karelian tradition, including the arctic territories are used. The principles underlying the creation of the List of Settlements of Traditional Karelian Residence and examples of using the List to determine the boundaries of economic activities of logging and fish farming enterprises are listed. application of the List is aimed at creating conditions for sustainable development of Karelians in the Republic of Karelia. One of the important factors is the cultural continuity of the way of life. Violation of the principle of gradualness when introducing innovations, the lack of an adaptation period to build a balance with traditions can lead to unpredictable consequences for society.

1 Introduction

The stated topic may cause confusion to an inexperienced reader. What is the meaning of the existence of such a List of Settlements of Traditional Karelian Residence? [1]

The notion "List of places of traditional residency and traditional economic activity of indigenous minorities of the Russian Federation" is one of the main concepts in the field of legislation related to the protection of the rights of indigenous minorities to their native habitat. The law "On guarantees of the rights of indigenous minorities of the RF" understands as the ancestral habitat "the historically established area within which indigenous minorities carry out their cultural and household activities and which influences their self-identification and way of life" [2]. The Constitution of the RF classifies the "-protection of the original habitat and the traditional way of life of small ethnic communities" as the joint jurisdiction of the Russian Federation and the subjects of the federation (Article 72, item "m"). Therefore, the List is approved by the Government of the RF on the basis of documents submitted by the authorities of the constituent entities of the RF, where indigenous minorities live.

The main source for the preparation of such materials by the RF subjects is census data on the settlement of indigenous numerically small peoples of Russia in the regions with justification of their traditional habitation in administrative districts, municipalities, and settlements. The consolidated list of places of traditional residence of indigenous peoples is approved by a decree of the Government of the Russian Federation [3]. Ethnological expertise at the federal level is entrusted to the Institute of Ethnology and Anthropology of the Russian Academy of Sciences. When forming the List of places of traditional residence of small indigenous peoples (hereinafter referred to as the Federal List), the criterion of their numbers in administrative units is not established; the substantiation of their primordial habitation in these territories is of decisive importance.

2 Research

The definition of territories of traditional residence of indigenous minorities is fundamental for the implementation of the collective rights of these peoples: from the protection of their ancestral habitat and control over industrial activities on their territories to claims for compensation for damage caused to the ancestral habitat as a result of industrial activities. Representatives of indigenous peoples who live in such territories are provided with a number of benefits and state support measures.

* Corresponding author: jalov@yandex.ru

Article 72, paragraph "m" of the RF Constitution speaks of the joint authority of the federal centre and its subjects to "protect the native habitat and traditional way of life of indigenous ethnic communities living in the territories of these subjects", without specifying that only the indigenous minorities of the Russian Federation included in the list established by the government of the Russian Federation may be considered as such. The Karelian people may be rightfully considered an indigenous ethnic community, whose native habitat is partially located in Karelia. While the total number of Karelian population in the country is 60.8 thousand people according to the 2010 census, 45.6 thousand of them live in Karelia, which is below the benchmark (50 thousand) established by the federal law "On guarantees of the rights of small indigenous RF peoples of Russia" to classify peoples as indigenous minorities of the RF.

An additional reason for the special legal regime in places of traditional residence of Karelians is the inclusion of six municipalities of Karelia into the Arctic zone of the Russian Federation (AZRF). In 2020, the law "On state support of entrepreneurial activity in the Arctic zone of the Russian Federation" [4] was adopted, and the Ministry for the Development of the Far East and the Arctic by order of the Russian Government prepared a standard of accountability for interaction between the residents of the Arctic zone of the Russian Federation and indigenous peoples. The authorities of Karelia, relying on the Karelian public initiative, may start developing a legal act "On the special legal regime of the places of traditional residence of Karelians in the Russian Arctic." Currently, the Ministry of the Russian Federation for the Development of the Far East and the Arctic is working on the "Program to support the economic activities of the small indigenous peoples of the North, Siberia and the Far East of the Russian Federation living in the Arctic zone of the Russian Federation (2021-2024) [5]. The Government of Karelia may come up with an initiative to extend its effect to the Karelian population of the regions included in the Arctic zone. In this regard, in 2020, the necessity arose to create a regional List of Settlements of Traditional Karelian Residence (hereinafter referred to as the STRK List or the List), which turned out to be not quite a trivial task. Below we will elaborate on the essence of the problems that have arose.

The current number of Karelians, the indigenous population of the Republic of Karelia, has significantly decreased. According to the 2010 census, their share was only 7.4%, while in the 1920s they accounted for 37.4%. Compared with the 1989 map of the settlement of the peoples of Karelia, (Figure 1), only three districts regions with predominantly Karelian population remained - Olonetsky, Pryazhinsky and Kalevalsky.

Karelians are characterized by dispersed living on the territory of the republic, which developed quite a long time ago. If we look at the ethnic map of the 1920s. (Figure 2), we can see that Karelian settlements gravitate to the western borders of the republic, which partly remains to this day.

The post-Soviet formation of three national regions (Kalevalsky, Olonetsky, Pryazhinsky) did not lead to a concentration of the Karelian population in these regions, and the dispersion remained.

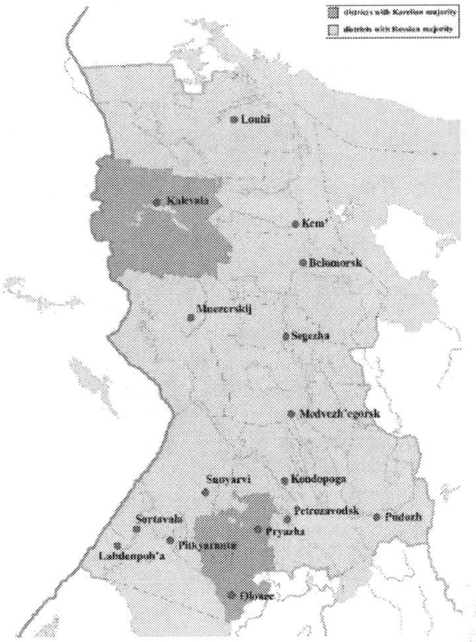

Fig. 1. Settlement of the peoples of Karelia defined by the ethnic majority (according to the 1989 census)

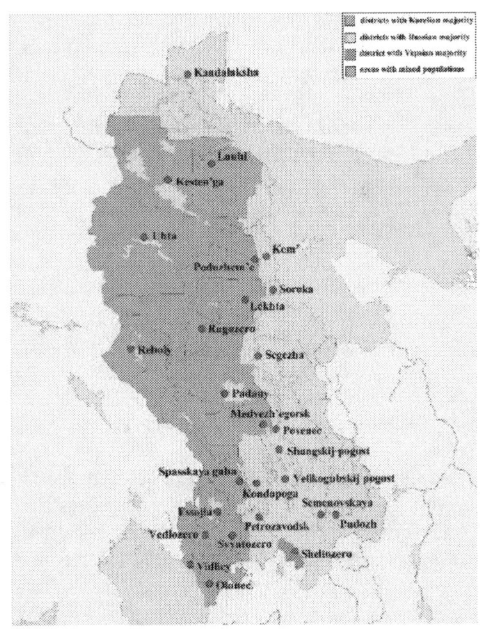

Fig. 2. Settlement of the peoples of Karelia defined the ethnic majority (according to the 1926 9 census)

The percentage of interethnic marriages among Karelians in the Soviet period was quite high. This was due to urbanization processes, resettlement (e.g., eviction of Karelians from border areas; evacuation during the Second World War; recruitment of workers to the timber industry enterprises in Karelia, etc.), unstable national and language policy, which experienced waves of internationalization and return to ethnicity [6].

The above-mentioned circumstances led to the fact that Karelians lost their linguistic competence. Their status as the titular ethnic group in the republic had to be defended. An example of this is the activities of the Congress of Karelians of the Republic of Karelia, and the fact that within the period of its 30-year existence (1990-2020), the issue of the state status of the Karelian language in the republic was not settled [7].

These conditions have formed real threats of loss of Karelians' ethnic identity. Research data preceding the 2021 population census show a high probability that the number of Karelians in the Russian Federation will be less than 50 thousand people. In this case, in accordance with the law "On guarantees of the rights of small indigenous peoples of the Russian Federation" of 1999, they would be entitled to the status of an indigenous small-numbered people. These peoples include those living in the territories of the traditional settlement of their ancestors, preserving their traditional way of life, economy and crafts [8] and distinguishing themselves as independent ethnic communities. Ethnos, once titular for the republic, will change its status to that of an indigenous minority, whose rights are subject to special protection in accordance with international and Russian legislation.

The preferences related to the status of an "indigenous minority" can be implemented provided that their allocation is targeted. This determines the need to create the List of Settlements of Traditional Karelian Residence.

Historical conditions of the formation of Karelian population in the republic, briefly described above, prove that solving the problem of creating the List of STRK is complicated. Along with the dispersion of residence, it is problematic to assess the current situation of Karelians settlement, since ethnicity is not currently documented and estimates of the ethnic composition can only be made on the basis of census data. Regulatory documents include the term "traditional residence", i.e., it is not about identifying the places of current COMPACT residence of Karelians, but about the existing TRADITION linking the place and the people.

These difficulties, as well as the practical significance of solving these issues, determined a wide range of historical sources used for compiling the List. Thanks to their informative value, the materials used in the zoning of the republic in the 1920s turned out to be the most useful for the solution of this problem. In Karelia, the preparations for the censuses of 1920 and 1926 were very thorough. The head of the republic E.A. Gylling direct participated in it. He established close ties with D.A. Zolotarev, who was responsible for the work of the European Department of the Commission for the Study of the Tribal Composition of the USSR Population. A unique reference book "The List of Populated Areas of the Karelian Autonomous Soviet Socialist Republic" was prepared based on the results of the census, which, in its turn, became the foundation for the project of zoning of Karelia and fully represented the ethnic composition of the regions [9].

In the process of its creation the results of the 1926 census materials were recalculated (since they had been collected according to the old administrative division – "uezd" and "volost") according to the newly formed in 1927 administrative districts, which allowed obtaining data comparable with the subsequent censuses.

The information in the reference book is structured by districts, village councils and settlements. Types of households, occupation, gender, and ethnicity were taken into account (only the major ones: Karelians, Finns and Russians). The List is unique as it reports the names of the settlements of the republic, both official and those used by the local population, with maximum completeness. The complexity of documenting the names of settlements in the so-called nesting type of settlement was also taken into account.

In total, 26 districts were identified. Twelve districts were considered as ethnically Karelian (Kestensky, Ukhtinsky, Kemiretsky, Rugozersky, Rebolsky, Tungudsky, Segozersky, Petrovsky, Syamozersky, Svyatozersky, Vidlitsky, Olonetsky), one Vepsian (Sheltozersky), eleven Russian (Pronezhsky, Povenetsky, Pudozhsky, Segezhsky, Soroksky, Kemsky, Kondopozhsky, Kandalakshsky, and two mixed (Loukhsky and Medvezhyegorsky). This historical source became the foundation for the formation of the STRK List, which was approved by the Order of the Government of the Republic of Karelia in July 2021.

On the basis of this document the settlements (places) where the share of the Karelian population exceeded the republic's average of 37.4% (1926) were identified. Those settlements were included in the 2021 STRK List, regardless of whether Karelians were still a significant group there. The 1928 reference book [10] included villages and isolated farmsteads, which for various historical reasons ceased to exist, which did not allow them to be reflected in the List-created on the basis of the currently existing settlements. However, given that their boundaries on modern cadastral plans also cover adjacent territories, in most cases the disappeared villages also happened to be included in the STRK number.

Thus, several principles were used when compiling the STRK List in the Republic of Karelia.

1. combination of principles of inclusion of both indigenous regions and specific places of traditional residence of Karelians

2. consideration of the information about the primordial residence of Karelians within specific settlements in the 1920s, when Karelia received its own statehood.

3 Conclusions

We believe that the approval of the List will become the initial measure aimed at creation of conditions for sustainable development of Karelians in the Republic of Karelia.

The List may be used to regulate the protection of the linguistic and cultural rights of minority peoples a (Constitution of the Russian Federation: Articles 68, 69, 71, 72); to protect human and civil rights and freedoms; to protect the rights of ethnic minorities (Article 72, paragraph "b"); to protect the original habitat and traditional way of life of ethnic minority communities (Art. 72, paragraph "m").

Inclusion of a number of Karelian municipalities in the Arctic zone of the Russian Federation, where STRKs are also allocated in accordance with the List, will enable entrepreneurial activities, regulated by Federal Law No. 193-FZ "On state support for entrepreneurial activity in the Arctic zone of the Russian Federation" dated July 13, 2020, to be conducted in these areas more responsibly and taking into account the interests of the local population. By now, the List has already become the legislative framework for delimitation of economic activities of logging and fish farming enterprises.

In conclusion, we note that although the connections of territory, tradition and identity are important, they change over time. Currently, borders are losing their former meaning. Everyone becomes a migrant in a virtual sense, exposing the instability of cultural values that have developed in a particular territory. Even though the List, considering the peculiarities of the regulatory situation in the Russian Federation, is intended to establish territorial and ethnic ties, it is important to understand its relativity.

Contemporary researchers note the fallacy of an approach, when identity is studied only in the context of a specific place where certain groups of the population happen to be [11]. When discussing this issue, the concepts of "hybrid identity" and "third space", arising from the mixing of two or more cultures, are increasingly used. In this case, a completely different quality and ratio of traditions and innovations may arise in the third space. The sustainable development of society depends on many circumstances. One important factor is the historical and cultural continuity of the way of life. Radical cultural change, violation of the principle of gradualism in the implementation of innovations, the lack of an adaptation period for their launch and building a balance with tradition can lead to unpredictable consequences for society.

The research was carried out within the framework of the grant of Russian Science Foundation, (RSF) https://rscf.ru/en/project/22-28-20318/.

References

1. Order of the Government of the Republic of Karelia 06/07/2021 417r-P Available online: https://gov.karelia.ru/news/15-06-2021-pravitelstvom-karelii-utverzhden-perechen-mest-traditsionnogo-prozhivaniya-karelov-v- respublike / (2021)
2. Federal Law "On guarantees of the rights of indigenous peoples of the Russian Federation" 05/03/1999 Collected legislation of the Russian Federation 18 2208
3. Order of the Government of the Russian Federation 08.05.2009 631-p Available online: https://docs.cntd.ru/document/902156317 (2009)
4. The standard of responsibility of Arctic residents to indigenous peoples 20.07.2020, *Collected Legislation of the Russian Federation*, **29**, 4503 Available online: http://minvr.gov.ru:443/press-center/news/28001 (2020)
5. E.I. Klementyev, *Language processes in Karelia on the example of Karelians, Vepsians, Finns* (2013) p 195
6. V.A. Kryazhkov, *State and Law*, **4** (2020)
7. V.A. Kryazhkov, *State and Law*, **5** (2018)
8. V.V. Sharok, et al., Dynamics of students' ideas about the work in the Arctic, *Vysshee Obrazovanie v Rossii*, 08693617, **30**, pp. 112-119, doi: 10.31992/0869-3617-2021-30-4-112-119 (2021)
9. B. Potapov, *Economy and statistics of Karelia*, **1-3**, pp. 5-8 (1926)
10. List of populated areas of the Karelian ASSR (1928)
11. H. Can, Biographische Konstruktionen im multikulturellen Bildungsprozess, *Individuelle Standortsicherung im globalisierten Alltag*, pp. 115-134 (2006)

Influence of the arctic climatic conditions of the Kola Peninsula on the reactivity of the sympathetic-adrenal and cardiovascular systems of junior high school children

Fyodor Scherbina[1,2], *Mikhail* Shelkov[3], *Alla* Trotsenko[4,*], *Anatoly* Scherbina[5], *Tatyana* Volkova[6] and *Peter* Zvyagintsev[7]

[1]Doctor of Sciences in Biology, professor, Murmansk State Technical University, 183010, Murmansk, Russian Federation
[2]Doctor of Sciences in Biology, professor, Murmansk Arctic State University, 183038, Murmansk, Russian Federation
[3]Candidate of Sciences in Pedagogics, professor, Murmansk State Technical University, 183010, Murmansk, Russian Federation
[4]Candidate of Sciences in Biology, Senior Lecturer, Murmansk Arctic State University, 183038, Murmansk, Russian Federation
[5]Candidate of Sciences in Pedagogics, Senior Lecturer, G.V. Plekhanov University of Economics, 117997, Moscow, Russian Federation
[6]Candidate of Sciences in Philosophy, Senior Lecturer, Murmansk State Technical University, 183010, Murmansk, Russian Federation
[7]Senior Lecturer, G.V. Plekhanov University of Economics, 117997, Moscow, Russian Federation

Abstract. The growth and development of children aged 7-10 years takes place under conditions of continuous learning, and the level of fatigue increases by the end of the school year. Environmental factors of the Kola Peninsula play a special role here. the process of adaptation of the organism takes place in all age and gender groups. The most obvious is the reactivity of the cardiovascular and sympathetic-adrenal systems to physical load, which enhances the correlation links between indicators of the two active systems. It happens in the following way: by the end of the academic year, in all age and gender groups before the polar day, a proven shift of the observed indicators is recorded. The results obtained provide a comprehensive picture of age-related adaptation changes and increase in the level of fatigue in the arctic climate conditions of the Kola Peninsula. Examinations of children aged 7-10 years showed that age dynamics of cardiovascular and sympathoadrenal systems indicators is non-linear and significantly depends on specific parts of the academic year. This may be due to the periodicity of the polar night, namely, as the functional capabilities of particular systems tend to increase with age, certain indicators have a wave-like character of changes.

1 Introduction

High latitudes are characterized by numerous extreme climatic and geographical, environmental factors that have a significant impact on the human body. The complex of northern environmental factors has a significant impact on the systems and mechanisms of homeostasis [8-14]. peculiarities of changes in the indicators of sympathoadrenal and cardiovascular systems in children living in the Arctic regions remain practically unstudied. From the point of view of physiology, the healthy growth and evolution of the child's body system can only occur in an environmentally clean and sound environment; so the question of researching the adaptive behavior of the child in a nearly extreme environment is very relevant [5].

The reaction to the increased level of information and physical activities throughout the academic process normally goes beyond the innate ("default") adaptation mechanisms, which requires that a new level of adaptation is studied - long-term adaptation in conditions of long-time exposure to school academic activity and special environmental factors [3]. It should be mentioned that in younger school age, when structural and functional changes in various body systems of the individual child's organism occur, including the impact of environmental factors, a special role here is played by the cardiovascular and neuroendocrine systems [1].

There is evidence that at the initial phases of schooling children often demonstrate the following reactions of the body that affect the rate of physical development: psychological stress, decrease hemoglobin levels in the blood, autoimmune disorders [2,4]. Besides, some data show that at the level of humoral regulation in primary school children may retain the tension of hormonal mechanisms [7]. Several authors indicate that studies of phase structures of cardiac cycles have revealed a tendency to unfavorable changes by the end of the school year. cholinergic effects increase with age, but for in younger schoolchildren the chronotropic activity of the heart is regulated mainly by the sympathoadrenal system (SAS) [5, 6]. It has been noted that by the end of the school year, decreased SAS activity is reported, as well as an increase in the influence of the parasympathetic department of the autonomous nervous system on the heart [2,7]. Moderate physical activity reveals the functional reserves of a

* Corresponding author: trotcenko2007@yandex.ru

number of systems, the nature of their implementation in the course of adaptation and the dependence on age and gender. The reactivity of the cardiovascular system (CVS) and SAS to a graded physical load in children 7-10 years old, in the dynamics of the school year, in the conditions of the Kola Arctic, has not yet been studied.

In view of the above, the aim of the study was defined as the study of CVS and SAS indicators in the children 7-10 years old and the changes in these indicators during the school year in the state of relative rest and under the influence of physical activity in the environmental conditions of the Kola Polar Region.

2 Research methods and organization

Externally healthy children of primary school age studying in a secondary school of the Kola Peninsula were examined. The study was conducted by age and gender groups three times a year: in October, February and May. A total of 20 boys and 20 girls were examined. Physical development was evaluated according to standard methods. Functional test was dosed for each individual – cycling on bicycle RITM ergometer: 1 W per 1 kg of body weight, pedal frequency – 60 rpm, duration - 3 minutes. The functional state of the CVS was studied by means of electrocardiography, and by the Korotkov sonic method. The excretion of catecholamines was determined fluorometrically [7]. The tests were carried out under laboratory conditions in a state of operative rest at the same time of the day in compliance with ethical standards (Helsinki Declaration and the European Community Directives 8/609 EC). Reliability of the obtained results was confirmed by – using SPSS 15.0. statistical program based on parametric and nonparametric data (t-Student's statistical test and Wilcoxon statistical criterion).

3 Research results and their discussion

the indicators of physical development change gradually with age from 7 to 11 years, becoming higher every year. Changes of CVS and SAS indicators with age have a wave-like character (see Figure 2). it is known that with age there is a decrease in heart rate (HR) and an increase in indicators of stroke volume (SV) and cardiac output (CO). However, as our studies have shown, age-related changes of these indicators are nonlinear.

By age 10, there is an increase in the heart rate in boys, with a decrease occurring at an earlier age. It is also noted that SV and CO change unevenly in children from 7 to 10 years of age, increasing sharply at the age of 8 in boys and at age 9 in girls, then decreasing, although this does not violate the general trend of age-related increases.

When comparing the indicators of the CVS and SAS in the age dynamics with the data of the end of the school year, a significant discrepancy is revealed. Thus, the indicators of the end of the school year reflect the adaptive shift in the indicators of the functional state of the CVS and SAS.

According to our data, by the end of the school year, there is a decrease in systolic (SBP) and pulse (PBP) pressure in all age-gender groups, with the exception of the boys of the first year of study. Shifts in diastolic blood pressure (DPB) do not reach a reliable value. From the beginning to the end of the school year, all age-gender groups also showed a decrease in history, more pronounced in the boys. By the end of the school year, the reliability of the annual dynamics of the HR was found in boys of the first and second years and girls of the third year of study.

By the end of the school year, the SV in boys of the third year and girls of the first, second and third years significantly increased, reaching reliable significance in these groups. But a credible increase in cardiac output occurred only among girls of the third year of study: from 3.64 to 4.41 l/min. The tendency to increased cardiac output was found in boys of the third year of study and girls of the first year of study; boys of the first and second grades demonstrated a slight decrease in CO: from 3.03 to 2.76 and from 4.15 to 4.04 l/min, respectively. This indicator practically did not change in the girls of the second year of study: from 3.19 l/min at the beginning of the school year to 3.18 l/min at the end of it. The pronounced discrepancy between the indicators of age groups and the annual dynamics in younger schoolchildren indicates a substantial influence of the educational load on the CVS at this stage of a child ontogenesis. The intensive nature of the CVS functioning is more pronounced at the end of the school year in boys of the third and girls of the first and third years of study.

Signs of fatigue are most pronounced in boys of the first year and girls of the second year of study.

During the school year there were changes in the CVS reactivity to the graduated exercise. In boys of all age groups, the growth in CO was higher at the end of the school year and was caused by an increase in both chrono- and inotropic effect of the heart. In girls, at the end of the school year, an increase in HR prevailed. The increase in SV was slightly lower than at the beginning of the school year; as a result, there was a slight increase in CO at the end of the school year when exercising. Changes in arterial blood pressure (ABP) in response to exercise both at the beginning and the end of the school year were reliably significant.

The increase in the chronotropic effect of the heart in response to stress at the end of the school year represents a non-conomic type of CVS response as a consequence of hypodynamia. Thus, the functioning of the CVS in primary school children of all age and gender groups under the influence of the graded physical activity at the end of the school year proceeds to a certain extent in a stressful mode. especially significant is not only to estimate the functionality of the system, but also to disclose its reserves, which is possible when studying the regulatory mechanisms of the CVS, in particular the SAS.

According to our data, by the end of the school year there was a decrease in catecholamine excretion (CA) in all age-gender groups, more pronounced in boys. The excretion of adrenaline (A) was significantly decreased in boys of all age groups and in girls of the second and third years of study; noradrenaline (NA) – in all age groups of boys and girls of the second and third years of study; dopamine (DA) – first and second grade girls; DOPA (D) – in second and third grade boys third grade girls.

Catecholamines (CA)	time of research	7 years old		8 years old		9 years old	
		1	2	1	2	1	2
Adrenalin	October	3.74±0.66	9.67±1.85*	5.02±0.51	11.76± 1.38*	4.22±0.39	8.03±1.03*
	February	4.18±0.64	7.83±0.78*	2.95±0.93	4.81±1.24*	4.87±0.6	8.57±1.29*
	May	3.43±0.4	5.1±0.55*	2.13±0.6	3.22±0.94	3.38±0.67	7.01±1.22*
Noradrenaline	October	8.29±1.25	18.43±1.42*	11.48±1.42	23.50±3.08*	11.42±1.32	18.88±2.89*
	February	17.9±3.43*	19.05±2.19*	9.64±3.02	10.10±2.65	10.19±0.85	17.9±3.43*
	May	8.92±0.93	11.18±1.46	6.84±1.32	7.35±1.46	7.41±1.52	18.02±2.98*
Dopamine	October	106.1±25.83	157.68±43.12	126.85±22.74	242.20±32.55	90.46±15.29	134.16±27.76
	February	124.21±28.2 8	137.9±19.59	122.23±29.0	232.42±73.58	66.72±16.61	88.32±22.12
	May	55.03±8.33	125.02±28.19 *	41.92±5.25	75.7 ± 23.8	84.04±10.47	13.14 ±10.64 *
DOPA	October	20.74± 4.11	32.2±4.99*	30.17±3.75	47.28 ± 4.97*	24.49± 1.83	4.09±3.18 *
	February	27.71±4.96	31.14±4.41	23.83±7.32	28.27 ± 3.01	26.55±1.82	35.43±3.33
	May	16.26±2.04	26.24±4.6*	20.5±3.79	17.18±2.4	15.74±2.39	27.16 ±3.25*

Fig. 1. Changes in the excretion of catecholamines (kg/min) under the influence of graded physical activity in primary school children during the school year (M±m).

Note: 1– excretion of CA before graded physical activity, 2 – after graded physical activity

* Reliability of changes in indicators shift under the influence of graded physical activity

The most significant decrease in CA excretion was found in second-grade girls and third-grade boys. The predominant decrease was observed in the final links of SAS – A and NA (please refer to Table 1). boys (first- and second-graders) and girls (second- and third-graders) showed a significant decrease in precursors of DA and D, which probably reflects the growing deficit of DA and D. Only in girls of the 3rd year, the decrease in the excretion of CA was moderate in all stages of the SAS. Decrease in SAS indicators by the end of the school year was typical for all age and gender groups (the result of educational process and environmental and social factors of the Arctic region). Decrease in CA excretion by the end of the school year was indicative of increased fatigue and was manifested by strenuous SAS functioning. The graded exercise revealed some functional reserves of the child's body. The reactivity of the SAS at the end of the school year indicates the lack of balance between the synthesis links (uneven decrease in the growth rate of CA excretion under the influence of the functional test).

Graded physical activity revealed the incompleteness of adaptive changes in the body during the first three years of study, as well as the imperfection of regulatory mechanisms in boys. Higher rates of cardiac output in response to physical activity at the end of the school year and their correlation not only with SV, but also with increased heart rate with decreased CA growth, indicates the presence of CVS reserves, increased sensitivity of the system to hormones and mediators (see Figure 2). At the same time, there is a strenuous functioning of the CVS and a decrease in the reserve capabilities of the SAS (as a result of increasing fatigue against the background of developing adaptation mechanisms). high reactivity of the CVS with decreased reactivity of the SAS to physical load is interpreted as an economization in the work of regulatory mechanisms, which serves as a favorable background. At the end of the school year, such economization reveals the reserves available in younger students. In girls, the growth of CA under the influence of the graded physical load at the end of the school year differs from the growth at the beginning of the school year. The increase of A was lower in the first, second and third grades, but in girls rose to 107.4% for girls in grade three, compared to 90.3% at the beginning of the school year.

The increase in NA in girls of the first and second grades decreased almost fivefold, and became twice as high in third grade. It is noteworthy that in the girls of the first grade at the end of the school year the growth of CA and NA sharply deceased; the excretion of DA increased just as sharply, and excretion of D was almost unchanged. The dynamics of the ratio coefficients of different SAS (CA) links indicates an increased level of use of D, but a decrease (A+NA)/DA (especially in the NA/DA link), while the A/NA indicator increased. Thus, there was a real vulnerability in the DA-NA link. The reactivity of the SAS in primary school students, both at the beginning and at the end of the school year, was at a fairly high level, however, at the beginning of the school year, the increase in A and NA prevailed over DA and D, while at the end of the school year, the increase in DA and D was prevailing over A and NA. This indicates the tension of the system, connection of the reserves, although the effectiveness was most evident in the enhancement of the chronotropic effect of the heart.

In general, s an uneconomical type of response to the presented exercise is observed. The high reactivity and coherence of the SAS links in the girls of the 3rd grade give reason to speak about the significant role of this system in regulation and provision of a high level of performance. All this is an evidence of achieving a certain level of adaptation to the conditions of the educational process in a sub-extreme environment.

Analysis of the correlative relationships between the studied parameters allowed concluding on the conjugacy in functioning of sympathoadrenal and cardiovascular systems.

All this leads to the following recommendations. When organizing educational work, sports training of primary school children, it is necessary to take into account the intense nature of CVS and SAS functioning in children of the first and second year of education, as well as the immaturity of these systems in children of the third grade.

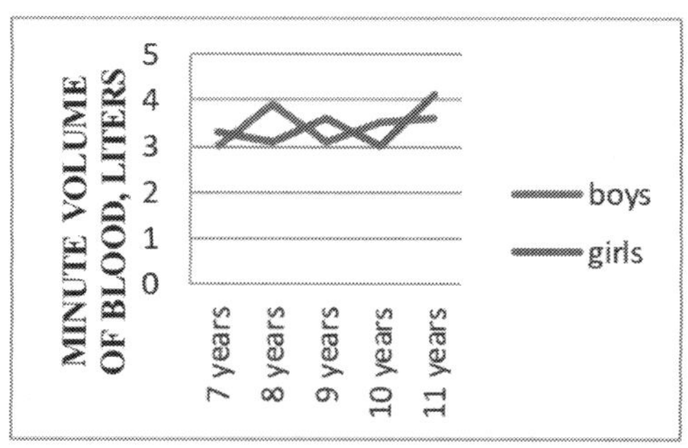

Fig. 2. Age dynamics of cardiovascular system indicators in boys and girls aged 7-10 years.

The regulation of graded load must correlate with the age and year of study: in the 1st year, a moderate and short-term graded load is recommended; in the 3rd year, a long graded load of moderate power is acceptable; in the 4th year, a moderate and short-term graded load at the beginning of the school year, in the middle of it – intensive and short-term graded load, at the end of -50% of PWC170 is an adequate load.

4 Conclusions

1. For younger school children it was found that by the end of the school year, prior to the polar day, catecholamine excretion significantly decreased in the relative resting state in all the groups studied, as well as it increased in response to a dosed physical load as a result of systematic exposure to the educational process.

2. Adaptation to physical exercises and learning activities for younger students is stressful due to increased fatigue levels.

3. the immaturity of the functional systems of younger schoolchildren may be due to the predominance of temporal functional dependencies between the indicators of the cardiovascular, sympathoadrenal systems and physical development. This pattern is a part of research results – by the end of the school year in all age and gender groups, the functional dependencies, both in the state of relative rest, and after physical activity, were weaker.

4. The results of research related to the functional state of the heart have shown that psychophysiological recreation and corrective activities should be provided, for children studying in the first 3 grades of primary schools in the arctic Kola Peninsula both during each school year and at the end of it.

References

1. A.A. Pseunok, *Uspekhi sovremennogo estestvoznaniya*, **8**, p. 26 (2007)
2. A.Sh. Arzikulov, *Molodoy Ucheny*, **20(124)**, p. 63 (2016)
3. O.E. Kondakov, et al., *Zhurnal Sibirskogo federalnogo universiteta*, **10(2)**, p. 312 (2017)
4. L.K. Buslavskaya, *Health-saving technologies in primary education* (Belgorod, IPK NIU BelGU, 2011)
5. N.A. Agadzhanyan, *Normal Physiology Textbook for medical students* (Moscow, Medical Information Agency, 2009)
6. M.M. Bezrukikh, *Pedagogical physiology textbook* (Moscow, Forum, 2013)
7. F.G. Sitdikov, et al., *Fiziologiya cheloveka*, **32(6)**, p. 22 (2006)
8. A.N. Nikanov, et al., *Ecologiya cheloveka*, 5, p. 11 (2019)
9. O.N. Popova and Yu.F. Scherbina, *Ecologiya cheloveka*, **5**, p. 3 (2012)
10. F.A. Scherbina, et al., *Human respiration in the Arctic during the seasons of light aperiodicity monography* (Arkhangelsk, 2018)
11. Yu.F. Scherbina, *Physiological characteristics of seasonal reactions of the respiratory system in young people under the conditions of the Kola Arctic region autoabstract dis cand boil. sciences* (Arkhangelsk, 2013)
12. A.B. Gudkov, et al., *Vestnik Severnogo (Arkticheskogo) Federalnogo Universiteta Seria Mediko-Biologicheskiye Nauki*, **1**, p. 33 (2013)
13. G. Gerasimova, et al., The Demographic Factor Impact on the Economics of the Arctic Region, *Resources*, **10**, 117. doi: 10.3390/resources10110117 (2021)
14. V.V. Sharok, et al., Dynamics of students' ideas about the work in the Arctic, Vysshee Obrazovanie v Rossii, 08693617, **30**, pp. 112-119, doi: 10.31992/0869-3617-2021-30-4-112-119 (2021)

Transformation of daily food practices of indigenous peoples of Taimyr through the prism of environmental change

Vasilisa Bobrova[1,*]

[1]Senior curator, Siberia Foundation, Museum of Anthropology and Ethnography, Peter the Great Russian Academy of Sciences, 199034, St. Petersburg, Russian Federation

Abstract. This paper presents the results of a trip to the Krasnoyarsk Krai in the spring of 2021. The work was carried out in the cities of Dudinka, Norilsk and in the settlements of Ust-Avam, Volochanka, Levinsky Peski of the Taimyr Dolgan-Nenets municipal district of the Krasnoyarsk Krai. The main purpose of the study was to examine the transformation of the food system of the residents of the above-mentioned settlements, as well as to consider the problem of delivery of goods and their transportation to these remote settlements, and the impact of these difficulties on the life and economy of local residents. Environmental changes are also an important factor in the formation of local food resources. Their influence plays one of the main roles in the self-sufficiency of the peninsula residents in food supply. The work was conducted with various categories of the population: people of different ages, men and women, hunters, fishermen, hunters, reindeer breeders, representatives of the administration, employees of trading companies and people employed in the sphere of educational field.

1 Introduction

Every year, the interest and close attention of the world community to the study and development of the Arctic region and the study of the indigenous peoples of the North is steadily increasing. Due to various recent events, special attention in the media agenda is given to Taimyr. This unique and picturesque peninsula is rich in minerals and inimitable human resources. Strongpoints of the Northern Sea Route, industry, production, mining are located and develop on its territory. However, among the global projects and primary issues that are important for the well-being of the state, we must, first of all, remember that this region is home to unique peoples - Dolgans, Nganasans, Nenets, Evenks, Enets. The peoples who have been living for centuries in harmony with the nature of Taimyr and have drawn from it the everything necessary for life. However, with the development of technology, the increase in production and other industries, the blow to the ecology of the region is becoming more and more powerful every year. One of the striking examples of recent years is the diesel fuel spill in Norilsk on May 29, 2020. It is important for researchers to analyze how this environmental disaster affected the local population in the year after the spill and how the situation will develop in the future. In March-April 2021, as part of the project on nutrition research in the Russian Arctic (RSF, project No. 19-78-10002), field work was carried out in Taimyr, namely in the heart of the Avamskaya tundra. The materials of these field studies are presented in the article and are intended to demonstrate the main difficulties in maintaining the everyday food practices of local residents and the transformation of traditional environmental management and nutrition within the framework of globalization processes.

2 Research

The main part of the expedition work was carried out in the settlements of Ust-Avam and Volochanka of the Taimyr Dolgan-Nenets Municipal District of the Krasnoyarsk Krai. As a brief historiographical reference, it should be noted that all the original ethnographic sources of the region under study include all sorts of information, such as: official dispatches of service people about the current state of administrative, military, economic affairs and events, yasak, salary, yavche and other books which recorded information on the number, structure of the Russian and aboriginal population, and also reflected economic processes, in particular, the payment of yasak, taxes. Various expeditions, including geological exploration expeditions, were carried out on the territory of the region. Of the ethnographic expeditions, it is necessary to note the materials of the expedition of A.F. Middendorf (1842-1845), who in his materials touched upon the issues of traditional nature management and food practices of the local population. In the 1970-1980s, in the territory, at that time of the Taimyr Dolgan-Nenets National District, the leading researcher of the MAE RAS G.N. Gracheva, Candidate of Historical Sciences, collected the unique field materials and the ethnographic objects that affect various

* Corresponding author: bobrovva.vasilek@yandex.ru

aspects from shamanism and shaman attributes to everyday life and mapping fishery grounds [1].

The first settlement is Ust-Avam in the Taimyr Dolgan-Nenets Municipal District of Krasnoyarsk Krai. Administratively, included in it is part of the urban settlement of Dudinka. It is located in the tundra zone within the North Siberian Lowlands on the right bank of the Avam River (left tributary of the Dudypta River), on the western slopes of the Maly Kamen Upland. The direct distance to the administrative center of Dudinka is 313 km. It was founded in 1935 on the 15th of July - the day of the village foundation. At that time, the first Red Nart plagues were organized on this territory. The population according to the 2010 All-Russian Census is 513 people. According to the official page of the settlement, 336 people live in it. Of these, 327 people are indigenous peoples: Dolgans (172), Nganasans (154), Nenets (1) [2]. Currently the village has 387 residents [3]. Ethnically, the population is represented by Dolgans, Nganasans, Russians. Representatives of small indigenous peoples of the Nort - Dolgans and Nganasans- in the village are approximately equal in numbers.

The second settlement is Volochanka settlement of the Taimyr Dolgan-Nenets Municipal District of the Krasnoyarsk Krai. Administratively, it is included in the urban settlement of Dudinka. It is located in the tundra zone in the southern part of the North Siberian Lowland on the left bank of the Kheta River in the tundra area. The distance to the town of Dudinka in a straight line is 410 km. To Ust-Avam in a straight line is 70 km. The first mention of the Volosyanka camp dates back to the 1640s. The settlement emerged as an intermediate on the crossing that connected the rivers Kheta and Pyasina. Old Volochanka was located a few kilometers from the new settlement. The modern village was founded in March 1931 as the center of the newly formed Avamsky district until its abolition in 1964. According to the 2010 All-Russian census, the population is 530 people [2]. The official page of the settlement lists 445 people [4]. There are 415 indigenous nationality: Dolgans (200), Nganasans (212), Nenets (3). As of January 1, 2021, 422 people lived here. The mortality and fertility rates are 1 to 1, there are 8 newborns for 8 dead people [3]. The increase in the population is due to those who are registered but live in Dudinka. Ethnically, the population is represented in the same way as in Ust-Avam.

The transporttaion situation is as follows. Year round, the villages can be reached by helicopter, mainly MI-8. Norilsk Avia is the monopolist of air transportation on these routes. The helicopter flies 2 times a week: Monday and Wednesday. The tickets are subsidized, so they are relatively inexpensive: Dudinka - Ust-Avam - 3971 rubles; Ust-Avam - Volochanka - 1047 rubles. The luggage is 15 kg. You have to pay extra for the rest of the cargo, but there is a restriction on the transportation of paid kilograms. In particular, the territorial department for the Volochanka settlement has disagreements with the carrier, due to which the transportation of paid cargo to the settlement is considerably complicated. Also, fishermen travel to Dudinka and Norilsk in caravans on snowmobiles to deliver over their catch to the town. The main modes of transport for the local population are: in winter - snowmobiles, it is worth noting that in Ust-Avam the "vehicle fleet" is being renewed, including after compensation payments after the diesel fuel spill at CHPP-3 in Norilsk, many have new Arctic Cat and Yamaha snowmobiles, while in Volochanka the Soviet Burans prevail. In summer, modes of transportation are boats, cutters, and cars.

The main commercial activities are fishing and hunting. They hunt hares, partridges, wolverines, wolves, bears and so on. The main commercial fish are broads, omul, whitefish, tugunok, pike and some more. In Soviet times, a branch of the Taimyr State Industrial Enterprise was located on the territory of Ust-Avam settlement. fish and carcasses of wild reindeer, extracted as a result of fishing, were handed over for processing to the Gospromkhoz. In Volochanka, the collective farms named after Stalin (among the specializations are horse breeding and gardening), Schmidt and Iskra were subsequently merged into the state farm Volochansky, which was engaged in hunting fur-bearing animals, fishing, and reindeer breeding. Livestock and fur farms existed on the basis of the state farm t. Currently, crop farming is virtually non-existent. Some grow green onions and potatoes in buckets. There is no cattle breeding. It is becoming more and more difficult to hunt wild reindeer, as they changes their migration routes and moves away from villages and outlets every year further. Some informants attribute this to geological prospecting expeditions, environmental degradation, and warming. In addition to the above, in the last few years, there have been fewer fish, while fish with mutations (growths, black meat, helminths) began to appear; it is worth noting that such fish began to appear before the 2020 spill, but locals expect long-term consequences this year when the water rises ... Many note that it is necessary to monitor this issue, because, even despite the fact that there is no direct pollution, harmful substances can be carried by living organisms: birds, fish, mammals and, according to residents, toxic substances got into the groundwater [3]. For the first time in three years, according to informants, the winter of 2020-2021 was the coldest and snowiest, which is why they expect floods and, accordingly, the consequences of a spill due to underground currents.

The nomadic way of life left this territory. The locals stopped reindeer husbandry in Soviettimes. Only the elderly remember nomadism and reindeer herding is remembered. In 2007, the Committee for the North and Indigenous Peoples of the RF Federation Council published data from the Federal Real Estate Cadastre Agency : "Established since the beginning of the 1930s, the structure of natural resource use and the concept of development of the North gave priority to the development of industry at the expense of traditional industries, which led to extensive foci of severe pollution and degradation of the natural environment, which led to the disturbance and retirement of the most agriculturally valuable lands. ... First of all, significant damage was caused to reindeer pastures ... The factor that significantly destabilizes the environmental situation in the territories of traditional nature management is the stressful impact of industrial facilities on reindeer pastures and hunting grounds, covering up to 40% of the area of traditional nature management "[5]. As part of the policy, a number of steps were taken that have negatively affected the indigenous population of the Arctic region, including the transfer of the nomadic population to sedentary position, a blow to family reindeer herding and "the conversion of the northern population to an average diet of 'mainland inhabitants'" [6]. All this resulted in the younger generations losing touch with traditional aspects of the primordial nature resource management. attempts are being made to revive reindeer husbandry, but time will tell how successful they will be.

Many societies, tribal communities, and family farms have been formed in the settlements. Within up to 100 km, each community has its own point. This is the name given to the places where the main fishing grounds are located. The formation of these points was influenced, in addition to the fact that during the Soviet period, houses were built there, by the fact that people stayed in these places during their nomadic wanderings. The main dwelling a such points was a beam or a typical Soviet house for 2-4 families. Currently, there are points where people live all year round. However, it should be noted that the composition of residents living at these points has changed significantly. Previously, families lived there, but now it is mainly men [3].

The main difficulties are associated with the delivery of food by winter road, by air and by water. This affects the final cost of goods, significantly increasing it. A large number of overdue items occur in late summer and fall, when there are difficulties with deliveries due to interruptions in the establishment of communication lines. for the period March-April there were no expired goods in stores. It should be also noted that the assortment in Volochanka stores is wider than in Ust-Avam (although two stores belong to the same entrepreneur). This is due, among other things, to slight differences in the routes of delivery. With the establishment of the winter road, fresh dairy products, fruit and vegetables are delivered to the shops. Due to the absence of a bakery in Ust-Avam, fresh bread and flour are transferred from Volochanka (parcels by snowmobiles, helicopters, boats). Some residents of Ust-Avam bake bread in small batches for sale. However, there is an acute shortage of fresh flour products in the village, in contrast to Volochanka, where the local bakery offers a relatively wide range of products on the shelves (bread, loaves, buns with jam / poppy seeds / sprinkles, etc.), in addition to cereal products. between the two villages, there is a circulation of food in both directions, as well as to points [3]. Regarding the first point: when there is no delivery, residents of the two settlements go to the neighboring village for food; goods are transported between Ust-Avam and Volochanka. The second point: the main products are brought to the point in the tundra, taken out for subsequent delivery to entrepreneurs/sale, the carcasses of wild reindeer, fish, game. As for the geography of exchange in settlements, the following schemes operate: between villages, between tundra and settlement, between settlement and town, between friends and relatives, between residents and entrepreneurs. Barter and debt exchange are also in effect.

They store food outside, in containers, sheds, glaciers, chests, refrigerators (for example, the grocery store in Volochanka sells refrigerators brought by the winter road). In this regard, the temporality of the products changes. Seasonal foods are available frozen all year round. For example, autumn berries (cloudberries, lingonberries, blueberries) in March or tugunok, which goes in August-September, in April.

Regarding questions related to changes in food taste and taste preferences, few informants reported a change in taste in traditional foods. Also, there is a slight tendency to modernize dishes, a synthesis of traditional and European cuisine [3].

3 Conclusion

Thus, the main difficulties of the settlements are related to transport infrastructure and accessibility. Additional helicopter flights are needed to increase the supply of basic consumer goods, as there is a demand for them. the lack of local food processing/production facilities also negatively affects the infrastructure and local residents. Self-sufficiency in food security is a key component in maintaining the traditional diet of local people. The basis of the population's diet is the meat of wild reindeer, fish, seasonal game, collecting of wild plants. It should be noted that traditional local food is becoming a local brand in the region. Those who come to these settlements eat traditional fish dishes (stroganina, sugudai, yukola, etc.), venison (jerky, liver/kidney stroganina, tongue, brain, etc.), berries/mushrooms/young larch cones. Last but not least, they collect young shoots of cones from larch trees and cover them with sugar, wait for the syrup to be produced. This is a local anti-cold remedy, quite fortified. Locals try to maintain food autonomy as much as possible through various strategic decisions: moving out on their own, various forms of exchange and the use of new technologies for storing food. However, due to environmental pollution, the expansion of mining, geological surveys, fires, oil spills and other factors, the ecosystem is changing, which leads to adjustments in the migration routes of wild reindeer, mutations in fish, and the gradual disappearance of various plants. The environmental consequences of the the Norilsk CHPP spill must be monitored over a long period of time to idetermine how they will affect the life of the local population and the traditional use of natural resources. Any changes in the ecosystem, especially such large-scale ones, could entail a significant decrease in the ability to maintain food autonomy and, accordingly, the inability of the population to make up for food shortages on its own.

RSF project "Food in Russian Arctic: resources, technologies and innovations" (№ 19-78-10002)

https://doi.org/10.1051/e3sconf/202337806011

References

1. G.N. Gracheva, Traditional outlook of Taimyr hunters: (Based on materials from the Nganasans of the 19th - early 20th centuries) (Leningrad, Science, 1983)
2. All-Russian population census 2010, **4**, Ethnic composition and language skills, citizenship 1 National composition of the population online: http://www.gks.ru/free_doc/new_site/perepis2010/croc/Documents/Vol4/pub-04-01.pdf (2010)
3. Field materials of the author collected within the framework of a grant from the Russian Science Foundation (project No. 19-78-10002) Expedition to the Taimyr Dolgan-Nenets municipal district Krasnoyarsk Territory March-April 2021 (2021)
4. About the village of Volochanka Dudinka Official site of the city online: http://www.gorod-dudinka.ru/volochanka/o-poselke (2021)
5. Materials for the Parliamentary Hearings in the Federation Council on the issue "Legal support of ethnological expertise as a prerequisite for the development of northern territories" (Moscow, Federation Council of the Federal Assembly of the Russian Federation, 2007)
6. O.A. Murashko and V.K. Dallmann, Transformation of the traditional lifestyles and nutrition of indigenous Nenets Autonomous District, *Ves. Moskov. Universiteta, S. XXIII Antrop.*, **4**, p. 20 (2011)

Paleoreconstruction of natural conditions as the basis for the rational use of natural resource management on the territory of the Neva Lowland

Roman Paranin[1,*], *Alina* Paranina[2,*], *Alexey* Grigoryev[3,*], *Lyubov* Larchenko[4,*] and *Margarita* Kiseleva[5,*]

[1] Master in Geography, graduate student, Herzen State Pedagogical University of Russia, 191186, St. Petersburg, Russian Federation
[2] Candidate of Geographical Sciences, Associate Professor, Herzen State Pedagogical University of Russia, 191186, St. Petersburg, Russian Federation
[3] Doctor of Geography, Professor, Herzen State Pedagogical University of Russia, 191186, St. Petersburg, Russian Federation
[4] Doctor of Geography, Professor, Herzen State Pedagogical University of Russia, 191186, St. Petersburg, Russian Federation
[5] PhD in Education, Associate Professor, Herzen State Pedagogical University of Russia, 191186, St. Petersburg, Russian Federation

Abstract. Modeling the evolution and dynamics of hydrological conditions of the Neva Lowland is of great theoretical and practical importance. The lowland is located in the Neva River basin, through which water from Europe's largest Lake Ladoga flows. From the Mesolithic to the present day, the river has been a source of drinking water and biological resources and an important transportation route. The location of the river at a major crossroads of transcontinental routes connecting West and East, North and South Eurasia, determined the metropolitan functions of the territory and the development of the multi-million city of St. Petersburg. The city's location on the Neva Lowland caused floods and gave rise to a number of environmental and geo-environmental problems. To optimize the use of natural resources, the cartographic models of the terrain are in great demand, on their basis it is possible to improve quantitative assessment and trends of waterlogging processes, submergence, risks of floods and disturbances in the water balance, exceeding the concentration of pollutants in water and bottom sediments. Reconstructions of hydrological conditions of the Neva Lowland in the Holocene are of interest for clarifying the evolution of natural and climatic conditions of the North-West Russian Plain. The aim of the article is to discuss the possibility of using GIS methods to study changes in hydrological conditions on the basis of actual data obtained so far in geography and archeology (geo-position and dating of objects of nature and material culture). The result of the work are maps that allow solving the problems of rational zoning of the territory, taking into account geological and environmental risks.

1 Introduction

Under conditions of global warming significant changes may occur north of 56°N, in ancient times this latitude was considered as the border of the Arctic [1]. This region is the most sensitive to global climate changes due to the lack of solar energy. Moreover, tectonic movements are significant here. Paleogeographic studies carried out in Northern Europe and North America have revealed traces of paleo-tsunamis and large-scale changes in hydrological conditions (water level fluctuations, slow and catastrophic floods). Such processes require detailed research and should be taken into account in the long-term planning of the development of the northern territories. catastrophic floods).

Possibilities of modern geographic methods and software for spatial analysis of potential environmental claims can be considered using the example of the Neva Lowland (Prinevskaya Lowland). The lowland is located at 60° latitude at the junction of the Baltic crystalline shield and the ancient Russian platform. The centerpiece of this territory is the Neva River, which carries water from Lake Ladoga, Europe's largest lake, to the Gulf of Finland in the Baltic Sea. The Neva River is used by the population as a source of drinking water, biological resources, as well as a water transport route. At the mouth of the Neva there is the northernmost city with a population of over five million, and the banks of the river are being actively built up. Building a dynamic GIS model and 3D model of the Neva Lowland will improve the assessment of waterlogging processes, submergence, as well as the risks of floods and excess concentration of pollutants in water and bottom sediments. This information is necessary for the optimization of modern nature management and long-term territorial planning.

The aim of the article is to discuss the possibility of using GIS methods and studying changes in hydrological

* Corresponding author: paranin.roman@yandex.ru, galina_paranina@mail.ru, neva8137@mail.ru, lubalar@mail.ru, ritakis@mail.ru

conditions based on the facts accumulated to date in history, archeology and earth sciences. To achieve this goal, the authors have analyzed the main sources of information – archives of natural and archaeological data, existing experience in the reconstruction of natural environments of territories using GIS. The authors' models of the hydrological situation of the Neva Lowland in Holocene (the boundaries of the Baltic and Ladoga transgression) are presented and their correlation with the elements of modern territorial and production structure of this territory is analyzed.

2 Methods of Research

The geographic database (DB) includes characteristics of natural and anthropogenic objects that preserve information about the past. The main source of information is the published paleogeographic data on the litho-, bio- and chronostratigraphy of bottom sediments of lakes and peat deposits of bogs in the region under study. Additional information may be obtained during expeditionary work on sampling in poorly studied areas of the territory (at different hypsometric levels). Comprehensive laboratory processing of samples includes radiocarbon dating, lithological, diatom, palynological, chironomid analyzes, as well as determination of the content of organic matter in the samples by the method of weight loss on ignition. To include new data in the database, the correlation of new data with paleogeographic information obtained from scientific publications is carried out. The results of the reconstruction of changes in the natural environment of the territory are presented in the form of paleogeographic maps showing the dynamics of hydrological conditions, changes in the types of lake sedimentation, changes in the main dominants of the vegetation cover and climatic characteristics.

ERS and GIS methods are used for the spatial analysis of paleogeographic data. To create a series of maps characterizing the main stages of the evolution of the hydrological conditions of the territory, the QGIS program is used. The program allows to work with the map layer by layer and analyze the spatial combinations of information presented on different layers, as well as to group the marking objects by their dates and qualitative differences.

The cartographic basis for modeling the hydrological conditions of the Neva Lowland territory was created on the basis of a series of 1:100000 scale maps. The boundaries of the maximum flooding of the Neva Lowland territory in the Holocene were drawn in accordance with the data of different authors: Litorina transgression of the Baltic Sea 10 m above the current world ocean level 7.6-7.2 kyr BP [2]; Ladoga transgression raising the level of Lake Ladoga to 20 m above sea level 3.5 kyr BP [3]. However, the dating of these events remains a matter of debate. There is a possibility that the rise in water levels in the region occurred at the same time.

3 Results and Discussion

3.1 Sources of information for creating a geographic database and reconstruction of the stages of evolution of the natural environment of the Neva Lowland in the Holocene

Archaeological materials on the Neva Lowland are very limited, due to the fact that ancient people chose for permanent settlements dry elevated places of the Karelian Isthmus and the Izhora plateau. Excavations at the promontory settlement at the mouth of the Okhta river (in the center of St. Petersburg) testify to the economic development of the river and the broad contacts of the population of these lands with the Baltic regions and the north of the Russian Plain about 5,000 years ago [4].

The position at the crossroads of transcontinental routes connecting West and East, North and South of Eurasia explains the metropolitan function and high development potential of this territory [5]. Today the territory is actively developing under the influence of the largest industrial and cultural center in Russia, sea and river port - the city of St. Petersburg [6, 7]. However, the development of a multi-million city is associated with a number of environmental problems (drinking water supply, pollution, flooding, pressure on the earth's surface), which require reliable models of the relief and hydrological conditions of the territory (taking into account their modern dynamics and evolution). The most suitable tool for solving such problems is GIS modeling, which is widely used in scientific research, in the dissemination of reference information and in management [8-10].

The works of A.V. Amantov, V.M. Anokhin, D.Yu. Bolshiyanov, N.N. Verzilin, D.A. Subetto, M.V. Shitov and other researchers [11-14] are devoted to the issue of the evolution of the nature of Lake Ladoga and its territory, the Karelian Isthmus and the Neva Lowland. However, the published data poorly correlate with each other, and their interpretation varies significantly. The main differences are related to the question of the history of the Neva River formation. The time and reasons for the formation of the Neva - whether it was the result of a revolutionary restructuring of the river network or a gradual transformation of a narrow sea channel into a river - are debated. It should be noted that macro-regional characteristics obtained in different parts of the Baltic Sea and similar conditions in North America [15-23], as well as the methods of general physical geography and geomorphology developed in recent decades [24], are still poorly involved in this discussion. The doctrine of rhythms in nature by A.V. Shnitnikov - E.V. Maksimov [25], which makes it possible to determine and predict the changes in humidification conditions of the global climate and a separate territory, has also not found application.

3.2 Capabilities of software tools for spatial analysis for solving theoretical issues and applied problems.

The paleogeographic database (DB) was created to reconstruct the stages of the development of the hydrological situation and to create prognostic models of the development of the territory. The elements of the database are the following information: 1) object name; 2) geographical coordinates of the object (latitude and longitude); 3) altitude above sea level; 4) altitude marks of lower and upper surfaces of bogs (or cores of lake bottom sediments); 5) dating and characteristics of the studied quaternary deposits; 6) archaeological data: names and location of sites, nature and age of finds. Placing the data on the map made it possible to highlight the areas where it is necessary to collect additional information. The augmented spatial model will be able to provide the zoning of the territory in accordance with the proposed hypothesis or search question.

Spatial analysis of these materials contributes to solving theoretical problems and practical problems of modern nature management, for example: 1) the problem of the origin of the Neva River and its name, 2) modern environmental zoning of the territory, and 3) drawing up long-term perspective plans for economic development. A few examples are considered further in the study.

3.2.1The issue of the origin of the Neva River

The formation of the Neva River is one of the key issues in the geography of the North-West of Russia since this river receives the flow of three large lakes –Ladoga, Onega and Lake Ilmen, as well as many rivers, including the Volkhov. archaeological studies show, the Neva basin has been inhabited by mankind for a long time - during almost the entire Holocene. However, on the banks of the modern Neva, the sites of ancient people are practically absent. The exception is the lower course of the river at the confluence of the Okhta river, where cultural layers of the Stone Age have been investigated. In the descriptions, written by ancient authors, it was noted that the Ocean Bay (Gulf of Finland of the Baltic Sea) continues as a waterway to the Caspian Sea. This suggests the transit position of the river and its role in the economic development of the region. However, scientifically, the prevailing hypothesis is the recent formation of the river as a result of a break of the waters of Lake Ladoga, which is also confirmed by legends of the Flood and the testimonies of ancient authors about catastrophic floods that set in motion many European peoples. The discussion about the origin of the Neva and the history of Lake Ladoga has been going on for over 100 years. According to various sources, the time of the breakthrough of the Ladoga waters into the Baltic along the channels of small rivers, the predecessors of the Neva, is estimated for the period from 3 thousand years BC up to 7th century AD, some authors attribute this event to the late Ice Age [26-28].

The results of GIS modeling of the maximum transgression boundaries of the Baltic Sea and Lake Ladoga are presented in Figures 1 and 2. Analysis of the images shows that the hypothesis of a catastrophic flooding breakthrough does not fully correlate with the obvious facts: 1) the maximum level of the Baltic Sea does not affect the structure of the hydrographic network - its main directions remain (towards the Neva riverbed); 2) the maximum level of the Ladoga transgression of about 20 m could not lead to a breakthrough of water through the watershed, since the excess water is taken away by the Vuoksa, which, even in historical memory, reversed the direction of flow - this phenomenon often occurred during the spring floods. The watershed breakthrough version makes no sense at all if we accept the refined model of the lake development, in which the maximum rise of water is up to 12 m above the sea level. At the same time, it can be seen that the Neva River in each of the models is much shorter in length than today (74 km). When the two models are superimposed, its length is reduced to 15-20 km, which does not allow us to speak of it as a river, but only as a short channel.

3.2.2 Discussion about the origin of the name of the Neva River

The name of the river is interpreted in different ways today. It can mean "new" (in Sanskrit and in many other languages - nava [29, 30]), then this confirms the youth of the river and the preservation in the hydronym of the memory of its formation (the emergence of the New River). On the other hand, Lake Ladoga had for a long time the same name as Nevo, the bowl of which began to fill with water 14 thousand years ago (the whole system "lake - river" was denoted by one name). The age of the lake does not agree with the ideas about the youth (newness) of the objects and allows to consider their connection with the root navis "ship, vessel" (Latin nāvigātiō "navigation", navigare "to sail a ship", English navy "fleet") [31]. Modern toponyms retain their navigational meaning: Lake Ladoga, the city of Lodeinoe Pole (Russian ladia – shallop, a boat of large displacement). Navigational conditions in Lake Ladoga are estimated as very difficult (over its surface, the wind is gathers strength, raising high waves). For this reason, a canal was built in the area of Southern Ladoga bypassing the shallow southern part of the water area in Southern Ladoga, by which the ships bypassed dangerous part of the way further to the south, north and east. Work on the creation of the water system, later called Mariinskaya, began under Peter I and lasted for over 100 years. During the laying of canals in the middle of the 19th century, cultural layers of the Stone Age settlements were uncovered (findings of A.A. Inostrantsev). All this confirms the transit function of the Neva as an important link in water communications.

3.2.3 Territory zoning for sustainable development

Due to global warming of the climate, flooding of territories caused by floods exceeding average multi-year levels has become more frequent. During floods, settlements located in river valleys are submerged [32]. It is possible to avoid such consequences by creating

https://doi.org/10.1051/e3sconf/202337806012

new sites based on the selection of relief elements exposed to similar phenomena in the past. Zoning taking into account natural environmental risks is especially important for areas of new development. In the absence of information stored in the historical memory, the only basis for assessment is the reconstruction of the evolution of hydrological conditions of the territories. Overlaying the Holocene flood boundaries on the satellite image of the Neva Lowland allows to single out zones of natural environmental risk and zones of sustainable development (Figure 3). Elevated relief elements can be recommended for implementation of promising long-term projects requiring large investments. On low-lying territories, it is necessary to limit construction and organize monitoring of such phenomena as waterlogging of the soil and the release of radon, development of flooding and submergence processes. Taking into account such zoning, anthropogenic risks should also be analyzed – flooding during dam closure, pollution of surface water and soil, etc.

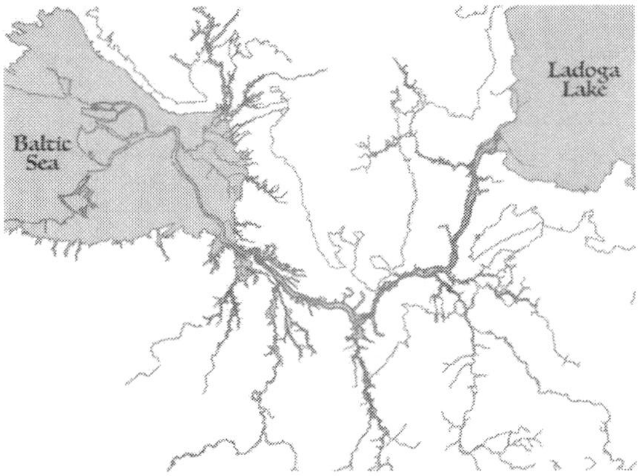

Fig. 1. Littorina transgression of the Baltic Sea (10 m above the current level)

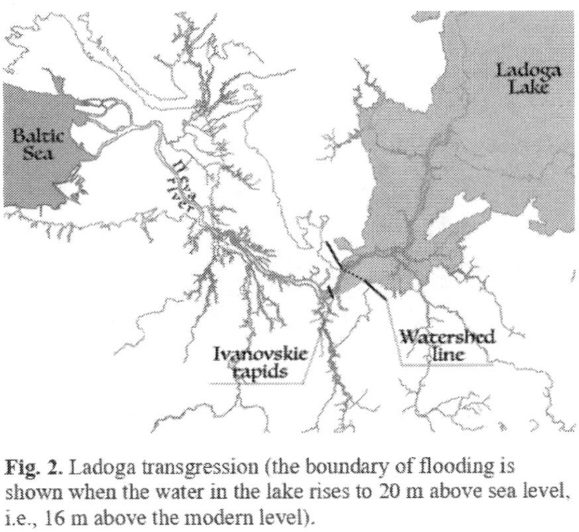

Fig. 2. Ladoga transgression (the boundary of flooding is shown when the water in the lake rises to 20 m above sea level, i.e., 16 m above the modern level).

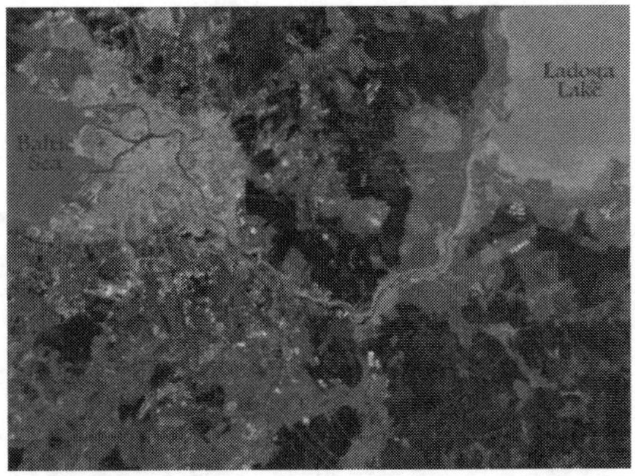

Fig. 3. Overlaying the Holocene flooding boundaries on the satellite image of the Neva Lowland, identifying zones of natural environmental risk and zones of sustainable development

4 Conclusions

The reliability of the forecast is determined by knowledge of the past, as well as the characteristics of the previous stages of development of the territorial system. It is especially important to assess the rate and direction of changes in natural processes in areas of prospective development, which include the Arctic. To reconstruct changes in the past, it is necessary to study archives of natural and anthropogenic data. Information about the development of the territory over hundreds and thousands of years is contained in such landscape characteristics as the height of river terraces, the composition and thickness of quaternary deposits, and the nature of organic inclusions.

At the present stage, sufficient data have been accumulated to start creating spatial paleogeographic reconstructions of the Neva Lowland. Such models will contribute to solving modern problems of nature management and can be taken into account in developing a strategy for sustainable development of the region in the context of global warming. Among the immediate scientific tasks of this area of research it is possible to allocate: 1. modeling the development of the Neva Lowland in the Holocene and creating a digital model of the region's relief at different stages of its existence, followed by calculations of surface levels in different historical periods using GIS methods; 2. clarification and obtaining new data on prehistoric settlements in the context of environmental changes; 3. establishing the time of formation of the Neva River based on the combined use of paleogeographic data of both natural and anthropogenic nature (archaeological data).

Fundamental studies of the spatiotemporal heterogeneity of the natural environment are useful for optimizing existing planning schemes for any territory.

The research was carried out with the financial support of the Ministry of Education of the Russian Federation (project No. FSZN-2020-0016)

References

1. V.S. Zhekulin, *Historical Geography: Subject and Methods (Istoricheskaja geografija: Predmet i metody)* (Leningrad, Nauka, 1982) p 224
2. A. Miettinen, et al., Palaeoenvironment of the Karelian Isthmus, the easternmost part of the Gulf of Finland, during the Littorina Sea phase of the Baltic Sea history, *Boreas*, **36**, pp. 1-18 (2007)
3. A.V. Amantov, et al., Problems of the Holocene development of the Southern Ladoga region. (Problemy golocenovogo razvitija Juzhnogo Priladozh'ja), *Regionalnaya Geologiya i Metallogeniya*, **65**, pp. 37-4 (2016)
4. M. Kulkova, et al., Chronology of neolithic-early metal age sites at the Okhta river mouth, *Radiocarbon*, **54(3)** (2012)
5. V.I. Paranin, *History of the Barbarians (Istorija Varvarov)* (St. Peterburg: Russkoe geograficheskoe obshhestvo, 1998) p. 283
6. L. Larchenko, et al., Industrial Innovation Clusters of St. Petersburg: Problems and Development Prospects, *I. Sci. Conf. on Innovat. in Digital Econ.: proc. St. Petersburg 24-25 Oct. 2019*, 3374655 (2019)
7. V.D. Sukhorukov, et al., The Russian Arctic World as a "Cultural Circle" and an Object of Development, *IOP Conf. Ser.: Earth and Environ. Sci.*, **816(1)**, 01201 (2021)
8. H-L. Habicht, et al., GIS-based multiproxy coastline reconstruction of the eastern Gulf of Riga, Baltic Sea, during the Stone Age, *Boreas*, **46**, pp. 83-99 (2017)
9. M.V. Minina, et al., Formation of the PALEOLADOGA database for paleolimnological research (Formirovanie bazy dannyh PALEOLADOGA dlja paleolimnologicheskih issledovanij), *Geograficheskij vestnik*, **2(45)**, pp. 18-28 (2018)
10. J. Muenchow, et al., Reviewing qualitative GIS research - Toward a wider usage of open-source GIS and reproducible research practices, *Geography Compass*, doi: 10.1111/gec3.12441 (2019)
11. A.L. Aleksandrovskij, et al., New data on the transgression of Lake Ladoga, the formation of the Neva River and the agricultural development of the north-west of Russia (Novye dannye otnositel'no transgressii Ladozhskogo ozera, obrazovanija reki Nevy i zemledel'cheskogo osvoenija severo-zapada Rossii), *Doklady RAN*, **424(5)**, pp. 682-687 (2009)
12. N.N. Verzilin, et al., Manifestation of the history of the Neva River in the granulometric state of Late Glacial-Holocene sediments (Projavlenie istorii r. Nevy v granulometricheskom sostave pozdnelednikovyh-golocenovyh otlozhenij, *Vestnik SPbGU) Ser. 7. Geologija Geografija*, **1**, pp. 64-72 (2012)
13. D.A. Subetto et al. Catastrophic changes of the Karelian Isthmus hydrographic network in the Late Glacial – Holocene: paleoseismological origin From past to present – Late Pleictocene, last degaciation and modern glaciers in the centre of Northern Fennoscandia Excursion guide and abstracts of the INQUA, *Peribaltic Working Group Meeting and Excursion Rovaniemi: Geological Survey of Finland*, pp. 156-157 (2017)
14. A.O. Aksenov and D.Ju. Bol'shijanov, New Information on the Late Holocene Ladoga Transgression Based on Field Research Data (Novye svedenija o pozdnegolocenovoj ladozhskoj transgressii po dannym polevyh issledovanij), Rel'ef i chetvertichnye obrazovanija Arktiki, *Subarktiki i Severo-Zapada Rossii*, **6**, pp. 184-19 (2019)
15. V.R. Krossa, et al., Late Holocene Baltic Sea outflow changes reconstructed using C37:4 content from marine cores, *Boreas*, **44**, pp. 81–93, doi: 10.1111/bor.12093 (2015)
16. M. Bartels, et al., Wahlenbergfjord, eastern Svalbard: a glacier-surrounded fjord reflecting regional hydrographic variability during the Holocene?, *Boreas*, **47**, pp. 1003-1021, doi: 10.1111/bor.12325 (2018)
17. M. Czymzik, et al., Differential North Atlantic control of winter hydroclimate in late Holocene varved sediments of Lake Kortejarvi, eastern Finland, *Boreas*, **47**, pp. 926-937, doi: 10.1111/bor.12315 (2018)
18. C.F.M. Lewis and B.J. Todd, The Early Lake Ontario barrier beach: evidence for sea level about 12.8-12.5 cal.ka BP beneath western Lake Ontario in eastern North America, *Boreas*, **48**, pp. 195-214, doi: 10.1111/bor.12351 (2019)
19. S.L. Norris, et al., Geomorphic, sedimentary and hydraulic reconstruction of a glacial lake outburst flood in northern Alberta, Canada, *Boreas*, **48**, pp. 1006-1018, doi: 10.1111/bor.12403 92019)
20. J.L. Peters and T.A. Brennand, Palaeogeographical reconstruction and hydrology of glacial Lake Purcell during MIS 2 and its potential impact on the Channeled Scabland, USA, *Boreas*, **49**, pp. 461-476, doi: 10.1111/bor.12434 (2020)
21. A. Rosentau, et al., Holocene relative shore level changes and Stone Age hunter-gatherers in Hiiumaa Island, eastern Baltic Sea, *Boreas*, **49**, pp. 783-798, doi: 10.1111/bor.12452 (2020)
22. S. Engels, The influence of Holocene forest dynamics on the chironomid fauna of a boreal lake (Flocktjarn, northeast Sweden), *Boreas*, **50**, pp. 519-534, doi: 10.1111/bor.12497 (2021)
23. J.P. Lunkka, et al., Deglaciation dynamics of the Scandinavian Ice Sheet in the Salpausselka zone, southern Finland, *Boreas*, **50**, pp. 404-418, doi: 10.1111/bor.1250 (2021)
24. A.N. Lastochkin, *System-morphological basis of the Earth sciences (Geotopology, structural geography and general theory of geosystems (Sistemno-morfologicheskoe osnovanie nauk o Zemle) (Geotopologija, strukturnaja geografija i obshhaja*

teorija geosistem) (St. Petersburg, SPbGU, 2002) p. 762

25. E.V. Maksimov, *Learning about rhythms in nature (Uchenie o ritmah v prirode)* (St. Petersburg, SPbGU, 1992) p. 123
26. U. Ailio, The geographical development of Lake Ladoga in the post-glacial period and its connection with Stone Age settlement (Die Geographische entwicklung des Ladogasees in postglazialer zeit und ihre beziehung zur steinzeitlichen besiedelung), *Helsinki: Bulletin de la Comission geologique de Finlande*, 45, p. 159 (1915)
27. A.A. Nikonov, The birth of the Neva, stormy and hidden (Rozhdenie Nevy burnoe i potaennoe), *Obshhestvo Sreda Razvitie*, 1(101), pp. 212-229 (2009)
28. N.N. Verzilin and G.I. Klejmenova, On the problem of assessing changes in the water level of Lake Ladoga in the late and postglacial periods (K probleme ocenki izmenenij urovnja vod Ladozhskogo ozera v pozdne- i poslelednikov'e), *Vestnik SPbGU Ser. 7*, **4**, pp. 55-70 (2006)
29. *New Sanskrit-Russian dictionary: more than 81300 words (Novyj sanskritsko-russkij slovar': bolee 81300 slov) by S V Efimovskij* (Moskva, Torus Press, 2012) p. 1168
30. A.A. Grigoryev, et al., Prehistoric stone objects of cultural heritage as a resource for the development of tourism in the Russian Arctic, *IOP Conference Series: Earth and Environmental Science*, **539**, 012093 (2020)
31. A. Paranina and R. Paranin, The Research of Northern Labyrinths as Navigation Network Elements I. J., *TransNav*, **10(3)**, pp. 451-456 (2016)
32. V.A. Matveeva, et al., Manganese Pollution in Mining-Influenced Rivers and Lakes: Current State and Forecast under Climate Change in the Russian Arctic, *Water (Switzerland)*, **21100255400(14)**, doi: 10.3390/w14071091 (2022)

Prospects for the use of Arctic routes in the context of climate change and the implementation of the concept of sustainable development of the Arctic zone

Valentina Gogoleva[1,*], *Natalia* Kirsanova[2], *Tatyana* Zyabkina[3] and *Tatyana* Dolgopolova[4]

[1]Candidate of Pedagogical Sciences, Associate Professor Graduate School of Legal Studies, Forensic Enquiry and Technical Expert Examination, Peter the Great St. Petersburg Polytechnic University, 195251, St. Petersburg, Russian Federation
[2] Assistant professor, Graduate School of Legal Studies, Forensic Enquiry and Technical Expert Examination, Peter the Great St. Petersburg Polytechnic University, 195251, St. Petersburg, Russian Federation
[3]Assistant professor, Graduate School of Legal Studies, Forensic Enquiry and Technical Expert Examination, Peter the Great St. Petersburg Polytechnic University, 195251, St. Petersburg, Russian Federation
[4] Candidate of Legal Sciences, Associate Professor Department of General Theoretical Legal Disciplines, North-Western Branch of the Russian State University of Justice, 197046, St. Petersburg, Russian Federation

Abstract. The urgent need to solve environmental problems and meet the needs of people requires finding alternative ways to deliver goods in the most environmentally friendly and safe way. In addition, the climatic changes occurring on our planet open up additional prospects for the use of the Arctic for economic purposes, in particular, the use of the Northern Sea Route for cargo transportation. In the modern world economy, sea cargo transportation accounts for about 80% of all cargo transportation. The main cargo flow is carried out through the Southern Sea Route (SSR), along which goods are transported from Europe to Asia. For quite a long period of time, shipowners and logistics companies did not perceive the Arctic region as an alternative to traffic from Asia to Europe or North America, while navigation in the Arctic was much more than just transport shipping. At the same time, the situation has changed significantly in the 21stcentury. The interest in this region among the states of the Arctic zone, as well as other countries, is connected not only with the presence there of large reserves of mineral resources, solving geopolitical problems, but also the search for alternative sea routes for cargo transportation between Europe, North America and Asia. In connection with the blockade of the Suez Canal in March 2021, the question of the need to redirect cargo transportation to alternative routes – the Northern Sea Route (NSR), as part of the Northeast Passage, and the Northwest Passage became acute. The Suez Canal is the busiest and longest sea artery. In addition, this area is unsafe due to frequent attacks by pirates. Due to the geographical features of the SSR - the presence of sufficiently narrow places along the route - under the condition of increasing traffic, does not contribute to the free passage of ships. Meanwhile, the active melting of the Arctic ice in the foreseeable future will allow navigation along Arctic routes at higher latitudes and without the escort of icebreaker transport. All of the above allows us to consider navigation in the Arctic latitudes as an actual alternative to the Southern Route, which will serve as the beginning of sustainable development of the Arctic region.

1 Introduction

The UN Global Climate Conference was held in Glasgow from October 31 to November 12, 2021. The purpose of the meeting with thousands of participants from all over the world was to approve a plan for the transition to a global economy with zero greenhouse gas emissions by 2050. Only its achievement will make it possible to contain global warming at the current level. The problem is that the commitments already made around the world to reduce emissions are categorically insufficient. And in order to achieve the goal, huge funds, efforts and sacrifices will be required from the whole world. The whole world will need to pay trillions of dollars and make huge organizational efforts to make the transition "organized" - otherwise the world risks being stuck halfway to a clean future without sufficient energy sources and with huge prices for them. Probably, political cataclysms await the world in the process of transition: from geopolitical revolutionary shifts to social discontent of large groups of people.

But even if the goal is successfully achieved, from today, the new clean world of 2050 does not seem comfortable: in winter it will be colder in houses, in summer it will be hotter, driving cars (of course, electric) will be allowed rarely, and flights on airplanes (on

* Corresponding author: 79217420164@yandex.ru

hydrogen fuel) on vacation will not be available to everyone. However, if warming cannot be stopped, the world will probably be even less comfortable.

The main task of countries to meet the needs of citizens for the necessary goods is their timely delivery in the most environmentally friendly way. existing international channels are permanently overloaded. And the recent crisis in the Suez Canal has shown the need for alternative routes.

In our article, we propose an informed solution combining ecological approaches to the development of effective directions of traffic flows.

Until recently, navigation in the Arctic region was hampered by difficult ice conditions. At the same time, a number of scientists believe that in the next 50 years, the predominant traffic flows will pass through the Arctic routes. Geographically, two transport routes have developed in the Arctic space: the Northern Sea Route (NSR), also called the Northeast Passage (NEP), passing along the northern shores of the Russian Federation, and the Northwest Passage (NWP), passing through the Canadian Arctic [1]. These polar routes between Asia, North America and Europe are currently the most promising from the point of view of commercial use. The Northwest Passage has a length of approximately 900 nautical miles and runs from the Atlantic Ocean to the Pacific Ocean along the route: the Labrador Sea and Hudson Strait, then through the Bering Strait or the Beaufort Sea. The length of the NSR is somewhat longer and is about three thousand nautical miles. However, despite this, the use of the NSR as an alternative route seems to be the most promising in the conditions of geopolitical and climatic changes, since it has a number of significant advantages.

Let us turn briefly to the legal side of the issue of securing sovereignty over the Arctic sectors.

The Russian Empire, and later the USSR and the Russian Federation, apart from hydrographic, navigational and technical equipment, also took actions to organizational and legal formalization of the Northern Sea Route, concerning the sea customs strip (1910), arrangement of radio stations and navigation signs on the coast (1911), Decree No. 1606 and Resolution of the Council of People's Commissars of the USSR No. 1873 dated 17.12.1932 established the Main Directorate of the Northern Sea Route; the Government of the Russian Federation adopted a Resolution of 19.06.1994 "On measures to improve the management of the Northern Sea Route" and others [1]. Currently , the organization of the navigation along the NSR is carried out by the following regulatory legal acts of the Russian Federation and international treaties : "UN Convention on the Law of the Sea" 1982 [8], "Convention on International Rules for the Prevention of Collisions of Ships at Sea, 1972" [9], Resolution N MSC.46(65) of the International Maritime Organization "Approval of Amendments to the International Convention for the Safety of Life at Sea 1974" [10], "International Convention for the Prevention of Pollution from Ships 1973" [11], etc.

It should be noted that none of the existing world agreements has defined the legal status of the NWP. At the same time, the NWP is subject to the sovereignty of Canada, since the route passes within its territorial waters. Concurrently, there has been a dispute between Canada and the United States over the status of the NWP since the late 1960s, which is still unresolved. The State policy of Canada considers the NWF under the jurisdiction of Canada, since it defines waterways through the Canadian Arctic archipelago, its internal waters and, accordingly, the passage of transit vessels through it requires special permission from the Canadian authorities. In 1970, Canada expanded its territorial waters to 12 nautical miles, passed the Arctic aquatic Pollution Prevention Act, empwering it to apply strict measures against potentially dangerous vessels navigating in the Arctic, and established a control zone of 100 nautical miles. The USA, motivated by the violation freedom of navigation, categorically protests against this law [3]. It should be noted that this position of the United States is shared by the European Union [4].

Tensions over the legal status of the NWP between the United States and Canada persisted until January 1988. In order to resolve the differences, the Agreement on Cooperation in the Arctic was signed, regulating the relations between the parties in exploring the Arctic region and navigating the parties' icebreaking vessels in their Arctic zones. At the same time, the United States has not recognize Canada's position regarding the extension of its sovereignty over the Canadian Arctic Archipelago, but agreed to obtain permits from Canadian authorities for the transit passage of its vessels through Canadian waters [5]. Nevertheless, the dispute over the status of the NWP has not been resolved to date. It is worth noting that at present the NWP is currently a system of interconnected individual short passages in the Canadian Arctic Archipelago and is sufficiently developed in terms of organizing navigation and search and rescue activities, but for commercial purposes the Northwest Passage is still underused [6]. Until the middle of the twentieth century, the NWP was not of commercial interest for the passage of cargo ships from due to the presence of sufficient ice cover and the complexity of the geographical landscape. Meanwhile, the ongoing climatic changes in the world in general, and in the Arctic region in particular, contributing to the reduction of the ice cover first in the southern and then in the northern part of the NWF, have helped to pay attention to the NWF as an alternative sea trade route. According to experts, climate change opens up the possibility of almost year-round navigation on the NWP for the transportation of goods from Japan and China to Europe or the east coast of the United States. Firstly, the NWP is several thousand nautical miles shorter than the Suez and Panama Canals. Secondly, the Government of Canada has high hopes for commercial use of NWP, since mining is increasing in the Canadian Arctic region, which will entail an increase in maritime traffic. However, other experts believe that the NWP will not be able to become an effective and economically profitable trade route, since it is less favorable for navigation for several reasons: the presence of a large number of islands in the region that contribute to ice blockage, the lack of deep-water ports, and the aforementioned disagreement over the legal status and access to the

 https://doi.org/10.1051/e3sconf/202337806014

passage between Canada and the United States together with the European Union. Recently, Canada has strengthened its military presence in the region; in 2009, the Canadian Parliament voted almost unanimously to change the name of the strait to the "Canadian Northwest Passage" [4].

It should be emphasized that Russia's position regarding the status of the Northeast Passage (its main part is the Northern Sea Route) regarding the extension of sovereignty largely coincides with Canada's position. At the same time, the United States and other countries consider the NSR as an international route for the transit of foreign vessels. However, it should be noted that the overwhelming majority of researchers believe that in the future, with the expansion of the navigation capabilities of the NSR and NWP, navigation in higher latitudes will require the world community to develop new international norms and rules regarding the regulation of navigation in the Arctic region [7].

Despite a number of advantages, currently the use of the NWP in terms of the commercial transit route is somewhat inferior to the NSR. In particular, due to its geographical location, the NSR reduces the length of the sea route from Europe to the countries of the Asia-Pacific region by up to 50%. According to experts, as a result of global warming, by 2030 it will be possible to navigate the NSR in ice-free water for 6 months, and by 2050 – all year round without the escort of icebreakers [12]. This will allow the NSR to become the most commercially attractive route, including for the transportation of transit cargo.

The ongoing climate changes make it possible to confidently speak about the almost year-round use of the NSR and NWP as a single Arctic marine transport artery, which will contribute to the sustainable development of the Arctic zones.

2 Main Body

It should be noted that the comparison of the NSR and the NWP does not seem quite correct, since they cannot be regarded as alternatives to each other, since the NSR is a short route between Europe and the Asia-Pacific region, which is an alternative to the Suez Canal, and the NWP connects the countries of the Asia-Pacific region with the east coast of the USA, which is an alternative to the Panama Canal.

At the same time, when assessing the prospects of using the NSR and NWP for the movement of transit cargo, both positive and negative factors are identified, and the similarity of both routes is determined.

When comparing the Northwest Passage and the Northern Sea Route, it is obvious that both transport communications are not clearly defined routes, but sea passages consisting of numerous straits and various waterways.

The similarity of both routes is also seen in the fact that both Canada and the Russian Federation consider the NWP and the NSR, respectively, as their internal waters, in contrast to the position of the U.S. and some EU countries, which consider them international straits with the possibility of open and free (without additional notification) transit passage of foreign vessels. At the same time, such a position of the United States and other countries contradicts the position of the International Court of Justice [13, 14] according to which the definition of international straits requires compliance with certain functional criteria, in particular, whether navigation through a particular strait can be carried out without the help and assistance of the coastal state, without maintenance of the strait by this state, providing emergency rescue measures by the coastal state, if necessary, etc. In the case of the NWP, a sufficient number of registered transits cannot meet the above-mentioned functional criterion established by the UN International Court of Justice. It is also difficult to imagine the possibility of ships passing through the NSR without the provision of icebreakers by Russia. In addition, a certain stumbling block in regulation of the legal status of the NSR and NWP is the desire of foreign states to conduct their vessels through them without special notification and permission of the authorities of neighboring countries.

Thus, the dependence of foreign vessels' navigation on the support and assistance of coastal states is evident, and Canada and Russia quite rightly refer both sea routes to their inland waters. In addition, both the NSR for Russia and the NWP for Canada serve, first of all, the national interests, are used as an internal transport route.

Secondly, the obvious advantage of both sea arteries is that using the NWF, the route between Asia and Northwestern Europe is reduced by 30%, and the route between Asia and the East Coast of the United States is reduced by 20%, and the NSR is 2.5 times shorter than the route through the Suez Canal. Reducing the length of the route significantly reduces transportation costs, the cost of the ship crew salaries, etc. A significant advantage of using the NSR and NWP is their safety, because harsh climatic conditions naturally exclude pirate attack, which otherwise leads to loss of cargo or cargo and vessel. Also, safety of these routes is determined by the absence of military conflicts in the active (firing) phase.

An undoubted positive aspect of using the NSR and NWP is the reduction of harmful emissions into the atmosphere by reducing fuel consumption due to the shortening of the route and the absence of "traffic jams" on it, which directly corresponds to the "green agenda" - as one of the strategies for sustainable development in the face of global climate change.

In our opinion, the positive points in terms of competition between the commercial use of the NSR and the NWP include the absence of overlap between the interests of Russia and Canada in the use of Arctic transportation arteries, since the NWP is an alternative to the Panama Canal, and the NSR is an alternative to the Suez Canal.

The disadvantages today include the limited duration of navigation due to the ice conditions on the NSR and NWP and the need to use auxiliary fleet - icebreakers, which can increase the financial costs of cargo transportation. In addition, to pass the NSR and NWP it is necessary to use vessels of appropriate class with ice-resistant hull design ; icing of the ship's hull can affect the operability of the equipment; low temperatures and long periods of daylight and darkness affect the well-being and operability of the ship crew, which requires appropriate personnel training.

In favor of the priority of the NSR, it should be noted that the intensity of changes in the ice cover of the Arctic region, associated with global warming, is most noticeable in the Eastern hemisphere. With approximately the same duration of "clear water" navigation at present - about three to five months, the use of both Arctic routes in the remaining time requires the involvement of icebreaker fleet vessels for piloting. The Russian Federation currently, unlike Canada, has a sufficient nuclear and diesel icebreaking fleet, which already allows the use of the NSR practically year-round.

Also in favor of the NSR it should be noted that in conditions of global warming, intensive melting of the Arctic ice, geographical features, the prospects for active use of the NWF are associated with environmental risks related to possible man-made disasters. The greatest threat is represented by accidental spills of oil and petroleum products from ships or leaks from broken pipelines. Meanwhile, the strategy for the development of the Arctic zone of Russia until 2035, approved by the President of the Russian Federation Vladimir Putin in the fall of 2020, among other things, is focused at minimizes such risks.

As practice has shown, the last (and far from the only) incident in the Suez Canal is a "thin" place in the global economy, which allows us to intensify the work on the search and use of alternative logistics routes in addition to the NSR and NWP. The traditional and well known route for sailors is around Africa via the Cape of Good Hope. However, such a route adds transportation costs and increases by 10,000 km. One of the best alternative routes can be considered the Chinese "One Belt, One Road" - the transportation of goods to Europe by land, transiting through Russia, which is currently actively promoted by the PRC. According to some publications, in the first two months of 2021, China sent more than 2,000 trains with goods to Europe, twice as much as in the same period last year. Even before the closure of the Suez Canal in March 2021, tariffs for maritime transportation had significantly increased (several times), which became comparable to tariffs for rail transport tariffs. At the same time, as practice shows, rail transport is quite efficient and very competitive for the transportation of a certain category of goods – consumer goods, electronics, various components. The most exotic alternative is a return to the ancient plan of building a double of the Suez Canal - a canal through Israel using nuclear charges with a total capacity of 1.04 Gt. Another alternative route is the Transpolar Sea Route passing through the North Pole, connecting Europe and Northeast Asia. However, the use of this transport artery is possible only in decades, when the ice conditions in the region will radically change. And in the near foreseeable future, this route will require not only powerful icebreakers, but also vessels of the appropriate class.

Thus, taking into account the above, despite the availability of alternative routes, including land routes, the Northern Sea Route opens up new opportunities for cargo transportation, and with the increase in energy trade, the use of existing short trade routes for economic activity is a matter of time.

3 Conclusion

Using modern approaches to solving environmental problems and preventing the onset of energy collapse due to the lack of traditional energy sources, the expansion of the use of the Northern Sea Route for the delivery of hydrocarbons through alternative routes and other goods needed by the population of different countries, in our opinion, seems very promising and effective.

Summing up, we can confidently say that the Northern Sea Route can compete not only with the Southern Sea Passage, but also with alternative transport routes, but only on condition of innovative development of the NSR (availability of high-level coastal infrastructure, including land, services), equipping the route with efficient ports, developed railway network, in the future - connecting mainland and port infrastructure into a unified logistics network; modernization of existing ports and construction of new efficient transshipment hubs, formation of support development zones; taking into account the different levels of depth both in ports and in certain sections of the NSR, carry out measures to deepen the seabed to increase the number of passes of high-tonnage vessels; attract both private investors and foreign partners to implement projects; update and improve the organization of the icebreaker fleet; ensure the safety of vessel passage through the NSR (providing anti-terrorist protection, improving and developing the network of rescue and coordination centers, ensuring continuous monitoring of the location of vessels, ice, hydrometeorological and navigation conditions; improving navigation and hydrographic support and communication systems in order to ensure the ice wiring of vessels); strengthening radio and radar surveillance; modernization of the polar air fleet; in connection with the increase in economic activity, strengthening of measures to ensure environmental protection, to intensify work to eliminate environmental damage already caused.

Thus, the analysis allows us to conclude that the use of the Northern Sea Route is promising, which will lead to the economic development of the circumpolar regions and the regions of the Far North, allow Russia to increase the potential of international maritime cargo transportation, expand the boundaries of trade cooperation [15].

This strategy is based on the Russian concept of solving environmental problems and preventing climate change processes due to greenhouse gas emissions and other toxic substances.

References

1. Mordvinova T B et al. 2017 Polar law: monograph (Moscow: Justitia) p 398
2. S. Roach, Excessive Maritime claims. US Naval War Callege, *International Law Studies*, **66**, p. 207 (1997)
3. " United Nations Convention on the Law of the Sea "(UNCLOS) (concluded in Montego Bay on 10.12.1982) / / "Collection of Legislation of the Russian Federation", 01.12.1997, N 48, Article 5493
4. "Convention on the International Regulations for the Prevention of Collisions at Sea, 1972" (COLREG) (Together with < Location and technical characteristics of lights and signs, sound-signalling devices>, "Additional signals for vessels engaged in fishing close to each other", " Distress signals") (Concluded in London on 20.10.1972) Available online: http://www.pravo.gov.ru (1972)
5. Resolution N MSC. 46 (65) of the International Maritime Organization "Approval of Amendments to the International Convention for the Safety of Life at Sea, 1974" (Adopted 16.05.1995), *Bulletin of International Treaties appendix*, **1(2)**, pp. 13-14 (2011)
6. "International Convention for the Prevention of Pollution from Ships, 1973" (MARPOL) (ed. of 26.10.2018) (Together with <Protocol I on the provisions concerning the reporting of incidents involving the Discharge of Harmful Substances>, <Protocol II on Arbitration>, <Rules for the Control of Pollution in the Transport of Toxic Liquid Substances in Bulk, prevention of Pollution by Harmful Substances transported by Sea in packaging, sewage from Ships, Garbage from Ships, the air environment>, <Guidelines for the Distribution of Toxic liquid substances by category>, "Criteria for the determination of harmful substances in packaging", <Lists of Petroleum Products, Liquid Substances>, <Certificate Forms IOPP, IAPP, IEE, logs of oil and cargo operations, operations with garbage, the register, the certificate of compliance>, < Test cycles and weight coefficients>, "Criteria and procedures for the designation of emission control areas", "Type approval and limits of operation of ship incinerators", <Information included in the bill of lading for the supply of bunker fuel>, <Procedure for checking liquid fuel>, "Information to be sent to the IMO database on fuel consumption by ships") (Concluded in London 02.11.1973) Available online: http://www.pravo.gov.ru (1973)
7. D.R. Rothwell, The Canadian - U.S. Northwest Passage Dispute: A Reassessment, *Cornell International Law Journal*, **26(2)**, pp. 331 – 372 (1993)
8. I.P. Dudykina, Foreign researchers on the legal aspects of climate change in the Arctic, *Russian Law Journal*, **2**, pp. 178-188 (2017)
9. M. Byers, The legal status of the North-West Passage and the Arctic sovereignty of Canada: past, present, and desired future, *Episode 25. International relations and world politics*, **2** (2011)
10. Problems and prospects of the Arctic region development. Study-method. Materials No. 6/2017 / D. D. Maksimova; Russian Council for International Affairs (RIAC) (Moscow, NP RIAC, 2017) p. 56
11. R. Atlant, The Security Implication of Climate Change in the Arctic Ocean, *Environmental Security in the Arctic Ocean*, (2013)
12. M.Yu. Zelenkov, Transport and logistics system of the Northern Sea Route: prospects, problems and ways to solve them, *Arktika: ekologiya i ekonomika*, **4(36)**, pp.131-140, doi: 10.25283/2223-4594-2019-4-131-140 (2019)
13. "The Statute of the International Court of Justice" (Adopted in San Francisco on 26.06.1945) / Current international law, **1** (Moscow, Moscow Independent Institute of International Law, 1996) pp. 797 – 811.
14. International public law. Collection of documents vol 1 (Moscow, BEK, 1996) pp. 13-14, pp. 404-414
15. I.G. Gerasimova, I.S. Oblova, E.I. Golovina, The Demographic Factor Impact on the Economics of the Arctic Region, *Resources*, **10**, 117, doi: 10.3390/resources10110117 (2021)

https://doi.org/10.1051/e3sconf/202337806015

Analysis of the hydrodynamic modeling market under the environmental conservation concept

Artem Ivanov[1], *Marina* Kruk[2,*], and *Vladimir* Litvin[3]

[1]Student, Department of Development and Operation of Oil and Gas Fields, St. Petersburg Mining University, St. Petersburg, Russia
[2]Associate professor, Department of Economics, Organization and Management, St. Petersburg Mining University, St. Petersburg, Russia
[3]OOO Gazpromneft-Snabzhenie, St. Petersburg, Russia

Abstract. The paper presents a comprehensive analysis of the global software market for modeling technological processes in the field of oil and gas based on data from independent analytical agencies. The work covers various aspects of the industry through a study of key competitors. The paper also provides a better understanding of the meaning of modeling software implementation. The assessment of the potential of the domestic market (TAM-SAM-SOM method) was carried out and the expediency of introducing Russian software for hydrodynamic modeling of high-viscosity oil was substantiated.

1 Introduction

Today, the high-tech market enters many sectors of the economy, including the oil and gas business. Despite the challenges the world faced due to the pandemic, the market will only grow in the near future.

As part of the topic under consideration, it is worth noting that special attention within the country's fuel and energy industry is also focused on promising hydrocarbon reserves in the Arctic. This area is a challenge requiring the application of the most advanced scientific solutions, both theoretical and practical. It is in this regard, the possibility of creation and application of domestic software solution for hydrodynamic modeling of high-viscosity oils is able to guarantee the safety, stability and efficiency of works in this direction, as well as improve the environmental safety of their production. Highly specialized software for solving problems of hydrodynamic modeling under various conditions of occurrence and filtration of oil will allow large oil companies to significantly reduce costs by increasing the accuracy of field processes modeling.

The current global market of information and communication technologies, according to various estimates, is about 269.88 trillion. rubles (3.69 trillion dollars). The global lockdown situation in 2020 led to a 4.3% drop in the global economy, which is 2.5 times more than during the global financial crisis of 2009. The contraction of the global economy had a significant impact on energy markets and led to the 4.3% drop in global energy demand, mainly due to carbon-intensive electricity generation, transport, and tertiary sector [1]. According to Gartner [2], it is projected that after a decline of 2.7% in 2020 due to the coronavirus pandemic, the global IT services market will show stable growth of around 6% per year in the next few years. It is expected that an increase in the activity of field development around the world and an increase in the costs of oil and gas companies will contribute to the growth of global software for modeling technological processes in the oil and gas market. In addition, the growing penetration of IT solutions is one of the key growth drivers for the global oil and gas process modeling software market. The Russian market in 2020, according to the IDC analytical company, grew by 14% to 1.833 trillion. rubles ($25 billion) [2].

Scientific and technological progress in the energy sector is currently living at a tremendous momentum. According to some experts, the degradation of competencies in this industry will practically catch up with the IT sphere by 2030 and will have a lifespan of about 2–3 years. In this regard, the launch of the new personnel training with ''job-ready'' skills in the digital power grid complex should have started quite a long time ago. It is important to note that the digitalization should take place simultaneously with the modernization of the laboratory base, with the ad-hoc formation of the required number of students (technicians, engineers) by teachers having the situational knowledge in dedicated universities [3].

Despite the growth of the domestic IT services market, the fuel and energy complex remains highly dependent on imports of foreign software. In 2019, an RF Presidential Decree approved a new Doctrine of Energy Security of Russia. was It takes into account the latest changes in the international situation, actualizes the challenges and risks of energy security, directions and tasks of the state policy. The provisions of the Doctrine have been included in the Energy Strategy of Russia for the period up to 2035. Within the framework of these tasks, a reasonable question arises about the necessity to increase the competitiveness of the Russian

* Corresponding author: marina_kruk@mail.ru

software solutions for a number of field tasks, including the prospective development of the deposits with high-viscosity oils. Due to the shortage of heavy grades of oil (the parties to the agreement on regulation of oil production (OPEC+) mainly reduce the production of heavy oil), the role of Russian companies in ensuring global energy security has increased. By replacing Iranian and Venezuelan heavy crude, companies supplying Urals crude to the world market benefit. [4] Refinery efficiency is very sensitive when it comes to changes in consumer demand. For instance, the OPEC+ agreement on the reduction of oil production has affected oil consumption, the COVID-19 pandemic has reduced global demand for petroleum products (kerosene, gasoline, diesel fuel, etc.), and finally has affected oil and petroleum product prices. [5] Nevertheless, taking into account the implemented concept of sustainable development, it does not always take into account the specifics of the commodity sector of the world economy, which includes both "traditional" hydrocarbon carriers, and developed "green energy". The main challenges concern access to investment resources, as well as the impact of discriminatory policies that hinder the full implementation of strategic initiatives to find a path to sustainable development. [6] Some experts argue that in the case of Russia, the energy sector will develop according to the optimistic scenario with a significant share of conservative elements. The transition to the high-tech scenario requires a number of changes in the process of forming technological reserves, as well as elimination of institutional and infrastructural risks. [7] The purpose of the development of the Russian fossil fuel energy industry, on the one hand, is to strengthen and maintain its position in the global socio-economic development, and on the other hand, to strengthen and maintain its position in the global energy sector, as well as to maintain a reliable quality and economic supply of energy and raw materials, taking into account the principles of energy conservation and energy efficiency. [8] Like any developing system, there is both advantages and disadvantages in the modern Russian corporate relations model. A number of unresolved problems remain hindering it. These include the weakness of the institutional structure of the corporate sector, the presence of significant administrative barriers, the uncertainty of state policy, etc. An important problem is the uncertainty and imperfection of the regulatory framework governing the organization of corporate entities: legal regulation is fragmentary and superficial; methods of coordinating the work of industrial complexes with federal and municipal administrative and economic departments are not fully developed. [9]

Process modeling in the oil and gas industry is used for the design, development, analysis and optimization of technical processes. The use of machine learning technologies in the oil and gas industry has increased significantly in recent years. An important task in the development of the oil and gas industry in the coming years is to improve the efficiency of oil and gas production and well drilling, and the main impetus for the introduction of machine learning methods was the drop in oil prices. Oil and gas companies have focused on resource efficiency, optimizing production processes [10]. The demand for the introduction of software for modeling technological processes at oil and gas facilities is growing, because it allows to effectively design and analyze each process during exploration and increase oil and gas production, and allows to reduce the risk of errors and increase productivity [11]. For example, a Bayesian neural network was used in the work of Al-Yami and Schubert [12]. The method used made it possible to create a system for making expert decisions during drilling. This method can be implemented to train young engineers. The system can also provide advice at all stages of well construction. These recommendations may relate to well completion, monitoring of drilling and cementing of wells, selection of drilling fluids, etc. [11] Analysis of Data Bridge market research shows that the average annual growth rate of oil and gas modeling market will be 4.8% in the forecast period of 2021-2028 and the market is expected to reach 108.43 billion rubles (1.48 billion dollars) by 2028. The growing introduction of innovative and advanced technologies to improve production efficiency is a major factor driving the growth of the oil and gas process modeling market.

Growing demand for oil and gas, combined with the growth and expansion of the oil and gas industry, will drive the demand for modeling of oil and gas production processes. Increasing interest in the research and development area for investment in innovative technologies such as 3D technologies will create additional lucrative opportunities for the growth of the oil and gas process modeling market. Growing demand for automation systems in the oil and gas industry will also play a decisive role in the growth of the market [13]. The North American region has the largest market share worldwide, followed by Europe and the Asia-Pacific region. The USA and Canada dominate the North American market. The growth of the market in North America is due to technological advances and the wider adoption of software applications for modeling in small and medium-sized enterprises. In addition, the region has a well-developed infrastructure that allows for faster implementation of advanced technologies.

It is expected that during the forecast period, the Asia-Pacific region will see ample opportunities for growth in the manufacturing and IT sectors. The presence of a large population, developing infrastructure and technology are the main factors contributing to the growth of the modeling software market in the region. The governments of India and China are accelerating the development of the manufacturing sector to develop and test vehicles, IT-related products and solutions, as well as defense equipment. High market growth is also expected due to technological advances and the use of modeling and analysis technologies.

It is expected that the global market of IT solutions for modeling will grow at a rate of ~16% per year [14] during 2017-2023 (Figure 1).

 https://doi.org/10.1051/e3sconf/202337806015

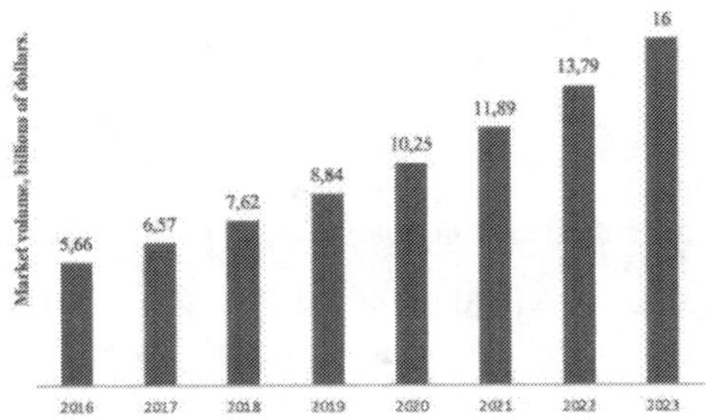

Fig. 1. Global modeling software market (Source: MRFR Analysis.

According to the Markets&Markets analytical agency, it is expected that the volume of the global modeling software market will grow from 928.86 billion rubles ($12.7 billion) in 2020 to 1.97 trillion rubles ($26.9 billion) by 2026 with a cumulative annual growth rate (CAGR) of 13.2% during the forecast period [15].

According to the mentioned data of a number of analytical agencies (Table 1), we can see that the annual growth of the whole market of IT-services in modeling in the coming years will average 14.6%, while the growth of the modeling market in the oil and gas sphere is only 4.8% [13]. Thus, according to this estimate, the market share of modeling in the oil and gas sector will decrease from 8.89% in 2020 to 3.4% in 2023 of the total market, but will grow in volume from 74.6 billion rubles (1.02 billion dollars) in 2020 to 85.57 billion rubles (1.17 billion dollars) in 2023. This is due to the widespread introduction of modeling in various industries due to the global digitalization of most technological processes. The main factors driving the modeling software market are the growing demand for an efficient solution to reduce production and training costs, as well as the expansion of modeling capabilities in advanced technologies such as digital twins, AR/VR and 3D printing. In order to be competitive in the market, enterprises are paying more and more attention to research activities.

Table 1. The volume of modeling and simulation markets in the oil and gas industry (billion dollars, excluding interest)

Source	CAGR, %	2020	2021	2022	2023	2024	2025	2026
MRFR Analysis	16	10.3	11.9	13.8	16	-	-	-
Markets and Markets	13.2	12.7	14.4	16.3	18.4	20.9	23.6	26.9
Data Bridge	4.8	1.02	1.07	1.12	1.17	1.23	1.29	1.35

The main players represented in the oil and gas process modeling market are Aspen Technology Inc., Honeywell International Inc., KBC., AVEVA Group plc, Process Systems Enterprise Limited, Chemstations Inc., KONGSBERG, GSE Systems, Inc., Schlumberger Limited., ProSim, Schneider Electric, Bryan Research & Engineering, LLC, Halliburton., CULGI B. V., SimuTech Group, Simul8 Corp., ANSYS, Inc., ESI Group, MOSIMTEC, LLC, Hexagon AB and TietoEVRY and domestic players [13].

2 The importance of the fuel and energy sector for Russia

Russia's energy sector, the basis of which is the fuel and energy complex, makes a significant contribution to national security and socioeconomic development of the country. Therefore, special attention should be paid to it, since the fuel and energy complex plays a key role in the formation of the country's budget revenues. Its share in fixed capital investment is about one-third of all investments in fixed capital of the country, in the structure of federal budget revenues it is about 40%, and in Russian exports (in value terms) - more than half.

Today special attention within the country's fuel and energy complex is paid to promising hydrocarbon reserves in the Arctic. This is due to the fact that the Arctic Zone of the Russian Federation (AZRF) is a zone of strategic interests of Russia, where 12-15% of the country's GDP is created, providing about a quarter of the export of Russian natural resources [16]. The Arctic zone has the potential of a hydrocarbon resource base of about 13 billion tons of oil and 86 trillion m^3 of gas. Their development is also a driving force for progress in the economic sector and a lifeline for the energy transition [17]. At the same time, the development of the hydrocarbon resource base on the Arctic shelf has considerable limitations, and all the significant advances in recent years in the preparation of new reserves have led to the rediscovery of offshore extensions of onshore fields [18]. The effective development of these fields provides not only material benefits, but also stimulates the development of science, because working with oil fields on the shelf, especially with large taxable oil reserves, has always required advanced solutions in related specialized disciplines. It is important not only to study individual problems of Arctic zone development, but also to form scientific knowledge about the Arctic as a unique integrative object of economic activity. The proposed main directions of scientific research approximately and comprehensively describe the problem of Arctic zone development, will help to solve the problem of improving the efficiency of economic activities in the Arctic zone of the Russian Federation [19]. The creation of an independent internal hydrodynamic simulator, which will also be applicable to Arctic sediments, can guarantee the safety, stability and efficiency of works in this direction. The potential expected effect for the mining areas from the implementation of the projects of the backbone zone will be as follows: development of production facilities; positive impact on investment attractiveness; reduction of the burden on the environment; increased accessibility of transport in remote areas; support of the interest of the Russian Federation in the presence in the Arctic; improvement of living conditions of the population; strengthening economic security of the macro-region [20]. Today, the volume of the domestic software market in the entire fuel and energy complex is 286 billion rubles ($3.91 billion) [21]. Russia has successfully implemented a number of projects to introduce Russian digital developments in the oil and gas industry. Thus, the tNavigator software product from Rock Flow Dynamics LLC, with the grant support of the Skolkovo Foundation, is successfully used in more than 200 companies around the world. The leading module of this product is a hydrodynamic simulator used to model the processes of field development. Another example is the Prime system of Yandex Terra LLC. This is full-featured Russian software for 2D/3D/4D/3C/4C seismic data interpretation that meets all the requirements of modern technical specifications, and also includes a number of innovative algorithms for solving geophysical problems. Among the domestic solutions in the field of geological modeling are Geoplat-Pro (GridPointDynamic company), the tNavigator software package module "Designer of Geology" (Rock Flow Dynamics LLC), as well as the "Sphere.Geology" module (the STC RIT Delta company) [22].

However, domestic IT companies offering their own solutions often face a number of difficulties, which can be divided into three groups.

Firstly, Russian companies representing traditional sectors of the economy, as a rule, demonstrate low interest in the introduction of digital technologies. One of the key tasks in this area remains the development of effective mechanisms to stimulate demand for digital solutions from companies.

Secondly, the lack of a necessary regulatory framework for effective regulation of the digital economy of Russia remains a barrier. Here, the task of the state is to create the clearest and most transparent rules of the game for all market participants. Active work in this area is being carried out within the framework of the corresponding federal project - "Regulation of the digital environment".

Thirdly, it is the low level of competitiveness of Russian digital solutions and, as a result, the high dependence of our market on foreign products.

3 Demand satisfaction level

According to Russian analysts, the dependence on foreign products in the oil and gas industry for 2020 is 80-90%, and the share of exports of ICT services in total Russian exports in the first half of 2019 was 8.4% [23].

Russian companies' exports of computer services grew much faster in the first half of 2020 than in 2019. According to Russoft, in the first six months of 2020 it grew by 14% compared to the first half of 2019 —to 175.53 billion rubles ($2.4 billion), whereas in the first half of last year the growth was 10.5% compared to the same period of 2018. And this is despite the fact that total exports of services during the same period decreased by 21% to 1.7 trillion. rubles ($23.3 billion) [24].

A number of difficulties arise in connection with the use of foreign solutions:

1. Foreign geological modeling software products use various interpolation and extrapolation algorithms, which are usually not common enough, which often leads to inexplicable results – for example, unjustified replacement of a reservoir by a non-collector.
2. The absence of a Russified shell of foreign products, which, on the one hand, complicates their use and, on the other hand, directly contradicts Russian legislation.
3. A significant workload of trainers with various options, which dictates the need for long-term training (which suppliers may be interested in).

In addition to the challenges mentioned, we may add the following [25]:

1. High dependence on the existing institutional environment – the state has to consider creating an independent regulator.
2. The need to develop additional procedures for monitoring capital expenditures.
3. Increased burden on the regulatory body to collect more information for better analysis.

Some large oil and gas companies deliberately refuse to use market simulators and prefer to create their own software tools, because, in their opinion, they are more trustworthy. Even inferior to the licensed analogues offered on the market, mainly by the services offered, they turn out to be quite a powerful computational tool in the hands of their authors. For example, work on the creation and improvement of hydrodynamic simulators in Severnipigaz, a branch of VNIIGAZ, has been going on for more than 20 years. This was caused first by the lack of reliable software tools on the market, then by a lack of funds in the branch to purchase them. All this resulted in the creation of the Proteus software package, which has been repeatedly tested at the oil and gas condensate fields of the Timan-Pechora province [26].

To date, among the popular simulators (Table 2) there also are Russian ones; however, for hydrodynamic modeling of high-viscosity oils on the market there are no domestic analogues to the most common CMG STARS simulator, which was created by the Canadian company CMG Software.

Table 2. Comparison of the most common simulators

Company	Revenue for 2020, million rubles	Profit for 2020, million rubles	The main activity of the company
tNavigator	958,1	116	Research and development in the field of natural and technical sciences
Schlumberger	26 213	6 100	Manufacture of other instruments, sensors, instruments and instruments for measurement, control and testing

Roxar	104,5	-83,2	Computer software development, consulting services in the field and other related services
CMG STARS	5 085	1 739,7	Computer software development, consulting services in the field and other related services
Surgutneftegaz (TechScheme)	1 062 200	729 600	Oil production
Rosneft (RN-KIM)	5 750 000	181 000	Oil production

4 Methods and methodology

Methods of critical and comparative analysis methods were used for the analysis. To determine the final market segment as a potential consumer of the developed software product, the TAM-SAM-SOM method was modified (Table 3). The method of estimation and calculation of it was improved for specific given conditions.

Table 3. Semantic content of indicators for assessing the market potential for the introduction of domestic software for hydrodynamic modeling of high-viscosity oils

Parameter	Meaning	Quantitative assessment
TAM (Total Addressable Market)	Is the product in demand?	The volume of the global market for the production of high-viscosity oils
SAM (Served/Serviceable Available Market)	How many companies can buy such an IT product?	The volume of the domestic market of IT solutions for high-viscosity oils
SOM (Serviceable & Obtainable Market)	The part of SAM that you can take	The volume of the domestic IT solutions market, which can be displaced due to the advantages of the new simulator

TAM:

$$Ae_{\gamma} = Q_d \cdot 365 \cdot Cp \cdot \gamma, \qquad (1)$$

where:
Q_d – average daily oil consumption
Cp – average cost of oil production in Russia
γ – the share of highviscosity oil in global reserves
Ae_{γ} – high – viscosity oil production costs

SAM:

$$V_m^R = \frac{(\frac{P \cdot \eta}{n} - \sum R) \cdot \gamma_R}{30} \qquad (2)$$

where:
P – economic effect due to digitalization;
n – duration of economic effect;
η – share of software for hydrodynamic modeling;
V_m – the volume of the hydrodynamic modeling market;
ΣR – total revenue of the main competitor companies;
γ_R – Russian share of high – viscosity oil reserves;
V_m^R – volume of the Russian hydrodynamic modeling market.

SOM:

$$V'^{R}_{m} = V_m^R \cdot 0{,}242 \qquad (3)$$

where:
V'^{R}_{m} – Serviceable & *Obtainable Market.*

5 Results

Table 4. Initial data for the calculation [26, 27, 28, 29, 30, 31, 32, 33, 34, [35]

Parameter	Dimension	Sign	Value
Average daily oil consumption	million barrels/d	Q_d	107,3
Average cost of oil production in Russia	USD /barrel	Cp	11
The share of high-viscosity oil in global reserves	%	γ	21
Economic effect due to digitalization	USD billion	P	1000
Duration of economic effect	years	n	5
Share of software for hydrodynamic modeling	%	η	5
The volume of the hydrodynamic modeling market	USD billion s/year	V_m	10
Total revenue of the main competitor companies	USD billion /year	$\sum R$	0.44574
Russian share of high-viscosity oil reserves	%	γ_R	22
Volume of the Russian hydrodynamic modeling market	USD billion /year	V_m^R	0.07
The really achievable volume of the domestic market	Usd billion /year	V'^R_m	0.017

To estimate TAM, we need to calculate the entire volume of the oil and gas market, which is accounted for by high-viscosity oil. To do this, we use OPEC data on daily oil consumption (107.3 million barrels per day) [27] and the average cost of production of one barrel of oil in Russia in 2020 ($11/barrel) [28]. Let us assume that the share of high-viscosity oil in the global recoverable reserves is 21% [29].

Since the paper calculates the feasibility of creating domestic software for hydrodynamic modeling of high-viscosity oil, we will proceed from the average cost of 1 barrel of Russian oil.

As a result of calculations, the total volume of the target market (TAM) amounted to 90.47 USD billion /year or 6,616.87 billion rubles/year.

When calculating Served/Serviceable Available Market (SAM), we take into account that, according to experts of the World Economic Forum, the global oil and gas industry due to digitalization can achieve an economic effect of $1 trillion by 2025. Let us assume that digitalization of processes will allow oil and gas companies to generate a total of $200 billion annually [27, 32]. However, these amounts do not reflect the "percentage of IT solutions in the cost of 1 barrel of oil". An accurate model helps to reduce costs and prevent possible complications, therefore mining is getting faster and cheaper.

According to the most conservative estimate, let us assume that hydrodynamic modeling of high-viscosity oils gives only 5% of the "gain" [30,31]. We get that thanks to digitalization, we will receive USD 200 billion, but only 5% of it will be considered the market volume of hydrodynamic modeling.

From Table 2, we take into account that the main competitors in the market of hydrodynamic simulators are tNavigator, Roxar and CMG STARS, Schlumberger; we subtract the annual revenue of these companies for a rough estimate. The total revenue of the companies for 2020 amounted to $ 0.44574 billion.

Based on the fact that Russia accounts for 22% of all high-viscosity oil in the world [29, 34], it is possible to make a formula for calculating the available volume of the domestic market. The duration of field development should also be taken into account. For a rough estimate, we may assume that the development period of an oil field is 30 years, then we get the value of the volume of the served available market (SAM) is equal to $ 0.07 billion /year or 5.8 billion rubles/year.

To calculate Serviceable & Obtainable Market (SOM), we take into account that companies such as Rosneft and Surgutneftegaz have their own hydrodynamic simulators, and the volume of the market they occupy, according to various calculations, can account for 34.8% and 10.8%, respectively [32, 35-38]. Let us assume the worst-case scenario in which all major players in the Russian oil and gas market have their own hydrodynamic simulators, then the market share that these simulators can occupy, will be 24.2%. Thus, the resulting value of the Serviceable & Obtainable Market volume was $0.017 billion/year or 1.23 billion rubles/year.

6 Conclusion

Thus, we can say that under the conditions of the implementation of the strategy of digitalization of the fuel and energy complex and the doctrine of digital security of the energy complex, it is advisable to introduce new special software for modeling high-viscosity oils. An economic analysis was carried out, on the basis of which it is clear that such a software solution will be able to occupy its niche of 1.23 billion rubles/year in the case of its low price and high quality.
Such a solution will increase competition in the Russian market and will enable many companies to expand their activities due to the concomitant increase in the accuracy of development and reduction of expenses. The accuracy of the development, in its turn, will minimize possible costs and meet the key criteria and requirements of today's environmental agenda. It is also important that the implementation of such a highly specialized Russian software solution can be more accurate due to the fact that all algorithms will rely on the experience of developing domestic deposits, and more operational support from the software manufacturer is also possible.

References

1. N. Romasheva, D. Dmitrieva, Energy Resources Exploitation in the Russian Arctic: Challenges and Prospects for the Sustainable Development of the Ecosystem. *Energies,* **14**, 8300, https://doi.org/10.3390/en14248300 (2021)
2. Gartner and TAdviser online database. Available online: https://www.tadviser.ru/index.php/Статья:ИКТ_(мировой_рынок)(2021)
3. M.Yu. Shabalov, Yu.L. Zhukovskiy, A.D. Buldysko, B. Gil, V.V. Starshaia, The influence of technological changes in energy efficiency on the infrastructure deterioration in the energy sector. *Energy Reports*, **7**, pp. 2664-2680, https://doi.org/10.1016/j.egyr.2021.05.001 (2021).
4. V.L. Ulanov, E.Yu. Ulanova, Impact of External Factors on National Energy Security. *Journal of Mining Institute*, **238**, pp. 474-480, doi: v10.31897/PMI.2019.4.474 (2019)
5. I. Filatova, L. Nikolaichuk, D. Zakayev, I. Ilin, Public-Private Partnership as a Tool of Sustainable Development in the Oil-Refining Sector: Russian Case, *Sustainability*, **13**, p. 5153, https://doi.org/10.3390/su13095153 (2021(
6. V.S. Litvinenko, P.S. Tsvetkov, M.V. Dvoynikov, G.V. Buslaev, W. Eichlseder, Barriers to implementation of hydrogen initiatives in the context of global energy sustainable development, *Journal of Mining Institute*, **244**, pp. 428-438, doi:10.31897/PMI.2020.4.5 (2020)
7. A. Kopteva, L. Kalimullin, P.Tcvetkov, Soares, A. Prospects and Obstacles for Green Hydrogen Production in Russia, *Energies*, **14**, p. 718, https://doi.org/10.3390/en14030718 (2021)
8. Y.L. Zhukovskiy, D.E. Batueva, A.D. Buldysko, B. Gil, V.V. Starshaya, Fossil Energy in the Framework of Sustainable Development: Analysis of Prospects and Development of Forecast Scenarios, *Energies*, **14**, 5268. https://doi.org/10.3390/en14175268 (2021)
9. A.A. Lapinskas, Modern Specificity of Corporate Relationson the Example of Vertical-Integrated Company, *Journal of Mining Institute*, **233**, 561-570. DOI:10.31897/PMI.2018.5.561 (2018)
10. Sh. Islamov, A. Grigoriev, G.Yu. Beloglazov, S. Savchenkov, O.T. Gudmestad, Research risk factors in monitoring well drilling – A case study using machine learning methods, *Symmetry*, 7(13), pp. 1-19. (2021)
11. Y.A. Borisevich, G.V. Tomilov, *The problem of developing fields with high-viscosity oil* (Tomsk, *National Research Tomsk Polytechnic University, 2015*) pp. 69-70
12. A.S.H. Al-yami, J. Schubert, Systems and Methods for Expert Systems for Well Completion using Bayesian Decision Models (BDNs), *Drilling Fluids Types, and Well Types*. Available online: https://hdl.handle.net/1969.1/177120 (2021)
13. Data Bridge Market Research online database. Available online: https://www.databridgemarketresearch.com/reports/global-oil-and-gas-process-simulation-software-market (2021)
14. Market Research Future online database, Global Simulation Software Market Research Report: By

Component (Software, Service). Available online: https://www.marketresearchfuture.com/reports/simulation-software-market-5769 (2021)
15. Markets and Markets online database; Simulation Software Market by Component - Global Forecast to 2026. Available online: https://www.marketsandmarkets.com/PressReleases/simulation-software.asp (2021)
16. M.B. Alekseeva, V.F. Bogachev, M.A. Gorenburgov, Systemic Diagnostics of the Arctic Industry Development Strategy, *Journal of Mining Institute*, **238**, pp. 450-458, doi: 10.31897/PMI.2019.4.450 (2019)
17. Y. Zhukovskiy, P. Tsvetkov, A. Buldysko, Y. Malkova, A. Stoyanova, A. Kozhenkova, Scenario Modeling of Sustainable Development of Energy Supply in the Arctic, *Resources*, **10**, 124, https://doi.org/10.3390/resources10120124 (2021)
18. O. Prischepa, Y. Nefedov, V. Nikiforova, Arctic Shelf Oil and Gas Prospects from Lower-Middle Paleozoic Sediments of the Timan–Pechora Oil and Gas Province Based on the Results of a Regional Study, *Resources*, **11**, 3, https://doi.org/10.3390/resources11010003 (2022)
19. S.A. Agarkov, A.V. Kozlov, V. Fedoseeva, A.B. Teslya, Major Trends in Efficiency Upgrading of the Economic Activity in the Arctic Zone of Russian Federation, *Zapiski Gornogo Instituta*, **230**, pp. 209-216, doi:10.25515/PMI.2018.2.209 (2018)
20. S.A. Lipina, L.K. Bocharova, L.A. Belyaevskaya-Plotnik, Analysis of Government Support Tools for Mining Companies in the Russian Arctic Zone, *Zapiski Gornogo Instituta*, **230**, pp. 217-222, DOI:10.25515/PMI.2018.2.217 (2018)
21. Information and analytical publication "Neftegaz", Import substitution in the oil and gas industry. Available online: http://oilandgasforum.ru/data/files/web14_1.pdf; (2020)
22. Information and analytical publication "Neftegaz"; Integrated energy Security and digitalization of the fuel and energy sector; Available online: https://www.neftegaz-expo.ru/common/img/uploaded/exhibitions/neftegaz2020/img/digest/Neftegaz_Digest_2020.17(24).pdf, (2020)
23. A. S. Gavris, V. P. Kosyakov, A. Y. Batalov, The concept of effective design of the development of hydrocarbon deposits, *Software solutions, Oilfield business*, **11**, pp. 75-85 (2015)
24. RBC; Exports of Russian IT services accelerated despite the pandemic. Available online: https://www.rbc.ru/spb_sz/17/11/2020/5fb3da859a794794f80a9d6c (2020)
25. I. Filatova, M. Shabalov, L. Nikolaichuk, Regulation methods in natural monopoly markets case of russian gas network companies, *International Journal of Engineering Research and Technology*, **5(12)**, pp. 624-630 (2019)
26. A.V. Nazarov, Modeling of oil and gas field development: state, problems, prospects, *Rassokhin readings : Materials of the seminar dedicated to the memory of Professor*, pp. 37-40, (2009)
27. Vedomosti. Technology will help oil companies earn another $1 trillion in five years. Available online: https://www.vedomosti.ru/partner/articles/2020/10/30/845034-haitek-nefti (2021)
28. A.A. Shchepalov, Heavy oils, gas hydrates and other promising sources of hydrocarbon raw materials: An educational and methodological manual, *Nizhny Novgorod: Nizhny Novgorod State University*, (2012)
29. Roscongress Insider; Russia and its regions in international trade in ICT services. Available online: https://roscongress.org/upload/medialibrary/ae6/ICT_May-2020.pdf; [Accessed 26th September 2021]
30. Rosbalt; OPEC slightly adjusted its estimate of oil demand this year. Available online: https://www.rosbalt.ru/business/2021/06/10/1906104.html, (2021)
31. A.Sh. Miyassarov, Improving the development of high-viscosity oil deposits with the use of resource-saving technology: Abstract of the Dissertation of the Candidate of Technical Sciences (Ufa, 2015)
32. G.G. Modorskaya, et al., External factors affecting the cost of oil production, *Skif. Student science issues*, **12(52)**, pp. 95-105, (2020)
33. BCS Express; How much does oil production cost in Russia. Available online: https://www.rosbalt.ru/business/2021/06/10/1906104.html (2021)
34. N.Y. Bashkirtseva, High-viscosity oils and natural oils, *Bulletin of Kazan Technological University*, **19**, 296-299. (2014)
35. Business Profile Group | MGI Worldwide; Analytical research. Oil production and refining market in Russia: Life after OPEC+ (2020). Available online: https://delprof.ru/upload/iblock/df8/DelProf_Analitika_Rynok-pererabotchikov-nefteproduktov.pdf (2020)
36. Comnews: Rosneft has sold 200 commercial licenses for the use of RN-GRID software. Available online: https://www.comnews.ru/content/204756/2020-02-27/2020-w09/rosneft-prodala-200-kommercheskikh-licenziy-ispolzovanie-rn-grid (2021)
37. M. Kruk, A. Semenov, S. Pudovkina, System of indicators allowing to assess the level of public awareness about sequestration technologies in the Russian Federation, *International Multidisciplinary Scientific GeoConference Surveying Geology and Mining Ecology Management, SGEM*, **20(6.2)**, pp. 395–403 (2020)
38. E.B. Mazakov, K.V. Matrokhina, V.Y. Trofimets, Traffic management at the enterprises of the mineral industry. Advances in raw material industries for sustainable development goals, *CRC Press*, pp. 397–405 (2021)

Financial potential as a basis for ensuring environmental and economic security of the Arctic regions

Mikhail Ulchenko[1,*] and *Nikita* Ponomarev[2]

[1] PhD (Economics), Associate Professor, Leading Researcher, Luzin Institute for Economic Studies – Subdivision of the Federal Research Centre «Kola Science Centre of the Russian Academy of Sciences», 184209, Apatity, Russian Federation

[2] Graduate student, Luzin Institute for Economic Studies – Subdivision of the Federal Research Centre «Kola Science Centre of the Russian Academy of Sciences», 184209, Apatity, Russian Federation

Abstract. Large-scale industrial development of the Arctic territories is a steady course of development that the world economy has been following for several decades. This development is facilitated by global warming, which creates prerequisites for the freeing more and more territories from ice, depletion of resources, especially hydrocarbons, as well as the emergence of new technologies that allow extraction of minerals in difficult conditions for economic activity in the Arctic. Given the vulnerability of the Arctic natural systems, the issues of environmental and economic security, as a result of such development, come to the fore. The study, analyzes the financial capabilities of the Arctic regions and concludes that they have significant financial potential, which, if used rationally, should contribute to ensuring economic and environmental security. The paper proposes the authors' interpretation of the concept of financial potential of the region. The review of domestic and foreign scientific literature showed that many works are devoted to topical issues of ensuring environmental and economic security of the Arctic regions, but the impact of financial potential on these categories is currently undefined. The paper also analyzes the financial potential of economic entities operating in the Arctic zone of the Russian Federation. It has been revealed that of the nine regions of the Russian Federation classified as Arctic, only one - the Krasnoyarsk Territory - has the value of the indicator characterizing the share of break-even enterprises, which exceeds the average Russian values. At the same time, when assessing the coefficient of provision with tax payments, in seven out of nine regions the value of this coefficient exceeds the average Russian values.

1 Introduction

The financial potential of a region is the key to ensuring sustainable economic development, which means the creation of conditions under which all modern needs of the population are met without compromising the needs of future generations. Obviously, the greater the financial potential of the region, the higher the probability of ensuring its environmental and economic security. At the same time, the issues of economic and environmental security, which are among the components of the concept of "national security", are not only inextricably linked, but are also decisive in the development of the Arctic spaces [1]. The active development of the Arctic, observed in the past few decades, is due to several reasons. Firstly, the exhaustion of finite natural resources in more southern climatic zones. Secondly, the emergence and development of new technologies that allow exploration and extraction of minerals in the difficult natural and climatic conditions of the Arctic [2, 3]. Thirdly, the emergence of a more favorable conditions for economic activity, due to the release of ever larger spaces from snow and ice cover. And the presence of huge proven reserves of hydrocarbons - oil and natural gas, as well as non-ferrous metals - gold, platinum, aluminum, copper and other minerals, only fuels interest in this macro-region from not only the Arctic states [4, 5]. At the same time, the vulnerability of natural systems of the Arctic, not resistant to external influences and characterized by an extremely low rate of self-purification, suggests that the development of these territories and water areas will be carried out in compliance with environmental safety requirements [6].

2 Overview of the main sources of research and publications

Despite the large number of scientific studies devoted to the study of environmental and economic security, the issues related to its financial component, if considered, then quite superficially. At the same time, finances in the modern economic world are a key component of the success of the development of any system. The lack of financial resources nullifies all the efforts of state

* Corresponding author: ulchenko23@rambler.ru

authorities aimed at creating conditions conducive to sustainable economic development of territories. However, not only the success of economic development of the country and/or regions depends on the availability of financial resources, the lack of funds challenges other types of security, such as military, medical, social, energy, food, environmental, which is especially important for the regions of the Russian North and the Arctic [1, 7, 8].

The issues related to the financial security of the regions as a key component of economic security are considered in sufficient detail in the work of R.V. Badylevich *Assessment of state financial support for the development of the Arctic zone of the Russian Federation* [7]. The article, based on official statistics, analyzes the main sources of replenishment of the budget of the Russian Arctic regions. Based on the analysis, two groups of regions were identified, depending on the level of financial independence of replenishment of the regional budget. The study also concludes that the state policy with regard to the Russian Arctic regions is gradually shifting from direct financing to creating favorable conditions for potential investors. We draw attention to the article by I.V. Nikulkina - *Strategic approach to the formation and implementation of state financial policy in the Arctic zone of the Russian Federation: methodological aspect* [9]. The main attention is paid to the study of the state financial policy in relation to the Russian Arctic territories. According to the author, the allocation of these regions into a separate category and the implementation of a special financial policy in relation to them by the state is a reasonable position of the federal authorities. The high importance of financial security and financial potential of the regions in ensuring economic and environmental security is also noted by Sh. Tabata in his article *Financial flows between the federal budget and the Arctic regions of Russia* [10]. The author notes a fairly high financial potential and financial sustainability of the Arctic regions, believing that most of them can be economically independent from the federal center. At the same time, the creation and maintenance of conditions characterized as attractive for private investors is a mandatory function of the state. The issues of environmental security in the Arctic regions of Russia are considered in sufficient detail in a joint article by I.A. Rodionova and S.A. Lipina *Ensuring environmental safety: public administration of the Arctic region* [3]. As part of this work, the authors identified the main sources of environmental pollution among the Arctic regions and concluded that the highest level of pollution is observed in those places where industrial enterprises that develop and process extracted minerals are located.

3 Assessment of the financial potential of the Arctic regions

The financial potential of a region takes into account all the resources that the subject has - these are the financial resources that are potentially possible to attract, and those resources that the regional authorities actually have today to ensure the sustainable development of the subordinate territory.

In the context of the sanctions policy of Western countries, as well as the spread of Sars-Cov-2 infection, the capabilities of almost all Russian regions, in terms of raising funds from outside sources, are seriously limited. That is why the issues of assessing the financial potential of the Arctic regions are of particular relevance and allow us to judge their ability to independently attract domestic financial resources.

disproportions in the development of the regions of the Russian Federation, observed for several decades, are primarily due to a significant difference in the level of financial provision of territorial systems with financial resources. At the same time, the state policy towards "needy" regions, which is commonly called the policy of equalization, has led to two extremely negative trends: firstly, it contributed to the growing dependence of the regions, in terms of financial assistance, from the federal authorities; secondly, it discouraged them from searching for new mechanisms for attracting financial resources within the regions. As a result, the problem of strengthening the financial potential of the regions, which existed before, remains unresolved at the present time.

In such conditions, one of the key tasks of the regional authorities is to create conditions under which the regions will become financially independent. In order to understand how achievable such a condition is, it is necessary to analyze the financial potential of a group of regions and/or an individual region.

When assessing the financial potential of several regions, it is possible not only to correlate their financial capabilities, but also to obtain data on their competitiveness, as well as reactions to certain management decisions made earlier.

Since the objects of research are the Arctic regions of the Russian Federation, it is necessary to determine which territorial systems can be attributed to the Arctic. Guided by the legal act of the President of the Russian Federation No. 296 of May 2, 2014 (edited on May 13, 2019 and June 27, 2017), the following regions are assigned to the Arctic zone of the Russian Federation: Arkhangelsk Region - partially, Krasnoyarsk Territory - partially, Murmansk Region - completely, Nenets Autonomous District - completely, Republic of Karelia - partially, Komi Republic - partially, Sakha Republic - partially, Chukotka Autonomous District - completely, Yamalo-Nenets Autonomous District - completely [11, 12, 13].

Table 1. Budget and tax potential of the regions of the Arctic zone of the Russian Federation [14]

Regions	Expense coverage ratio	The coefficient of concentration of own income	The coefficient of provision of the population with tax payments	The ratio of the deficit/surplus of the consolidated budget to the amount of expenditures
Arkhangelsk Region	0.76	0.73	68.3	0.04
Krasnoyarsk Territory	1.01	0.89	97.7	0.13
Murmansk Region	0.89	0.86	113.3	0.03
Nenets Autonomous Okrug	0.85	0.65	292.2	0.01
Republic of Karelia	0.61	0.6	51.7	0.02
Komi Republic	0.94	0.91	104.6	0.04
Sakha Republic	0.64	0.64	136.1	-0.01
Chukotka Autonomous District	0.28	0.28	307.5	0.006
Yamalo-Nenets Autonomous District	1.1	0.96	437.2	0.14
Russian Federation	0.81	0.81	69.2	0.0003

Despite the common features characteristic of the Arctic regions - low population density, unfavorable climatic conditions for living and working, significant reserves of natural resources, vulnerability of the environment, they all have different financial potential. Some of these regions need to attract financial resources from higher-level budgets, others do not. Tables 1-2 present data reflecting individual elements of the financial potential of the Arctic regions. Thus, Table 1 presents the data determining the tax and budget potential of the analyzed territorial systems of the Arctic.

The analysis showed that five of the nine Arctic regions have the cost coverage ratio exceeding the national average (0.81) - the clear leaders are the Yamalo-Nenets Autonomous District with a ratio value of 1.1 and the Krasnoyarsk Territory - 1.01. More modest ratio values were recorded in the Komi Republic, the Murmansk Region and the Nenets Autonomous District - 0.94, 0.89 and 0.85, respectively. In fact, we can say that the designated territorial systems have a greater share of their own funds (revenues) in total expenditures than the national average. The remaining regions can be characterized as territorial systems, which are characterized by low budget occupancy due to their own revenues.

Similar results were obtained when calculating the coefficient reflecting the share of own revenues of the Russian Arctic territorial systems in the total budget revenues - the coefficient of concentration of own revenues. The clear leaders here also remained the Yamalo-Nenets Autonomous District – 0.96 and the Komi Republic - 0.91. More modest, but above the national average values were recorded in the Krasnoyarsk Territory, Murmansk Region and Nenets Autonomous District - 0.89, 0.86 and 0.85, respectively. At the same time, the Chukotka Autonomous District became the most dependent on federal aid – 0.28. From the data in Table 1, it also becomes evident that in 7 out of 9 Arctic regions the coefficient of tax payments provision exceeds the average Russian values (69.2). The lowest values of the coefficient, according to the analysis, are noted in the Republic of Karelia and the Arkhangelsk region - 51.7 and 68.3, respectively. The clear leaders are the Yamalo-Nenets Autonomous District, Chukotka Autonomous District and Nenets Autonomous District – 437.2, 307.5 and 292.2, respectively.

Table 2. Financial potential of the population of the regions of the Arctic zone of the Russian Federation [14]

Regions	The number of people with monetary incomes above the subsistence minimum, %	The ratio of deposits of individuals to total income	The ratio of per capita income to the subsistence minimum in the region	Decrease or increase in financial assets, rubles / person
Arkhangelsk Region	87.3	0.35	2.75	-305
Krasnoyarsk Territory	82.7	0.28	2.58	857
Murmansk Region	89.4	0.39	2.65	3937
Nenets Autonomous District	90.6	0.20	3.96	37198
Republic of Karelia	84.3	0.33	2.32	648
Komi Republic	84.5	0.35	2.56	4526
Sakha Republic	82.2	0.20	2.63	21082
Chukotka Autonomous District	91.5	0.25	3.77	3919
Yamalo-Nenets Autonomous District	94.4	0.20	5.28	29413
Russian Federation	87.7	0.36	3.32	529

The second table contains data that reflect the financial potential of the population living in the regions assigned to the Arctic zone of the Russian Federation. It is quite remarkable that when estimating such an indicator as "the ratio of deposits of individuals to the total income of the population", the value exceeding the national average (0.36) was achieved only in the Murmansk region – 0.39. This indicates that the population of the Arctic regions of Russia is either poorly interested in accumulating financial resources, or, more likely, simply does not have free funds that can be used for these purposes. At the same time, the value of this indicator in 3 out of 9 regions is close to the national average – the Arkhangelsk Region - 0.35, the Komi Republic - 0.35 and the Republic of Karelia - 0.33.

When assessing regions by such indicator as "the number of population with monetary income above the minimum subsistence level ", 4 out of 9 regions have values that exceed the national average (87.7). So, the clear leaders are Yamalo-Nenets Autonomous District, Chukotka Autonomous District and Nenets Autonomous District - 94.4, 91.5 and 90.6, respectively. The Murmansk region, albeit slightly, exceeds the average Russian value for this indicator – 89,4873. In other regions, the values of this indicator are lower than the national average.

Table 3. Financial potential of economic entities of the Arctic zone of the Russian Federation [14]

Regions	The coefficient characterizing the net financial result of enterprises, million rubles per 1 enterprise	The coefficient reflecting the share of break-even economic entities	A coefficient reflecting the average level of profitability of goods, services, and works sold by economic entities
Arkhangelsk Region	2.19	0.64	5.2
Krasnoyarsk Territory	16.5	0.7	15.2
Murmansk Region	5.3	0.62	8.7
Nenets Autonomous District	5.01	0.4	4.7
Republic of Karelia	3.14	0.57	20.1
Komi Republic	6.27	0.6	9.6
Sakha Republic	7.77	0.65	5.6
Chukotka Autonomous District	20.8	0.57	21.1
Yamalo-Nenets Autonomous District	117.4	0.61	17.1
Russian Federation	4.35	0.68	10.8

The third table contains data that determine the financial potential of economic entities operating in the Arctic regions of Russia. Thus, the Yamalo-Nenets Autonomous District is a clear leader among all the territorial systems of the Arctic by such an indicator as the coefficient reflecting the net financial performance of enterprises – 117.405 million rubles. In general, in seven of nine regions, the values of this indicator exceed the average Russian value (4.35): Chukotka Autonomous District - 20.8, Krasnoyarsk Territory - 16.5, Republic of Sakha (Yakutia) - 7.77, Nenets Autonomous District - 5.01, Murmansk Region - 5.3, Komi Republic - 6.27. Among the regions that can be considered lagging behind are the Republic of Karelia - 3.14 and the Arkhangelsk Region - 2.1. Also noteworthy is the indicator characterizing the share of break-even economic entities. It is interesting that only one of the nine Arctic regions, the Krasnoyarsk Territory, has a value of 0.701, exceeding the Russian average of 0.68.

4 Conclusion

The following conclusions were obtained in the framework of the study:

1. The analysis of works by domestic and foreign authors showed that many scientific studies are devoted to the topical issues of environmental and economic security of the Arctic regions, but the impact of financial potential on these categories has not been defined.

2. the study proved that the value of financial potential directly affects the capabilities of regional authorities to create conditions for ensuring environmental and economic security of subordinate territories.

3. The analysis of the indicators characterizing the financial and budgetary potential carried out in the work confirmed that most of the Arctic regions have significant financial capabilities to ensure the environmental and economic security of their territories.

The study was carried out within the framework of the RFBR grant 20-010-00776 A "Improving the state financial regulation of the development of the regions of the Arctic zone of the Russian Federation as a basis for ensuring the economic security of the Russian Arctic".

References

1. M. Ulchenko, et al., Specific features of ensuring economic security of coastal Arctic regions, *IOP Conf. Series: Earth and Environmental Science*, **302(1)**, 012104 (2019)
2. E.A. Samylovskaya, et al., Digital Technologies in Arctic Oil and Gas Resources Extraction: Global Trends and Russian Experience, Resources, 11(3), 29, doi: 10.3390/resources11030029 (2022)
3. V.S. Litvinenko, The Role of Hydrocarbons in the Global Energy Agenda: The Focus on Liquefied Natural Gas., *Resources*, **9(5)**, 59. (2020)
4. N.V. Govorova, et al., Russian Arctic: problems of international cooperation, *Contemporary Europe-Sovremennaya Evropa*, **1**, pp. 156-159 (2018)
5. K. V. Matrokhina., V.Y. Trofimets, E.B. Mazakov, A.B. Makhovikov and M.M. Khaykin, Development of methodology for scenario analysis of investment projects of enterprises of the mineral resource complex, *Journal of Mining Institute*, **259**, pp. 112-124. doi: 10.31897/PMI.2023.3 (2023)
6. I.A. Rodionova, et al., Ensuring environmental safety: public administration of the Arctic region, *Modern Problems of Science and Education*, **1(1)** (2015)
7. R.V. Badylevich, Assessment of the state financial support for the development of the Arctic zone of the Russian Federation, *Fundamental research*, **9**, pp.11-19 (2020)
8. P.S. Tcvetkov, et al., The Changing Role of CO2 in the Transition to a Circular Economy: Review of Carbon Sequestration Projects, *Sustainability*, **11(20)**, 5834; doi: 10.3390/su11205834 (2019)
9. I.V. Nikulkina, Strategic approach to the formation and implementation of state financial policy in the Arctic zone of the Russian Federation: methodological aspect, *Economics and entrepreneurship*, **9(74)**, pp. 205-210 (2016)
10. Tabata Sh 2019 Financial flows between the federal budget and the Arctic regions of Russia, Region: Economics and Sociology 3(103) 3-25
11. Decree of the President of the Russian Federation №296 dated 02.05.2014 "On land territories of the Arctic zone of the Russian Federation. Available online: http://www.kremlin.ru/acts/bank/38377 (2014)
12. Decree of the President of the Russian Federation №287 dated 27.06.2017 "On amendments to the Decree of the President of the Russian Federation dated may 2, 2014 №296". Available online: http://www.kremlin.ru/acts/bank/42021 (2014)
13. Decree of the President of the Russian Federation №220 dated 13.05.2019 "On amendments to the Decree of the President of the Russian Federation dated may 2, 2014 №296". Available online: http://www.kremlin.ru/acts/bank/44231 (2014)
14. Federal state statistics service. Available online: http://www.gks.ru (2021)

https://doi.org/10.1051/e3sconf/202337806018

Strategies for Large Oil and Gas Companies Operating to Support Sustainable Development and Environmental Safety in the Russian Arctic

Ishel Bianco[1], *Alexei* Fadeev[2,*], and *Alexander* Illinsky[3]

[1]Peter the Great St. Petersburg Polytechnic University 195251, St. Petersburg, Russian Federation
[2]Doctor of Economic Sciences, Professor, Peter the Great St. Petersburg Polytechnic University 195251, St. Petersburg, Russian Federation
[3]Doctor of Economic Sciences, Professor, Peter the Great St. Petersburg Polytechnic University 195251, St. Petersburg, Russian Federation

Abstract. IAs conventional oil and gas sources become less productive, unconventional hydrocarbon sources become more attractive due to new technologies and become the target of expanded exploration. Oil and gas development in the Arctic is challenging but promising, the largest reserves of Oil and Gas in the Arctic are on the Arctic Shelf in Russia. Most Arctic communities, especially indigenous communities, depend on fishing, hunting, and gathering, this requires the preservation of environmental health. In this article we tried to delineate a strategy for large oil and gas companies to diversify risk by proposing a form of social entrepreneurship that works for environmentally vulnerable regions with low infrastructure or skilled labor. With this framework, large oil and gas companies could effectively and reliably support smaller-scale enterprises preserving environmental health and creating an innovative business ecosystem that promotes a Arctic. Our model could also lead to local and indigenous entrepreneurship and a more equitable distribution of wealth in the Arctic.

1 Introduction

The Russian Arctic is experiencing dramatic environmental and social changes because of global warming and extractive operations such as hydrocarbon production. Russia's Arctic Regions have a low population density, low levels of infrastructure and generally low GDP per capita. The Arctic population depends on the exploitation of natural resources, with indigenous regions strongly depending on subsistence fishing. Social conditions are generally lower than in mainland Russia. Some Arctic regions have high unemployment, low wages, and poor health conditions.

This study attempts to illustrate some of the challenges faced by oil and gas development in the Russian Arctic, and offers an investment framework for large oil and gas companies in the Arctic that can maximize social welfare while preserving renewable resources. This framework is created for the Russian Arctic in contrast to Scandinavian models, which apply to higher-income population with less cultural diversity and population structure.

2 Methodology

The literature used for this paper includes articles of Dr. Fadeev [1] and Dr. Ilinsky [2] on sustainability, as well as more specific material cited for its environmental and social importance, such as the articles by Unguryeanu and Chaschin et al. [3, 4]. The holistic nature of our study means that besides theoretical principles of management we must examine environmental, biological and social data. Overall, the authors agree with the need for more involvement of stakeholders, but highlight the contributions of new technologies on the analysis of human impact on the environment and long-term socio-environmental consequences. For this purpose we used quantitative and qualitative methods, and examined interdisciplinary research by Arctic scientists and experts based on publications of the Arctic council and Nordic Council of Ministers. Quantitative analyses were obtained from several sources, such as state, national, Arctic Council and Large Marine Ecosystem framework suggested by One Shared Ocean, collected in PAME publications.

Most of our parameters are objective, but in the formulation of our framework we use some qualitative research, this research combines several parameters and is reflected in the Human Development Index.

In calculating the social contribution (including the environmental component of social welfare in the Arctic) of oil and gas companies, we have analyzed several social components, such as infrastructure, education and tax revenue, which we measure in investments when possible, but in areas with low

* Corresponding author: alexfadeev79@gmail.com

transparency or conversion of investment to projects. For example, some of the roads built in the Arctic region translates into savings of 10 dollars per kilometer [1]. Most of these factors have been gathered in the Human Development Index [1, 2].

3 Results

Managing oil and gas resources in vulnerable regions like the Arctic requires a more holistic approach to operations that include consideration of environmental factors, environmental and socialist factors, human populations, and cultural dynamics. Oil and gas operations need to prioritize environmental factors since most of the coastal population depends on local natural resources, including a larger-than-usual percentage of fish protein [2]. Small spills have global implications because of the connections between Arctic countries, which are exacerbated by security considerations. In hhe strategy of Russia for development of the Arctic, human development is a priority along with environmental security. Russian oil and gas companies should integrate these environmental principles from the design stage of sites and the evaluation of licenses.

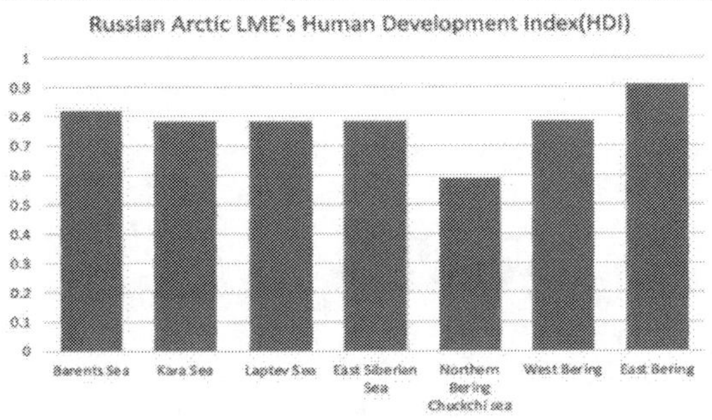

Fig. 1. Russian LME's Human Development Indexes. Calculating social contribution of oil and gas companies. Source: PAME

We analyzed several social components, such as infrastructure which we measure in investments when possible but in areas with low transparency or conversion of investment to projects. For example, some of the roads built in the Arctic region, translates into savings of 10 dollars per kilometer [3].

Some of the Arctic regions suffer from low wages and from low job opportunities [3]. To calculate income, we calculate income growth divided by prices growth in the Arctic region. The impacts on health can be calculated by the percentage of days lost or days missed from school because of health reasons, it can also be calculated by hospital expense per patient in the region.

Satisfaction with life is a subjective factor and is best calculated by questionnaire, indirectly it can be verified by mobility, but emigration may not reflect dissatisfaction with current life conditions if low income or low level of savings may keep the population from moving to other regions. life satisfaction can also cause declining birth rates and drug problems, which further reduce income and social development [4].

The social mobility factor can be related to education and social investment per child or student. However, education is not the only factor affecting mobility; family capital, passed down to the generation's can provide a similar advantage [4,1].

3.1. Environmental Service Preservation

Since oil and gas are non-renewable resources that largely affect renewable resources by diminishing their quality, quantity, and availability, the balance of production of oil and gas must be examined in the range of diminishing returns to a resource that is more durable and has limits where its renewability is compromised. Damage to fisheries or to forestry due to oil spills can affect the structure and long-term viability of environmental services. Poor populations are more than 45% more dependent on ecosystem services than urban populations, and therefore more vulnerable to the effects of climate change [5].

Although the consequences of oil and gas exploration are dramatic, there are technologies that can mitigate all of them, and many of them have been developed by small and medium-sized enterprises. Some of those technologies have been tested in warmer environments, such as biological solvents and some filtration systems. The cost of some of these technologies has declined due to competition and market effects, however, they show a slower rate of development when prices of oil and gas are lower, and so we can predict a post-Covid recession and subsequent decline in demand and contraction of the oil and gas service industry. The big oil and gas companies should invest in these technologies to reduce further liabilities and create more affordable investment systems. In Norway, large oil companies provide loans and grants to smaller enterprises to spread risks. Russian oil and gas companies should be incentivized to do so as well [3].

Some technologies, such as enhanced oil recovery, reduce the amount of space and resourced used for oil extraction and therefore two of the biggest environmental risk variables. Unfortunately, some of the chemicals and processes employed in these techniques, such as sound, can affect some organisms, such as whales and dolphins. Acids can also dissolve shells of crustaceans and corals. Other biocompatible chemicals such as lubricants or phthalates can bioaccumulate, affecting fisheries and agricultural productivity [2]. Production debris can also reduce both predation and reproductive success by affecting the communication of organisms leading to lower reproductive success or predation. Seismic activity has been shown to alter the behavior of whales and dolphins. Changes in nutrients can facilitate the establishment of invasive species, and accelerate eutrophication of waterways, reducing survival of fish and crustaceans [6]. These effects will affect the success fishing, hunting, and agriculture.

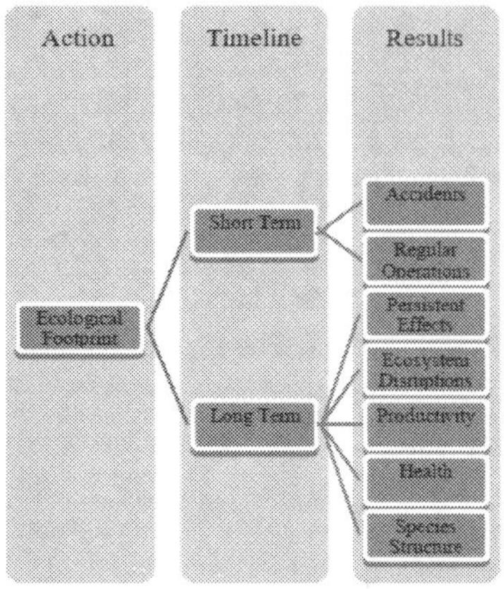

Fig. 2. Long and short- term environmental effects, Source: authors)

3.2 Framework for Arctic Social Investment

The introduction of digital technology, machine learning and remote operations can cut costs due to predictive maintenance and better risk assessment. These technologies help in both ways, reducing the costs of catastrophic events, and improving the quality of processes by ensuring optimal conditions operational conditions for all components and better use of labor force. Digital technologies can further reduce the number of accidents that affect human health and design systems that better manage the hazards of the workplace [6].

Using the LME framework, we observe that more rural areas have lower GDP and greater dependence on fish sources. A more damaged environment with higher risks requires more extensive management initiatives. Large oil and gas companies can implement initiatives that are more significant in its impact than smaller companies, as in the example of the production reduction agreements in central Texas. Large companies also can also pass on the high costs of innovation and competitive challenges to smaller companies and later either absorb them or continue partial ownership. Socially and environmentally, the process is complex [7].

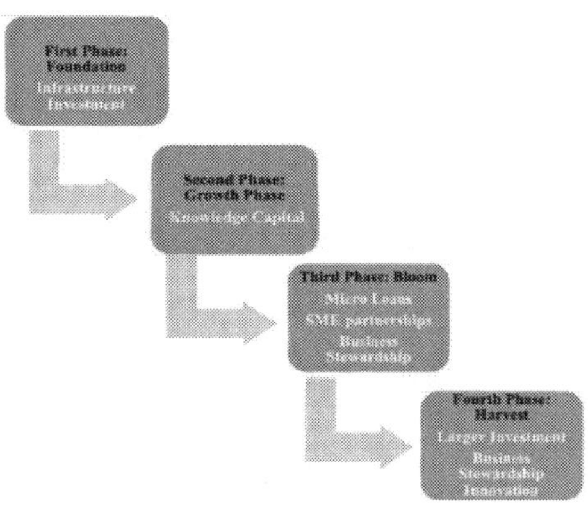

Fig. 3. Arctic social investment framework for large companies; source: authors

The first phase of investment should create conditions that can make the population capable of entrepreneurial functions, such as engineering education, skills development, languages, computer skills.

The second "Growth" phase intensifies education and professional investments, empowering an already skilled population with small grants, knowledge training, and further, investments that increase, usually without return. This phase can be accomplished with the support of government.

The third phase or "Blossom" involves the beginning of a strategy of building service companies that are still managed and advised by larger companies and management consultants who scan the most prospective companies, which are mostly owned by the larger company and still dependto some extent on the experience and reputation of the larger company, as well as guarantors of loans. Returns are low. Most investments at this stage are still considered social enterprises.

Phase 4 or "Harvest" creates well-funded clusters of companies that have already shown the ability to generate profits and focus on technology and services, with higher wages in this phase and possible returns on investment. Large oil and gas companies can then choose whether to continue as joint ventures or separate ventures, and receive patents created jointly, but without confidentiality agreements that can thwart competitiveness. The benefits are economic in nature and create secondary social benefits.

We believe that in some areas, such as Sakhalin and Yamal, Gazprom and Yamal LNG are between the first and second phases, and Rosneft is in the first stage of investment in its Arctic enterprises [2, 3, 7]. This is a good point to analyze their social investment strategy according to our framework.

We propose a diversified social investment strategy for major oil and gas companies to counterbalance some of the inefficiencies that have made rent revenues and local taxes ineffective in reaching poor and disadvantaged communities, particularly those located far from urban centers, which is a common condition for Russian Arctic communities, and for remediation of structural inequalities, particularly among indigenous communities. Indigenous communities suffer from higher rates of mortality, cardiovascular disease, injuries, and suicide. The most significant factor influencing poor health is indigenous ancestry. In the Russian Arctic, 40 different tribes are registered, 98,651 people in 2010, representing only 5.5% of the general population of the Arctic, and an increase of almost 1% [8, 1].

to optimize sustainable development in the Arctic with a focus on the prosperity of local populations, particularly indigenous and vulnerable groups, we chose a business stewardship system based on the concepts of social entrepreneurship, which is better suited to the structure of the Russian energy industry and may also be suitable for other countries that once had large state companies, or still have large state companies, that are vertically integrated and represent a large share of the domestic and export market. An energy environment with few big players tend to concentrate capital and

E3S Web of Conferences **378**, 06018 (2023) https://doi.org/10.1051/e3sconf/202337806018
IIRPCMIA 2021

weaken competition, resulting in a low development of the service industry; however, in some cases, like in Norway, a large company can spread risk by investing in smaller companies that specialize in services and produce innovative technologies, which in the case of Norway was also product result of government support [2, 10].

Some community investment models create a central tribal or local government, but if these communities have limited technological expertise, this investment will not lead to innovative products or services, and they may not support a robust service industry.

A new business model created specifically for the Russian Arctic may be more adequate to address the social problems of the region and to integrate and administer knowledge capital and avoid the Dutch curse of countries with a large base of natural resources that do not distribute benefits among the population. Since Russia's mineral rights are owned by the state, the model of Alaska or Canada may not be appropriate, and besides, the infrastructure of Russia is not as developed as in Norway. the concentration of capital in Russia also benefits from the stability of investments and the fluidity of capital resources, which can support the creation of a strong service industry even at times when oil prices oil may be low. Stability of investment is important for a population that does not generally have a large social network, as in Scandinavian countries. Russian entrepreneurs also do not have the financial opportunities available to North Americans and Western Europeans.

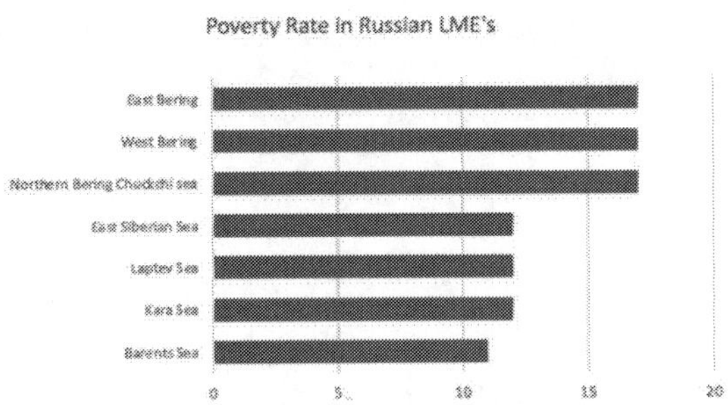

Fig. 4. Poverty rate in different LME's. Source: PAME

Structural inequality is aggravated by concentration of opportunity and capital, which become a loop that is extremely hard to overcome, as poverty leads to poor nutrition, poor health, lack of motivation and addictions [4]. There are several places where there are good opportunities to break this loop. For poor families, this can mean dependence on some social services that provide skills for parents and an advantage in education for children. These services may be just enough to survive but not enough to thrive, unlike middle-class families who have more support. Mentors and volunteers can overcome lower level of education of parents, less abundance of skills for better jobs, and other inequalities arising from the transfer of intellectual, emotional, and financial capital [8]. Gazprom and Rosneft already offer mentoring and skills development initiatives, but they can still be improved to intervene promptly at the most significant stages of development. This means early education and technological assistance to remote communities to ensure that they have access to quality education and training, this payment can also be non-financial, but can be translated into tax credit for the first steward company. Business stewardships may be most effective where there is no history of entrepreneurship or where there may be limited access to capital and training [7].

Table 1. Summary of contribution of benefits. Source: authors: PAME

Benefits	Charity	Social Responsibility	Small Business Investment and Microloans	Business Stewardship
Personal Development	Neutral, dependency	Reparative	Empowering, Enabling	Empowering, Enabling
Sustainability	Low, seasonal	As long as trigger exist	Long term if guided	Long term
Reinvestment in community	None	Possible but low	Long term if successful	Long term, preparing for transitions.
Risk	Misrepresentation	Misrepresentation	Medium dependent on control	Low by diversification
Returns	None	None	Medium	Low
Permanence	Ephemeral	<1 year	Average >5 years	Longer than Business Cycle
Immigration	Encourages	Neutral Local Action	Neutral if Native Focused	Discourages by nurturing Local Business

Social indicators show the advantages of business stewardship as they fill the capital and knowledge gaps that arise from charitable investments, the perceived cosmetic value of social responsibility and short-term or compensatory outcomes and the weakness of unsupported microloans.

Many authors agree that charitable support alone does not address long-term vulnerabilities of the community and can be low or seasonal and fail to develop the abilities of the local population [9,3,6]. They do not empower psychologically or emotionally, and they can lead to underdevelopment of abilities in the population. Strategic investments, such as educational programs for oil and gas professionals, produce immediate results, enabling the population to access jobs and are more participatory, but are still worth less than the ability of local people to become business owners and develop technologies that may be more suited to their environment and culture [2]. Participation is essential for the distribution of benefits and a more equitable distribution of wealth derived from the exploitation of natural resource [8].

Over the past 20 years, corporate social responsibility has expanded to the concept that business is an integral part of society and should be designed in a manner that is socially beneficial to society by creating more than profits, employment, or an increase in the tax base. This concept of social entrepreneurship not only absorbs the externalities usually addressed under social corporate responsibility, but also increases the social capital of the community by creating goods for the better functioning of society. Society, in its turn, returns investments in social goods, such as a better educated or more productive workforce. Investments in social capital can also return to companies in the form of reduced disruptive crime or higher productivity [9]. Currently, the lack of effectiveness of social investment by Shell has translated into piracy and disruption of operations for Shell in Nigeria.

Table 2. Economic Indicators of Arctic LME's; source: PAME

LME	Poverty Rate	Population	Rural %	Fishing Revenue $
Barents Sea	11%	2,023,335	33%	556,441,114
Kara Sea	12%	276,868	43%	826,300
Laptev Sea	12%	31,013	100%	2,989,354
East Siberian Sea	12%	34,151	100%	1,344,326
Northern Bering Chuckchi sea	17%	56,490	100%	327,890,066
West Bering/ East Bering	17%	310,725	31%	71,489,6683

3.3 Business Stewardship as a solution for Russian Arctic Development

The concept of "business stewardship" is similar to that of social entrepreneurship, but it is adapted to the functions of large companies that now today take the place of former state-owned enterprises. Soviet companies may have invested on long-term projects that did not bring immediate profits, because they considered it their duty to support employment and industry in times of recession. Russian oil and gas companies tend to be large, operate on down- stream, upstream and midstream, and have the flexibility to absorb price shocks better than service companies. The business stewardship model can do a better social job of creating social benefits and still profit from each new business it steers until the time of maturity, when most processes can be delegated to the new enterprise. Our model of business stewardship in the Arctic satisfies long- and short-term sustainable development goals, improving the quality of life through investment in infrastructure and technology development, scientific leadership, and stakeholder involvement at all stages.

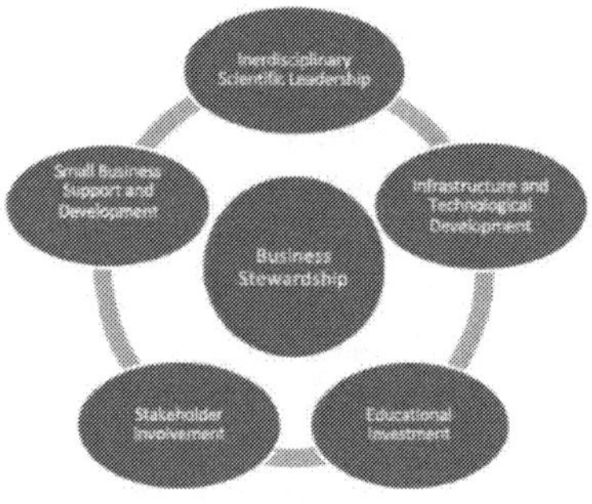

Figure 5. Business stewardship elements. Source: authors

Social enterprises in the UK lead in innovation at 65.8% compared to 42.5 SMEs and offer innovation processes at 47.9% compared to 19.2% of SMEs [10]. This hands-on l experience is the result of knowledge transfer and sustainability. It is similar to the process of an apprenticeship. Local and indigenous communities also transfer knowledge to the initial business steward. At the end of the term, results are evaluated and ownership can be transferred if goals and standards have been met. There may be stipulations that the initial investment can be paid out as seed money for new enterprises without interest and in shorter time frame [9]. This community investment model is more suitable for areas where the history of entrepreneurship has been short, or where a transition to other production modes or industries is needed. The parent company provides direction and expertise to drive innovation [10,13]. Business parks and incubators may have limited success where there are few elements to create a profit-conductive ecosystem. Another factor that has been shown to limit the success of business incubators is remoteness from learning and activity centers. In this way, business incubators can discriminate against vulnerable populations where transportation is challenging. Digital connections can also help coordinate contribution of local professionals [3].

4 Discussion

Preservation of the environment in vulnerable areas like the Arctic is essential to the long-term survival of local and indigenous communities. Some indicators of the environmental health needs to be updated more regularly because of global climate change, which is more acutely felt in the Arctic. Unfortunately, there is scarce data on the success of social entrepreneurship in former socialist countries. Most of the data come from developing nations in Asia or Latin America that lack access to education, science, and industrial development [7]. Russia has a high literacy rate and a large professional class, but it lacks financial tradition and reliable methods of facilitating credit to small entrepreneurs [9]. Some initiatives of the Russian government have tried to address his gap, but it remains an obstacle for most entrepreneurs, as does the low rate of savings which is affected by low wages [9].

Although some research has been done in the area of quantifying social benefits of environmental services, it needs to be reconsidered in light of. Arctic conditions [6, 13]. One measure could be the salaries of consultants or contractors who would do similar work. some time and service trading platforms, like Time Republik, have attempted to quantify this benefit as a tradable option. More research is needed on fair quantification of ecosystem services and their long-term losses, as well as the optimization of social benefits and transferring them to the community.

5 Conclusions

Russian Arctic regions depend on ecosystem services that account for a larger share than the average income of the most vulnerable populations, such as indigenous communities, therefore it is essential for oil and gas companies to respect the long-term quality of such ecosystems [8, 13].

The Russian Arctic is home to a small population that have some common features in its range. This population is dependent on environmentally vulnerable regions. poverty levels in most coastal areas are similar, and so is the dependence on natural resources and subsistence farming. Current industrial development is relatively low and largely dependent on extractive industries, especially of oil and gas. High levels of poverty tend to put pressure on animal and plant species through poaching and pollution [14].

Development in the Russian arctic requires not only investment in infrastructure, but also an understanding that the building infrastructure will take time and that investment cycles must focus not only on economic growth, but also on sustainable growth that creates well-being for local people while preserving the quality of their environmental resources.

Large oil and gas companies are heirs to the strength and resilience of the former Soviet state-owned companies. These companies could engage in investment strategies, in addition to charitable spending and corporate social responsibility, to promote local empowerment through entrepreneurship [13].

Our study explains the fundamentals needed in the Arctic for the development of successful entrepreneurship with a low economic footprint that preserves the integrity of renewable resources.

Our analysis suggests a strategy of business stewardship that has both social and profitable overtones and transfers knowledge along with ownership, minimizing risk for both start-up and mature companies and fostering innovation. Business stewardship will create prosperity that extends beyond the presence of oil and gas in the Arctic region.

Business innovation has to do with reducing environmental impact by reducing waste and integrating sustainable goals from the outset [15].

The final option for creating prosperity for oil and gas corporations is likely to be a combination of these different strategies, suited to each region and community with its individual characteristics.

The strategy of business stewardship promises to benefitlocal and indigenous communities by preserving their natural capital and distributing the wealth created by the exploitation of natural resources, as well as providing them with a higher level of participation on their own development.

References

1. Nordic Council of Ministers. Arctic Human Development Report: Regional Processes and Global Linkages. Norden. (2014)
2. A. Fadeev, et al., Innovative approaches to environmental management in the development of hydrocarbons in the Arctic shelf, *The Polar Journal*, doi: 10.1080/2154896X.2021.1889836M. (2021)
3. A. Ilynsky, *The Oil and Gas Industry of Russia: Problems and Priorities of Development. Monograph.* (St. Petersburg, Peter the Great St. Petersburg Polytechnic University, 2020)
4. T. Unguryanu, et al. Health risk perception of environmental factors in an urban population in Northwest Russia, *The European Journal of Public Health*, **20**, pp. 173-173 (2010)
5. V. Chashchin, et al. Health Risk Modifiers of Exposure to Persistent Pollutants among Indigenous Peoples of Chukotka, *International Journal of Environmental Research and Public Health*, **17**, 128, doi:10.3390/ijerph17010128. (2019)
6. T. Sorokina, et al., Impact of climatic effects on the environment and the economy of the Russian Arctic, *IOP Conference Series: Earth and Environmental Science,* 539, 012033, doi: 10.1088/1755-1315/539/1/012033 (2020)
7. A. Cherepovitsyn and O. Evseeva, Parameters of sustainable development: Case of Arctic liquefied natural gas projects, *Resources*, issue **21100808642(10)**, pp. 1-27, doi: 10.3390/resources10010001 (2021)
8. PAME, Large Marine Ecosystems, Arctic Ecosystems Available online : https://www.pame.is/projects/ecosystem-approach/arctic-large-marine-ecosystems-lme-s (2014)
9. F. Danielsen, N. D. Burgess, P. M. Jensen, K. Pirhofer-Walzl (2010). Environmental monitoring: the scale and speed of implementation varies according to the degree of people's involvement. *Journal of Applied Ecology*, Vol. 47, Issue 6
10. V. S Litvinenko, I. Bowbrick, I. Naumov, Z. Zaitseva, (2022) Global guidelines and requirements for professional competencies of natural resource extraction engineers: Implications for ESG principles and sustainable development goals , 2022, 338, 130530
11. M.S. Tysiachniouk, et al., Towards Understanding Benefit Sharing between Extractive Industries and Indigenous/Local Communities in the Arctic, *Resources*, **9**, 48, doi: 10.3390/resources9040048 (2020)
12. Department for Business, Energy and Industrial Strategy UK (2017) Social Enterprise: Market Trends 2017. Department of Media UK. Retrieved on June 17, 2021 Available online : https://assets.publishing.service.gov.uk/government/uploads/system/uploads/attachment_data/file/644266/MarketTrends2017report_final_sept2017.pdf (2021)
13. B. Urban, Evaluation of social enterprise outcomes and self-efficacy, *International Journal of Social Economics*, **42**, 163-178. 10.1108/IJSE-03-2013-0071. (2015)
14. I.G. Gerasimova, I.S. Oblova, E.I. Golovina, The Demographic Factor Impact on the Economics of the Arctic Region, Resources, 10, 117. doi: 10.3390/ resources10110117 (2021)
15. E.B. Mazakov, K.V. Matrokhina, V.Y. Trofimets Traffic management at the enterprises of the mineral industry, *Advances in raw material industries for*

https://doi.org/10.1051/e3sconf/202337806018

sustainable development goals (London, CRC Press, 2021) pp. 397–405

Environmental tourism potential of the Arctic regions of Russia

Natalia Roslyakova[1], *Ludmila* Dorofeyeva[2,*]

[1]Candidate of Economic Sciences, Senior Researcher, V.A. Trapeznikov Institute of Control Sciences, Russian Academy of Sciences, 117997, Moscow, Russian Federation
[2]Candidate of Economic Sciences, Associate Professor, North-west Institute of Management RANEPA, 199178, St. Petersburg, Russian Federation

Abstract. Tourism is an important field for international cooperation. The Arctic zone has a potential for the tourism activities development, the question remains on what policy principles should forming this industry be carried out. The goal of the work is to determine the advantages of the Arctic territories as a network cluster that takes into account modern security requirements and creates conditions for the development of territories based on tourist attractions. The work considers the best practices in the field of tourism clusters development in the world, Russian legislation aimed at activating the process of cluster formation in the field of tourism, key factors that determine the unique characteristics of the tourism sector of the Arctic. The main methods used were analytical generalization, comparative analysis, retrospective analysis of the legislative framework in the tourism sector of Russia. As a result, stating certain successes in the field of tourism development in the Arctic, the authors note the need for the development of the environment to attract the local population to forming cultural code of territory and note the key areas of digitalization that could contribute to improving the processes of tourist cluster formation in the Arctic territories.

1 Relevance

The year spent in long lockdowns has once again confirmed the importance of free movement, which acutely affects people's health and well-being. This made it clear that tourism: firstly, it promotes the balanced and harmonious development of the individual as a person, supplying them not only with recreational opportunities, but also opening up new horizons of nature and society, history and the modern world, as well as developing the world cultural heritage; secondly, by offering specific goods and services, tourism forms a special branch of national and global economy, and thirdly, it is a means of individuals to perceive themselves as an integral part of an emerging united humanity, thereby helping to shape global societal change, acting as one of the most important types of migration of people within and beyond their countries. Moreover, the classic mass routes have become in part inaccessible due to the danger of the spread of the cornovirus. And more isolated and somewhat less popular destinations have benefited from greater safety and sustainability. And the Arctic here can become a flagship destination for eco-tourism, which has been developing actively in recent years, increasing both economic the turnover and the number of people traveling around the world. This, in its turn, increases the anthropogenic load not only on the mass and well-developed tourist routes, but also forces the industry to look for new unique places for tourists and mechanisms for improving environmental friendliness and sustainability. In the past few years, the Arctic, which provides the opportunity to visit unique natural landscapes, has been a planetary centre of environmental tourism. This leads to the need to work out a strategy for the tourism development in the Arctic Zone, taking into account the crucial environmental role played by this region.

2 Review of approaches to tourism

Tourism, in its modern sense, is a fairly young phenomenon with deep historical roots. And as for any complex socio-economic and spiritual-cultural category with a variety of functions performed and a large number of forms of manifestation, there is no single canonical definition for the concept of "tourism". In the early interpretations, tourism was understood as a temporary movement of people for a certain period of time from a permanent place of residence. In the process of historical development, this term acquired new interpretations and various additions.

The tourism industry includes hotels, catering establishments, sightseeing firms, museums, cinemas, sports centers and other facilities. All of them are engaged in organizing the conditions for tourism and recreation. If we turn to the etymology of the word "tourism", then initially it came from the combination of the words "grand tour" or "great tour" - a familiarization

* Corresponding author: dorofeevalucy@gmail.com

trip. In the 17th and 18th centuries, such trips were made by representatives of the aristocracy to get to know different countries, cultures and nations. With the development of civil society, tourism became available to more and more people, and at the present stage has become a mass phenomenon.

According to the definition adopted by the United Nations (hereinafter - the UN) in 1954, "tourism is active recreation, affecting the health, physical development of man, associated with movement outside the permanent place of residence" [1,2].

In 1994, the UN World Tourism Organization (UNWTO) [3] approved a new expanded interpretation of the concept of "tourism", replacing the 1954 definition, which belongs to the UN Statistical Commission and was the first official and universally used one. Distinctive features of the new definition are the established "time limits" of tourism - " not exceeding one year in a row", as well as the specified possible purposes of tourists - "for recreation, business and other purposes" [4].

In order to streamline and systematize terminology in the field of tourism, as well as to make it easier to find the necessary definitions, at the beginning of the 21st century, the World Tourism Organization published the Thesaurus on Tourism and Leisure Activities [5] in three of the six official UN languages. This Thesaurus is a collection of terms and concepts directly related to tourism, tourism infrastructure and leisure activities [6].

Another definition is provided by the World Trade Organization (WTO), which interprets tourism as "the activities of people who travel and stay in places outside their usual environment for a period not exceeding a year for leisure, business or other purposes not associated with activities subject to remuneration for the place of stay "[7]..

According to the International Association Of Scientific and Practical Experts (AIEST) [8], in the relevant field, the concept of "tourism" refers not only to the economic sphere, but equally to the social one. In this context, tourism is considered as a set of connections and phenomena that arise in the process of movement of people on territory, different other than a permanent residence, with the purposes different from their work [5].

Turning to the consideration of tourism sphere in Russia, first of all, it should be noted that so far our country has not formed aconceptual apparatus in this area, exceptfor a number of concepts enshrined in Federal Law No. 132-FZ "On the foundations of Tourist Activity in the Russian Federation" (hereinafter 132-FZ) [9].

In Russia, tourism as a phenomenon in the modern sense (i.e., mass tourism) appeared during the period of the USSR. It was one of the main elements of social and cultural life of the population. Particular attention was paid to the development of health tourism in sanatoriums and rest homes, which was in line with the state policy of health protection of each citizen and the population of the state as a whole. Popularization of mass tourism among Soviet citizens took place in the context of "rational use of free time." It is important to note that at the turn of the 1970s and 1980s tourism in the USSR reached really high economic indicators compared to the world tourism industry, thanks to state management and, accordingly, to the absence of the risk of bankruptcy. After the collapse of the USSR, the transition to market principles of tourism organization in Russia led to a significant slowdown in the development of the industry. State regulation of the industry in that period was haphazard. In 1996, the first federal law in the field of tourism was issued, where it is interpreted as a sphere of services and becomes an object of state regulation in order to ensure constitutional rights of citizens to recreation and freedom of movement. In 2004, tourism came under the jurisdiction of the Ministry of Economic Development, and subsequently came under the jurisdiction of the Federal Agency for Tourism (Rosturizm). During this period, the key directions of state policy in the active development of tourism were not only to create comfortable conditions for tourists, but also to ensure travel safety.

the latest version of the Federal Law No.132-FZdefines tourism as "temporary departure of Russian citizens, foreign citizens, and stateless persons from their place of permanent residence for medical, health precautions, recreational, educational, sports, professional and business and other purposes without engaging in activities related to income generation from sources in the country of temporary stay"[9].

The Strategy for tourism development in the Russian Federation for the period up to 2035 continues in the post-pandemic era [10]. The implementation of the Strategy implies e comprehensive development of domestic and inbound tourism in the Russian Federation by creating conditions and promoting quality and competitive tourist product in the domestic and international tourist markets, strengthening the social role of tourism and ensuring the availability of tourist services, recreation and health for citizens of the Russian Federation. Regional strategies for tourism development are elaborated and adopted on the basis of this Strategy. The availability of regional strategies makes it possible to take into account a huge variety of types of tourism that may take place in Russia, and strategizing ensures identification and fixation of priorities relevant the region. In accordance with the available recreational potential of the country, the following main types of tourism have formed in Russia:

- cultural and educational tourism, which is characterized by a high level of travel costs
- active forms of tourism such as adventure and sports tourism, are becoming more and more popular in the world
- special forms of tourism, such as medical and health tourism, event tourism, eco-tourism, educational tourism and science tourism
- business tourism, which may include business meetings, conferences and various events.

The general state of the tourist market has been changing lately. This is due to the following factors, which, in turn, determine the positive and negative aspects. Negative factors include:

- aggravation of the political situation and complication of intercultural communication
- difficulties in obtaining visas
- a decrease in the level of safety of outbound tourism due to the spread of coronavirus infection
- unstable situation with the currency
- insufficient development of infrastructure for inbound tourism
- lack of qualified personnel in the industry
- limited access to financial risk insurance instruments for travel operators
- lack of effective investment incentives for the development of the tourist industry.

For Russia, as the largest country by area, it is extremely important to form and successfully operate tourism management bodies, both separately at the federal, regional and municipal levels, and in interconnection.

3 Formation of tourist potential in the Arctic zone

In modern Russian reality, unfortunately, there is no unified approach to defining the concept of "tourist potential", as evidenced by the analysis of tourism development programs in the regions. Most often, the tourist potential is understood as a certain number of objects used as an object of tourist attraction. This shortcoming can cause ineffective formation of development strategies for the Arctic regions and, as a result, ineffective use of budget funds, which, naturally, is a serious problem.

Before defining the concept of tourist potential, it is necessary to consider the main approaches to its interpretation. At the moment, there are three points of view on which further approaches to the understanding the tourist potential are based. The first approach is based on the assessment of supply in the tourism industry, the second - on the assessment of demand among tourists using tourist services, the third is a synthesis of the above approaches [11].

The first approach, which proceeds from the evaluation of a proposal, is found in the works of such authors as D. Spotts [12], B. Boers and C. Cottrell [13], A. Sazykin [14], D. Ushakov [15]. This approach is based on the view that tourist potential depends on two fundamental factors [11]. The first factor is the quantity and quality of tourist attractions. The second, in its turn, implies the development of the infrastructure of the tourism industry. This approach in modern Russian reality is the most frequently used by both researchers and representatives of state authorities. Under the first approach, tourist potential is based on both object and an subject" basis. The object basis, in this case, implies the presence of tourist attractors and infrastructure facilities. It should be noted that, if we follow this term, the value of tourist flow directly depends on the number of museums, tourist routes, hotels and other means of accommodation in the region. But, unfortunately, as seen in practice, this hypothesis does not always work. The creation of excessive infrastructure in the Arctic will lead to the destruction of the ecosystem, and the cost of visiting the Far North will remain high (due to the geographical remoteness) and will not allow the tourist to scale the tourist flow. Therefore, this approach cannot be used when considering the tourism potential of the Arctic.

The representatives of the second approach, which proceeds from the assessment of demand, are the following authors: J. Kripppendorf [16], M. Opperman [17], S. Jurowski and M. Olsen [18]. For example, in the works of J. Kripppendorf, the potential is viewed through the prism of the possibility to create the "final tourist product" and the ability to develop economically sound tourism (with an emphasis on the limitations and opportunities to use these objects to form the final tourist product) [11]. Nevertheless, from the practical point of view, this approach is considered more difficult to assess, but from the methodological point of view, it is more correct. The volume of the tourist flow is the basis for expressing tourist potential. But, unfortunately, the representatives of the second approach did not come to a consensus on what kind of tourist flow should be taken into account in this case (current, which is measured objectively, or predicted, the evaluation of which cannot avoid subjectivity). The only thing that the researchers of the second approach agree on is that the main critics of the tourist attractiveness of the region are tourists. The measure of the tourist potential of the region directly depends on them and their demand.

The pandemic showed a weak correlation between tourist potential and flow. The lockdown led to zero tourist flow, while the demand and desire to travel increased (with the slightest removal of restrictions, the cost of flights to a tourist destination increases many times over). The Arctic has tremendous touristm potential, while the current flow of tourists is low, primarily due to the cost of travel, as well as the need to develop regulations that allow travel in the Arctic zone without harming the ecosystem of the region.

In the process of forming the final approach, such researchers as M. Bulay and S. Yatu [19], L. Vereiro and J. Robeiro [20] and S. Formica [21] played an important role. It is this approach that synthesizes the first two [11]. In the third approach, h main property can be distinguished, which is that the tourist potential takes into account both demand and supply. In this case, the tourist offer of the territory, which includes attractors and infrastructure, will be the most important factor. However, no matter how important the tourist offer is, it directly depends on tourist demand (this may include the actual understanding of the brand of the territory, its history, the attractiveness of this territory for tourists in general). It should be added that in order to fully assess the tourist potential of the territory, it is necessary to understand that tourist supply and tourist demand have a direct impact on each other.

The above-mentioned possibility of the Arctic territories to form a tourist flow directly depends on the existing resources of the tourism industry [22]. These resources include:

• Tourist attractors

- Infrastructure in the tourism sector (including its development)
- Transport infrastructure and accessibility of the region.

4 Directions for the development of Arctic tourism

The previously identified factors lead to the need for state regulators to pay attention to the development of tourism in the Arctic Zone. The Federal Agency for Tourism (Rosturizm) [23] notes a rather weak tourist capacity of the Arctic territories of the Russian Federation compared to neighboring countries such as Norway and Sweden (the number of tourists visiting the Arctic in Russia is more than 50 times less than similar indicators in other Arctic countries).

This is due to several factors, which include poor infrastructural development, as well as the lack of a legislative framework governing tourism in the Arctic Zone. The formation of new reception infrastructure, reconstruction of ports is required, which will allow developing the most promising type of tourist travel for this region - cruise tourism. we will need active participation of the state in attracting cruise companies to work in Russia, the position of the federal center is important here. We should not forget about the commercial component of the business: the Arctic requires investment and individual work with investors, possibly, including a system of tax preferences for business. In general, we need to understand the fragility of the ecosystem of the Arctic zone and make any decisions taking into account the need to preserve the unique nature of the region. As the experience of other Arctic countries shows, Russia has great potential for the formation of a tourism cluster in the Arctic and is only at the very beginning of the journey [24].

In many ways, the direction of development of the Arctic is related to political decisions, today the region is perceived as a resource center, while other possible sectoral specializations are not given due importance. With developing tourism activities, especially in the Arctic, one should avoid the mistakes made by neighboring states, which include the inequitable distribution of the economic effect, when only the anthropogenic load remains on the territories, and all the profits are concentrated with tour operators. This can be avoided by organizing the work of small vessels, the development of reception infrastructure and the introduction of additional fees for environmental damage. Of course, it is necessary to provide incentives, which include preferential taxation at the initial stage of investment. People travel quite often, which obtaining an interesting unique tourist product can arouse considerable interest, therefore it should be remembered that in the Arctic not only the environment is fragile, but also the social organization of people living in this territory.

In the Arctic Zone there are protected zones that attract tourists, while it is not possible to organize tourism infrastructure on this territory, this is prohibited by current legislation. It is necessary to introduce into the legislative base the possibility of allocating municipal land plots on the borders of protected areas for the creation of mini-tourist clusters.

Even with the further development of tourism in the Arctic, it will not become widespread, therefore, so we should use the opportunities of cross-border cooperation using the experience of other Nordic countries to return investments. The Russian Federation has a unique opportunity to combine visits to eco-environmental trails and eco-parks with places of historical and cultural heritage, which is certainly a competitive advantage.

A unique place in the entire region is the Russian Arctic National Park [25]. With a fairly short season and a lack of the necessary number of equipped vessels capable of calling at Arctic ports, the park continues to attract tourists from all over the world. Today, the infrastructure formed back in the Soviet era is actively used, and new facilities are created to make tourist trips more comfortable. The main way to visit the park for tourists is cruise tourism, which, in its turn, is quite expensive for the average person; for example, at the moment, only 5% of the total number of tourists are citizens of the Russian Federation. development of small aviation can significantly reduce the cost of tours and intensify domestic tourism to the Arctic.

It is worth noting the importance of the development of interregional cooperation, especially in the context of the development of the Northern Sea Route [22], which can become not only a transit route, but also a tourist route. In this regard, the territory of Kamchatka can become a logistics center and a large tourist cluster for tourists wishing to travel in the Arctic. In 10 years it is planned to increase traffic capacity of Kamchatka ports by 10 times, which is facilitated both by development of infrastructure of the region (ports and airport) and by development of country's opportunities for navigating of ships along the Northern Sea Route (nuclear fleet).

The Russian Geographical Society (hereinafter RGS) [26] welcomes attention to the Arctic, educational activities are an important element in the development of a responsible attitude towards the country and the world in general. For the RGS, the main areas of activity in the Arctic Zone are educational and research activities. At the same time, tourism helps to build communication between all stakeholders - the population, businesses and the state. To formulate development strategies for the Arctic Zone, it is important to take into account the opinion of the business community for the formation of a full-fledged industry specialization.

The Arctic is actually quite different and its Russian component is the largest in the world. At the moment, Russia is not developing the Arctic from scratch, there is already a legislative base, which is certainly not enough to overcome all the problems facing today. Solutions are needed for the development of transport accessibility, taking into account the environment [27]. It is important to use the opportunities of cross-border cooperation in tourist activities, which makes it possible to form tourist reception centers in the territory, as well as to activate the local population for the development of folk crafts.

To develop cross-border cooperation, it is important to make changes in the country's visa policy.

5 Conclusions

The plan for the development of the Arctic as a center of eco-tourism requires a unified strategy with the definition of targets of the desired number of tourists and the required amount of infrastructure. At the same time, the opportunities of the Arctic zone can be used to develop environmental tourism in the northern regions. It is worth separating the application of the of state and private business forces, with the market's efforts concentrated on creating a tourist offer and increasing competition. The attention of the state should be aimed at attracting the local population to the creation of unique tourist products, taking into account the culture and traditions of the peoples of the Far North. In this process, we can turn to the institution of self-employment, which was rapidly introduced in all regions of Russia in July 2020 in order to reduce the negative impact of the crisis associated with the spread of the pandemic coronavirus.

In 2020, the Federal Agency for Tourism made significant efforts to create a separate non-resource sector of the economy in the Russian Federation, this has fully activated eco-environmental tourism in the middle belt of the country, but has not yet had a significant impact on the Arctic regions. Sustainable development and eco-tourism can become a source of investment for northern cities and give new meaning to spatial development. The pandemic has created unique conditions for the launch of domestic tourism, remote work allows small trips for 1-2 days; the developed national project on tourism "Tourism and hospitality industry" (Rostourism, Federal Agency for Tourism) [23] assumes many new support tool, which will affect the spatial development of the territory. Systemic and transparent management mechanisms for the industry will be created, the regulatory framework for the industrywill be improved taking into account environmental and social well-being norms, and training for the tourism sector will be organized. Specialists will receive the up-to-date digital and social knowledge and skills. The national project will make it possible to implement the potential of Russia in general and the Arctic in particular on the basis of:

- incorporatingthe priority of eco-environmental tourism and the active development of the territory, the emergence of a design code and the restoration of partially lost attractors
- taking into account modern threats associated with the spread of the coronavirus pandemic, creating conditions for ecology environment, social stability, public health and sustainability of the territories at whole
- increasing the safety of objects of the tourist industry facilities
- introduction of digital technologies in the tourism sector, in particular in the area of obtaining visas, booking accommodation and rental services, information support on routes, access to databases of the sights of the regional, online currency conversion
- elimination of administrative barriers to the creation of a tourism product, working on a one-stop-shop principle, with the inclusion of digital standards for the provision of public services.

The world is changing, people's needs are changing, the mechanisms of state influence are transforming, the sphere of tourism around the world is growing and developing, there are specific features of different types of tourism, in accordance with this, state management of environmental tourism should be improved in the context of constantly changing conditions of the internal and external environment.

References

1. United Nations Organization Convention on Customs Concessions for Tourists. Available online: http://docs.cntd.ru/document/1901570 (1954)
2. United Nations Additional Protocol to the Convention on Customs Concessions for Tourists, concerning the importation of tourism-related information documents and Available online: http://continent-online.com/Document/?doc_id=1009821 (1954)
3. World Tourism Organization (UNWTO) Basic Documents Tourists Available online: https://webunwto.s3.eu-west-1.amazonaws.com/s3fs-public/2019-11/130718basicdocumentsenweb.pdf (1993)
4. Concepts, definitions and classifications for tourism statistics ("Recommendations for Tourism Statistics" of the UN World Tourism Organization) Available online: https://www.unwto.org/ru (1993)
5. Tourest. Tourism Handbook Available online: http://tourest.ru/
6. E.L. Pisarevsky, *State and municipal management in the field of tourism: textbook 2nd ed.* (Moscow, Federal Agency for Tourism, 2017)
7. World Trade Organization (WTO) Available online: https://www.un.org/ru/wto/ (1995)
8. Worldwide network of tourism experts (AIEST) Available online: https://www.aiest.org/aiest-profile/profile/ (1951)
9. Federal Law of 24.11.1996 N 132-FZ (as amended on 09.03.2021) "On the Basics of Tourist Activity in the Russian Federation" Available online: http://www.consultant.ru/document/cons_doc_LAW_12462/, free (1996)
10. Strategy for the development of tourism in the Russian Federation for the period up to 2035 (approved by the Order of the Government of the Russian Federation dated September 20, 2019 No. 2129-r) Available online: https://tourism.gov.ru/upload/iblock (2019)
11. T.S. Krasnikova, The tourist potential of the territory and the need for its development, *Society: politics, economics, law,* **6** (2015)

12. D.M. Spotts, Regional analysis of tourism resources for marketing purposes, *Journal of Travel Research*, **35(3)**, r 3-15 (1997)
13. B. Boers and S. Cottrell, Sustainable Tourism infrastructure Planning: A GIS-Supported Approach, *Tourism Geographies*, 9(1), pp. 1–21 (2007)
14. A.M. Sazykin, Geographer's polemic notes on the problems of terminology in recreational geography and tourism Tourism in the Far East: business, investment strategies, education and ecology: materials of the region. *Scientific-practical conf. "Daltur-2001" (Vladivostok),* pp. 306–310. 7. (2002)
15. D.S. Ushakov, *Strategic planning in tourism* (Rostov n / D, 2007), p. 288
16. J. Krippendorf, Marketing im Fremdenverkehr. 2 aufl. (Bern; Frankfurt am Main, 1980), p. 168
17. M. Oppermann, Regional Aspects of Tourism in New Zealand, *Regional Studies*, 28(2), pp. 155-168 (1994)
18. C. Jurowski and M.D. Olsen, Trends in Tourist Attractions, *Journal of Travel & Tourism Marketing,* 4(1), pp. 71–96 (1995)
19. C. Iatu and M. Bulai, New approach in evaluating tourism attractiveness in the region of Moldavia (Romania), *International Journal of Energy and Environment,* **5(2)**, pp. 165–174 (2011)
20. J.C. Ribeiro and L.C. Vareiro, What influences a tourist to return to a cultural destination? *International Journal of Tourism Research*, **21(2)**, pp. 280-290 (2019)
21. S. Formica, Destination attractiveness as a function of supply and demand interaction. Unpublished doctoral dissertation (Blacksburg, 2000), p. 203
22. L.V. Dorofeeva et al., The transport relationship of the Arctic with the Far East, Siberia and the Urals and the issues of socio-economic development of territories, *IOP Conf. Series: Earth and Environmental Science,* **539**, 012083 (2020)
23. Federal Agency for Tourism (Rostourism) Available online: https://tourism.gov.ru (2021)
24. Roscongress Foundation information analytical platform Available online: https://roscongress.org/ (2020)
25. National Park "Russian Arctic" Available online: http://www.rus-arc.ru/ (2020)
26. Russian Geographical Society (RGO) Available online: https://www.rgo.ru/ru (2021)
27. L.V. Dorofeeva, *The essence and features of the infrastructural potential of regions Regional economy and development of territories Collection of scientific articles* (St. Petersburg, 2017), p. 183-189

https://doi.org/10.1051/e3sconf/202337806020

Development of tourism in the Arctic regions of the Russian Federation: the environmental aspect

Artur Kuchumov[1,*], *Galina* Karpova[2,*], *Elena* Pecheritsa[3,*] and *Marina* Voloshinova[4,*]

[1]Candidate of Sciences in Economics, Associate Professor, St. Petersburg State University of Economics, 191023, St. Petersburg, Russian Federation
[2]Doctor of Economics, Professor, St. Petersburg State University of Economics, 191023, St. Petersburg, Russian Federation
[3]Candidate of Sciences in Sociology, Associate Professor, St. Petersburg State University of Economics, 191023, St. Petersburg, Russian Federation
[4]Candidate of Sciences in Economics, Associate Professor, St. Petersburg State University of Economics, 191023, St. Petersburg, Russian Federation

Abstract. The article considers the opportunities for the development of tourism in the regions of the Arctic Zone of the Russian Federation in the environmental aspect. The authors examine the existing problems of the development of Arctic tourism in the Russian Federation, which include: low attractiveness of Arctic tourism for most Russian tourists, large investments with a long payback period, unacceptability of anthropogenic impact on the Arctic ecosystem, lack of modern passenger terminals, difficult transport accessibility of locations with specially protected natural areas. Analyzing the foreign experience of tourism development in the Arctic territories, the authors offer recommendations for the development of tourism in the Arctic zone of the Russian Federation. It also includes: support for indigenous peoples of the Arctic in order to increase employment of indigenous peoples while preserving their lifestyle, habits, and culture; assistance in cooperation and interaction with PR and marketing agencies, as well as design studios for creation of new souvenir lines; assistance in organizing excursions acquainting with the life of local residents, their traditional cuisine and local rituals, conducting educational seminars, organizing events for temporary residence in families of indigenous minorities; creation of information and tourist centers; development of events, ethnographic, and environmental types of tourism, intra-city tourism, addressing the lack of modern accommodation facilities and transport accessibility; assessment of environmental potential of specially protected natural areas in the Arctic Zone of the Russian Federation, so that the latter do not suffer from an influx of visitors; building of electric and hybrid polar expedition ships.

1 Introduction

The Arctic as a tourist route is not a popular destination among tourists. When choosing a destination for the next trip, Russian tourists limit themselves to the usual options: Moscow, St. Petersburg, Krasnodar region – or they choose from foreign destinations: Europe, Turkey, and Thailand. Despite the fact that the Arctic region has a unique and absolutely incredible flora and fauna, there is a stereotype about the eternal cold and endless ice, that the trip is long, and there is nothing to see. This is why tourism is a key part in the development strategy of the Arctic zone of Russia, which envisages the development of the Arctic until 2035 [1].

The development of tourism in the regions of the Arctic zone of the Russian Federation (AZ RF) is able to contribute to solving a number of social problems, such as low living standards of, unemployment, negative activity of young people in their spare time. For residents of the Arctic regions of the Russian Federation, tourism means increased employment, income, education and culture, and life satisfaction.

In accordance with the Decree of the President of the Russian Federation No 400 of 02.07.2021 "On the National Security Strategy of the Russian Federation", the objectives of ensuring environmental safety and rational environmental management are achieved through the implementation of state policy aimed, including, on solving environmental problems and rational use of natural resources of the AZ RF [2].

Cooperation in tourism development requires complex and strenuous management, joint operation of all tourism subjects, despite their different interests and goals. Due to the fragmented nature of tourism, individual tourism entities can rarely act autonomously. It is unlikely that any of the individual tourism entities can own or control all the relevant components necessary to offer a tourism product. Thus, the creation of a favorable political and operational framework for the production, creation and ultimate delivery of a tourism product depends on a collaborative effort.

* Corresponding author: arturspb1@yandex.ru, karpovaga@rambler.ru, helene8@yandex.ru, voloshinova@unecon.ru

https://doi.org/10.1051/e3sconf/202337806020

2 Materials and methods

Improving the competitiveness of the Russian tourism sector in the Arctic region will definitely lead to the acceleration of economic development of the Arctic territories of the Russian Federation [3].

The development of Arctic tourism is currently of great interest to Russian and foreign researchers.

For example, in [4], the authors examine trends in the development of Arctic expedition cruise tourism and current needs for Arctic research and monitoring, clarifying where exactly the expedition cruise tourism industry can have the greatest impact on providing data to the scientific community. The authors also compare the legal and regulatory context in Antarctica and in the Arctic and discuss how these differences might affect the widespread use of civil science.

The authors of [5] explore tourists' perceptions of the Arctic through the prism of the Instagram, a social network for photo sharing. The study focuses on the specific tourism experience of crossing the Arctic Circle in Rovaniemi, Finland, as in tourism it represents an "official" entry into the Arctic. The study also seeks to speculate on the future experience of crossing the polar circle with the development of new technologies, such as augmented and virtual reality.

The paper [6] discusses the external and internal factors that support or challenge the possible transformation of Arctic Sweden into a major ski destination in a changing climate. For instance, in Sweden (as in Norway, Russia, and Finland) there are conflicts in the sphere of land use when developing large ski resorts in the Arctic zones, especially when there are plans to build them on the territories of nature reserves and specially protected natural areas, which themselves are not only important attractions of natural and environmental tourism, but also traditional sources of livelihood for indigenous peoples of the Arctic territories.

Cooper E. A. [7]aims his study at exploring the potential of cultural centers as spaces that promote interaction between tourists and locals. This is done through the case study of the Katuak Cultural Center in Nuuk, Greenland. Combining the theories of cultural contact and accommodation, the article analyzes how locals and tourists use and perceive the cultural center. The author suggests ways to encourage positive interaction both at the local community level and at the level of regular visitors.

realities and opportunities of tourism in the Arctic are explored in [8]. The authors offer a descriptive overview of tourism statistics and its estimates.

The authors of these papers offer ideas on how to use the geographical, industrial and cultural identities of Arctic regions to present new "place innovations," configurations, content and communications that are attractive and accessible to both local people and tourists. The concept of "place innovation" helps to define the link between the identity of place identity and its main components: configuration, content and communication [9]. identity derives from the cultural, social, economic and geographical location of the tourist destination [10]. configuration includes the architecture, design and other physical components of a tourist destination. content includes commercial, public and non-profit services and activities that are offered at a particular tourist destination. With innovatively combined, these components can attract existing and potential residents, tourists, entrepreneurs, investors and other stakeholders. The authors studying territorial innovation highlight the need for parallel integration of the processes of urban planning, destination design, business promotion, cultural development, social cohesion, etc., suggest to surely take into account environmental aspects in the development of tourism in the Arctic territories [11-17, 23, 24].

The article used methods of scientific literature analysis, statistical data analysis, logical and content analysis, search, systematic approach, and comparison.

3 Results

There is a worldwide ranking of popularity of the Arctic among tourists, in which Russia ranks fifth. The annual global flow of tourists to the Arctic ranges from 10 to 11 million people. The annual flow of Russian tourists to the Arctic is just over 1 million people. And only half of these tourists are visiting for the purpose of recreation; the other half are business tourists, that is, those whose stay in the Arctic has nothing to do with recreation.

By the number of tourists in the Arctic, the Russian Federation is overtaken by Norway with 3 million people per year, as well as Iceland and the United States (Alaska).

As for the AZ RF, according to Rosstat data for 2019 (see Figure 1), the Krasnoyarsk region hosts the most tourists. More than 850,000 people come there. More than 300 000 tourists annually visit the Murmansk and Arkhangelsk regions. The third place is occupied by the Republic of Komi and the Yamalo-Nenets Autonomous Region, which host more than 200,000 tourists annually. The least number of tourists visit the Nenets Autonomous District - just over 13,000 people. Figure 1 lists some performance indicators of tourist organizations in the AZ RF.

Region	Number of collective accommodations (c.a.)	Places in c.a.	Overnight stays in c.a.	People housed in c.a.	Russian citizens housed in c.a.	Foreigners housed in c.a.
Archangelsk Region	186	14 052	1 180 457	359 794	347 028	12 766
Krasnoyarsk Region	420	34 776	3 520 570	861 541	808 051	53 490
Murmansk Region	163	8 139	1 004 277	319 580	263 791	55 789
Nenets Autonomous District	8	241	33 058	13 018	12 685	333
Komi Republic	128	7 310	923 185	233 373	228 144	5 229
Chukchi Autonomous District	48	984	99 816	27 838	27 426	412
Sakha Republic (Yakutia)	218	5 756	738 206	201 496	194 039	7 457
Yamalo-Nenets Autonomous District	103	4 269	394 690	208 076	203 763	4 313

Fig 1. Indicators of the economic activity of tourist organizations in the AZ RF for 2019. Source: Compiled by the authors based on the Rosstat data [18]

According to the Rosstat data, there are more than 1,200 collective accommodation facilities throughout the

Arctic zone of the Russian Federation, and more than 100 active tour operators. In recent years, before the coronavirus pandemic, there was a positive trend in the influx of tourists and, accordingly, revenues of tourism businesses.

If we consider the development of tourism in the AZ RF realistically, there is a lot of difficulties. On the one hand, the development of Arctic tourism is a priority for the state, and above all it is associated with the development of infrastructure at the expense of small and medium-sized businesses, in their turn, creating jobs for the local population, thereby replenishing the regional budget. On the other hand, the task of popularizing Arctic tourism is quite challenging, due to the fact that it requires huge investments, and their payback period is quite long, more than 10 years. And here we should also take into account that a tenfold increase in tourist traffic to the Arctic is impossible, due to the fact that we cannot allow any anthropogenic impact on the environment of the Arctic zone caused by large flows of tourists, which may negatively affect the environment of the Arctic.

Nevertheless, the Russian Government has a plan to increase the tourist flow to the Arctic to 3 million people per year, which is a permissible limit in terms of anthropogenic load on the ecosystem and should not violate the environmental principles of management in the Arctic zone. The figure of 3 million tourists a year should be reached within 15 years. An example of the work that begun is the detailed development of a strategic document regulating all tourist destinations in the region by the municipality of Taimyr and the Norilsk Development Agency, the main aim of which is to calculate the anthropogenic load on the ecosystem of the Taimyr Peninsula, for which purpose special tourist counters are installed in Norilsk airport. There are plans to install identical counters in the seaport of Dudinka, with the help og which it will become possible to calculate and forecast existing and foreseeable tourist flows.

The main problem of the Arctic region in terms of the organization of tourist flows is the lack of logistics infrastructure: a modern passenger cruise port, the availability of which would allow consolidating tourist flows. Also, we need to consider the difficulties faced by tourists in reaching the locations of specially protected natural areas: it is necessary to develop infrastructure for civil aviation, in particular heliporter landing sites facilitating flights from one archipelago to another.

The concept of sustainability in the context of tourism has been interpreted and rethought in various ways in recent decades, based on the interests of people or institutions. However, the danger of using different interpretations of sustainable tourism can lead not only to a certain degree of mistrust between stakeholders, but even to conflicts. The lack of consensus on the importance of sustainability in the context of modern tourism has become a serious trap in the development of sustainable tourism. The different interpretations are the result of different perceptions of tourism within two worldviews: on the one hand, the view of companies and their focus on business, on the other hand, the view of organizations and their attention to society and/or the environment. The first approach to sustainable tourism focuses on creating and managing businesses that can sustain themselves over a long period. From this perspective, the word "sustainable" is associated with supporting promotional programs that increase the number of tourists visiting tourist destinations. The second approach to sustainable tourism implies small-scale tourism, balanced with the social and natural environment, lowering the negative anthropogenic impact of tourism development. However, this view counteracts the concept of development of mass tourism and cannot alleviate the negative anthropogenic impact. Consequently, today in the Russian Federation, tourism, including in the Arctic territories, is considered as a tool for social and economic development. Tourism in this sense is used as a method of accumulating private and public revenues, which can later be used or invested in other development tools. From this point of view, tourism should not be small, since tourism is part of a broader policy aimed at creating a more sustainable society [15,19].

Experts warn that the growing tourist traffic in the Arctic increases the likelihood of a disaster such as an oil spill or sewage leak, that could wreak havoc in the pristine polar environment. According to experts, it is only a matter of time. For example, in 2004, a small cargo ship lost power while sailing through the Aleutian Islands of Alaska. When it ran aground, more than one million liters of fuel oil got into the water. Due to the remoteness, almost nothing could be salvaged [21]. Larger cruise ships carry much more fuel oil. The Arctic is subject to harsh and changeable weather conditions that complicate travel and pose hazardous to sailors. Sailing in high-latitude areas carries the risk of disruptions in of marine navigation and communication systems. In the event of an oil spill, accident, or equipment malfunction, the remoteness of the Arctic regions makes it almost impossible to respond effectively to emergencies [20]. Although there are arguments that such cruises increase awareness of climate change and the plight of endangered species such as polar bears, emissions from ships contribute to precisely these phenomena.

For a long time tourism has been considered a central principle of economic development plans in many countries, but it is dangerous to view tourism as just another Arctic development strategy, similar to mining or fishing, as this obscures the more direct link between tourist presence and activities and the livelihood and daily experience of local residents. With careful consideration and public participation and cooperation, tourism can be used to generate positive results through knowledge sharing and investment in nature conservation, but creating such mutually beneficial outcomes does not seem to be a priority for large companies. And while Arctic cities and towns, usually located in rural areas with harsh climates, sparse population, long distances, and scarce infrastructure, struggle to find innovative solutions to attract potential residents, tourists, entrepreneurs and investors, in a region where most of tourism is almost entirely focused on eco-friendly activities and cultural tours, visitors should be even more diligent to take into account the possible impact they have on the areas they are interested in. In other words, even the boldest desires cannot justify the harm they cause. As an example, we can consider Svalbard, one of the Arctic zones most severely affected by climate change to date. Steadily rising temperatures have caused more avalanches, the worst of which destroyed an entire street of residential buildings. Climate change is having a huge impact on the Arctic, and almost nothing on planet Earth is changing as quickly as the Arctic sea ice. satellite monitoring of ice levels began in 1979, and since then, every new report is frightening. Since 1980, temperatures in the Arctic have risen twice as much as in the rest of the world. 30 years later, both summer ice extent and thickness have almost halved, and nearly three-quarters of the ice has disappeared since 1980. In 2012, the ice area reached a record low level since monitoring began. One of the reasons for the rise in temperature in the Arctic is the self-sustaining loop, in which ice-covered areas are replaced by open water. This leads not only to less solar energy reflection, but also to more energy absorption by open water. Thin and cracked ice is also more susceptible to wind and currents, which bring warm water to the surface, boosting the melting effect. Increased ice melt not only contributes to local warming, but also affects ocean currents, leading to changes in our global weather systems and affecting climate beyond the Arctic. These effects are difficult to predict, and we do not know what impact they will have on overall sea-ice reduction in the future [22].

4 Conclusions

Thus, using the cluster campaign for the development of tourism in the regions of the Russian Federation, taking into account the environmental aspect, it becomes possible to recommend the implementation of the following measures.

To support the indigenous small-numbered peoples (ISP) residing within the AZ RF. Due to the fact that souvenir products made by the ISP in their current form are practically not in demand among tourists because more than half a century has passed since the time of their design and now they are not attractive for modern tourists and visitors to the Arctic territories, as they do not fit into their interior solutions and do not correspond to their aesthetic preferences, it is proposed to expand the range of souvenir products and traditional crafts of the ISP. In order to help the ISP in preparing souvenir products of appropriate design, as well as with marketing for their promotion, it is proposed to organize assistance in cooperation and interaction with PR and marketing agencies, as well as design studios to form new souvenir lines.

To develop event, environmental and ethnographic tourism. We propose organizing assistance to the ISP in the creation of modern ethno-settlements in terms of marketing, assistance with the organization of tours to get acquainted with the life of local residents, their traditional cuisine, participation in local ceremonies, and conducting educational seminars for the local population of the AZ RF preparing guidbooks on tourist routes, as well as organizing events for temporary residence in the families of ISP to get acquainted with their lifestyle.

At the same time, the main goal of the first two activities is to increase the employment of ISP as well as to preserve their lifestyle and culture.

To develop the intracity tourism in the cities of the AZ RF, we propose to develop intra-urban tourism. many cities in these regions lack unified structures uniting and integrating participants of the tourist cluster into a single conglomerate, so there is a need for the creation of information and tourism centers. Information and tourism centers should cooperate with the activities of museums, souvenir shops, catering enterprises (forming culinary trends specific to the ISP). Such measures should be aimed at improving the level of service for tourists in the territories of the AZ RF and should be one of the priorities in the activities of local administrations.

To solve the problem of the lack of modern accommodation and the development of transport infrastructure, it is proposed to attract construction organizations on mutually beneficial terms within the framework of public-private partnership for the emergence of new or reconstruction of old accommodation facilities, as well as for the construction of heliports that facilitate flights from one archipelago to another.

The potential of Arctic tourism, which is an important component of sustainable development of the AZ RF regions, consisting primarily in the unique nature of the Arctic, is used rather poorly. At the same time, there is a threat of degradation of fragile environmental system with the development of tourism in the Arctic. Two major limitations of Arctic tourism: environmental and logistical (poor accessibility, high cost), lead to marketing and management problems. Following the example of Scandinavian countries, it is necessary to assess the environmental capacity of specially protected natural areas in the AZ RF, so that the latter do not suffer from the influx of visitors, as is already partially happening in the popular Teriberka in the Murmansk region.

To neutralize the threats of fuel oil spills from cruise ships, construction of hybrid (equipped with batteries and hybrid power plants) polar expedition ships should be prioritized (following the example of Norway and France [21]).

References

1. Decree of the President of the Russian Federation of July 2, 2021 N 400 "On the National Security Strategy of the Russian Federation" Available online: http://www.consultant.ru/document/cons_doc_LAW

_389271/11f9b19337c1414c493bfd768cedffe7ff2cae88/ (2021)

2. F.G. Alberti and F. Belfanti, Do clusters create shared value? A social network analysis of the motor valley case, *Competitiveness Review*, **31(2)**, p. 326–350 (2020)
3. M. Dressler, Strategic profiling and the value of wine & tourism initiatives: Exploring strategic grouping of German wineries, International Journal of Wine Business Research, **29(4)**, pp. 484–502 (2017)
4. M.V. Ciasullo, R. Montera and R. Palumbo, *Online content responsiveness strategies in the hospitality context: exploratory insights and a research agenda Preprint TQM-12-2019-0299* (2021)
5. D. Chimirri, Collaborative configurations of tourism development: a Greenlandic example, *Journal of Tourism Futures*, **6(1)**, pp. 24–39 (2019)
6. A.R. Taylor , et al., Arctic expedition cruise tourism and citizen science: a vision for the future of polar tourism, *Journal of Tourism Futures*, **6(1)**, pp. 102–111 (2019)
7. A. Varnajot, Digital Rovaniemi: contemporary and future arctic tourist experiences, Journal of Tourism Futures, **6(1)**, pp. 6–23 (2019)
8. O.C. Demiroglu, et al., The last resort? Ski tourism and climate change in Arctic, *Sweden Journal of Tourism Futures*, **6(1)**, pp. 91–101 92019)
9. P.T. Maher, et al., Arctic tourism research: emerging scholars are moving the network forward, *Journal of Tourism Futures*, **6(1),** pp. 3–5 (2020)
10. E.A. Cooper, Cultural centres: a future for cultural Arctic tourism?, *Journal of Tourism Futures*, **6(1)**, pp. 57–69 (2019)
11. M.J. Enright, The Globalization of Competition and the Localization of Competitive Advantage: Policies Toward Regional Clustering, *The Globalization of Multinational Entreprise Activity and Economic Development ed N Hood and S Young* (London, Macmillan, 2000) pp. 303–331
12. J. Humphrey and H. Schmitz, How does insertion in global value chains affect upgrading in industrial clusters?, *Regional Studies*, **36**, pp. 1017–1027 (2002)
13. M.P. Voinarenko, *Clusters in the Institutional Economy* (St. Petersburg, ANO IPEV, 2013) p. 494
14. V.P. Tretiak, *Clusters of Businesses*, 2nd ed (Irkutsk, Balt. St. Econ. and Law Univ, 2006) p 219
15. M. Battaglia, et al., An innovative model to promote CSR among SMEs operating in industrial clusters: evidence from an EU project, *Corporate Social Responsibility and Environmental Management,* **16(3)**, pp. 133–141 (2010)
16. P. Collazzo-Yelpo and L. Kubelka, Shared value clusters in Austria, *Competitiveness Review*, **29(1)**, pp. 61–76 (2019)
17. K.S. Benidze, Cluster approach in the development of tourism, *Russian Entrepreneurship*, **18(6)**, pp. 895–908 (2017)
18. P. Maher et al., Arctic *Tourism: Realities & Possibilities Arctic Yearbook 2014* Available online: https://arcticyearbook.com/arctic-yearbook/2014 (2014)
19. M. Lindberg et al., Co-creative place innovation in an arctic town, *Journal of Place Management and Development*, **13(4)**, pp. 447–463 (2020)
20. L. Källström and J. Hultman, Place satisfaction revisited: residents' perceptions of "a good place to live", *Journal of Place Management and Development*, **12(3)**, pp. 274–290 (2019)
21. C. Kostenius, *Future Health Addressing Societal Challenges ed J Frishammar and Å Ericson* (Luleå, University of Technology, 2018) pp. 71–85
22. Co-Creating Attractive and Sustainable Urban Areas and Lifestyles: Exploring New Forms of Inclusive Urban Governance, *Nordregio* (Stockholm: Nordregio) p 53 (2016)
23. J.A. Yakovleva, et al., Sport and health aspects of adaptation in northern regions of Russia, *Teoriya i praktika fizicheskoy kultury=Theory and Practice of Physical Culture*, **4**, pp. 20-21 (2020)
24. V.V. Sharok, et al., Dynamics of students' ideas about the work in the Arctic, Vysshee Obrazovanie v Rossii, **08693617**, **30**, pp. 112-119, doi: 10.31992/0869-3617-2021-30-4-112-119 (2021)

Northern labyrinths: potential for interdisciplinary research and a resource for international tourism

Alina Paranina[1,*], *Roman* Paranin[2,*], *Alexey* Grigoryev[3,*], *Lyubov* Larchenko[4,*] and *Margarita* Kiseleva[5,*]

[1] Candidate of Geographical Sciences, Associate Professor, Herzen State Pedagogical University of Russia, 191186, St. Petersburg, Russian Federation
[2] Master in Geography, graduate student, Herzen State Pedagogical University of Russia, 191186, St. Petersburg, Russian Federation
[3] Doctor of Geography, Professor, Herzen State Pedagogical University of Russia, 191186, St. Petersburg, Russian Federation
[4] Doctor of Geography, Professor, Herzen State Pedagogical University of Russia, 191186, St. Petersburg, Russian Federation
[5] PhD in Education, Associate Professor, Herzen State Pedagogical University of Russia, 191186, St. Petersburg, Russian Federation

Abstract. The article presents the scientific rationale for the project of creating a geographic database on northern labyrinths - ancient stone structures and petroglyphs located on the coasts and islands of Northern Europe. The aim of the project is to conduct interdisciplinary research of labyrinths-gnomons (tools for orientation in space and time according to the Sun) as a source of information about ancient marine communications, development of navigation technologies, as well as coastal changes in the Arctic region and climatic rhythms of our planet. The research algorithm includes standard methods of complex geographical descriptions; solution of the equation based on the main trigonometric function; analysis of stable relationships in the system "object - landscape - geographic space - Earth - Universe". The relationship of the structure of labyrinths performing instrumental functions with the geographical latitude, as well as with the long-period rhythms of the planet, which are most clearly manifested in the Arctic region, is considered. The influence of navigation technologies on the physical development of geographic space and its modeling at the local (toponyms), regional (territorial systems and communications) and global levels (modern scientific picture of the world and ancient information models preserved in mythopoetic form) is shown. It is concluded that the creation of a regional geographic database on labyrinths will contribute to the preservation of cultural heritage sites, replenishment of the resources of international tourism in the Arctic region, the development of scientific communications uniting scientists from the northern countries.

1 Introduction

The northern labyrinths are mysterious prehistoric and medieval structures in the form of spirals. They are of great interest to scientists, independent researchers and tourists. Typically, the labyrinths are lined with wild stone or turf, and are located near water or on terraces. The largest number of labyrinths are concentrated on the islands and coasts of northern Europe. Especially many objects have survived on the shores of the Baltic Sea. Their highest density was noted on the Bolshoi Zayatsky Island in the White Sea - more than 30 objects on the area of 1.25 km2. Similar objects are situated in the Mediterranean and the Caucasus, they are found on all continents except Antarctica.

In culture, labyrinths and their images personify happiness - a harmonious human state associated with the fullness of human life, as well as with the understanding of the consistency of all parts of the world around humankind. In the urban environment, the sign is rethought with an exaggerated emphasis on certain semantic elements (for example, in Christian churches, an emphasis is placed on the "wandering of the soul" and the complexity of choosing the path of life). Artificial polarization of meanings made the interpretation of the sign contradictory. However, the geometry of the labyrinth retains the memory of its original instrumental function, using the orientation in geographic visual space and time.

The authors of the article offer a study of the labyrinths based on the methodology of the Russian school of cultural geography. The object of study of *cultural geography* is the geocultural space as the integrity of nature and culture. Culture is understood as a collective experience of adaptation (super-biological adaptation). The natural environment of the sites is studied using field and cameral methods of geography, landscape science, geomorphology, mathematical and evolutionary geography. The socio-cultural context of cultural heritage objects is considered taking into

* Corresponding author: galina_paranina@mail.ru, paranin.roman@yandex.ru, neva8137@mail.ru, lubalar@mail.ru, ritakis@mail.ru

account the data from archeology, anthropology, ethnography, cultural studies and historical geography.

In 2008 – 2010, with the help of the technology of sundialcalendars, the rational purpose of the northern labyrinths, located in the White Sea - on Bolshoy Zayatsky Island of the Solovetsky Archipelago, was determined. In 2010-2021, a navigational concept of information modeling of the world was developed, which explains the cosmism of the main national holidays, the integrity of traditional culture and consciousness [1, 2]. The concept was confirmed and developed in studies of rock carvings of the labyrinths located in the Caucasus. The navigational concept is consistent with the characteristics of earlier objects - the location of natural/man-made zoo- and anthropomorphic sculptures on ancient routes and the orientation of megalithic complexes [3, 4, 5].

The navigational functions of heritage sites open up new possibilities for exploring the past, because astronomical instruments reflect not only the age of the relief and catastrophic events, but also the angle of inclination of the Earth's axis, and associated with it mode of illumination, the speed of axial rotation, the magnitude of the gravity force, the rate of denudation and sedimentation. To extract this information contained in the structure of labyrinths and their position in the landscape, it is necessary to create a geographic database (hereinafter referred to as the DB).

To create a geographic database, a number of organizational problems must be solved. First of all, there is the problem of spatial accessibility associated with the location of objects in different states. In addition, the pandemic has created restrictions on direct contact. Under such conditions, the role of coordinated actions of research groups that work on the ground according to a single program increases.

When drawing up a research program and creating tourism projects, it is important to take into account the sensitivity of Arctic natural complexes to global change. It is due to the low biomass, low rate of biological circulation and permafrost. The current rise in global temperatures and an increase in anthropogenic load cause the most noticeable changes in all natural components and accelerated restructuring of native landscapes. These processes pose a potential threat to the objects of cultural heritage of the Arctic. The main factors in the preservation of labyrinths can be: 1) special status and role in the modern life of the indigenous population, 2) organization of interdisciplinary research on a natural-science basis, 3) inclusion of the objects in economic projects for the development of the recreation and tourism industry.

The aim of the article is to develop an algorithm for a complex study and involve sociologists, ethnographers, archaeologists, local historians, tour operators and interested representatives of other specialties from countries where ancient labyrinths and their images on the rocks (petroglyphs) have survived.

2 Methods of studying the labyrinths as ancient orienteering tools

The project of creating a geographic database on ancient northern labyrinths is based on cameral and field research methods. An important source of information is also scientific publications, which provide a versatile description of objects (unfortunately, the literature often does not contain reliable data on the geo-position of objects). For further paleogeographic reconstructions, only those objects for which the possibility of determining the noon, solstice and equinox days has been established can be selected.

The method of determining the instrumental functions of labyrinths is based on the correlation between the elements of their spatial structure and the parameters of the illumination mode (height and azimuth of the Sun). The main element of sundials and calendars is the gnomon (object giving the shadow). The height and location of the gnomon, at which the extreme positions of the shadow coincide with the diameters of the inner and outer arcs of the labyrinth, can be established using equation (1) based on the trigonometric identity ($H = \tan \alpha \cdot A$). Both sides of the equation give the same result - the height of the gnomon and describe the ratio of the angle of incidence of the sunlight (α) and the length of the shadow (A): the left one is for summer, the right one is for winter. To the distances measured from the center to the extreme arcs of the labyrinth (in this example - 1 m and 5.5 m), we must add a correction x in order to accurately determine its position in the center [6]:

$$\mathrm{tg}\,48{,}47\,(1 + x) = \mathrm{tg}\,4{,}97\,(5{,}5 + x) \qquad (1)$$

The greatest number of functions ("compass", clock, calendar) are performed by labyrinths symmetrical relative to the geographic meridian. In them, the diameter of the arcs in the northern direction marks the length of the noon shadow. In other objects, the boundaries of astronomical seasons and other dates are determined only by the azimuths of sunrise/sunset.

The universal algorithm for researching objects with instrumental functions includes the following stages: 1) application of standard methods of description (measurement, comparison); 2) characterizaion of the enclosing landscape (taking into account the evolution of the natural and climatic situation in the Holocene and geological and geomorphological features, including dominant systems in fracturing of rocks and in the extent of lineaments - linear tectonic structures that are expressed in the landscape); 3) astronomical and paleo-astronomical calculations of the calendar azimuths of sunrise/sunset and the Moon (taking into account the differences between the physical and astronomical horizons), height of the sundial-calendar gnomon and position of the midday shadow by seasons (for geographical coordinates of the object); 4) establishing correlations of the spatial characteristics of the object under study, the enclosing landscape and significant astronomical indicators recorded at a given point; 5) comparison of instrumental capabilities of the object with local and regional tasks of life support in different historical epochs (including analysis of the object location in the system of transport communications). Comprehensive studies of the primary purpose and use of ancient sites also include analysis of the cultural context (adjacent archaeological sites, toponymy, linguistics, mythology, sacred landscape and folk traditions) [7].

3 Geographical reasons for the diversity of labyrinth gnomons

The special geographical location of the Arctic region predetermines the increased dynamism of natural conditions and the display of this diversity in astronomical instruments created on the basis of universal solar technology. priorities of orientation by the Sun in the Arctic are associated with the phenomena of polar days and nights, which occur during the summer period of dynamic economic activity. The most significant calendar dates in extreme climatic conditions are equinoxes, dividing the year into two contrasting parts - warm and cold. The severity of nature, variability of all landscape components and special (more convenient) conditions for astronomical observations well explain the high level of navigational knowledge noted for the peoples of Northern Europe by ancient authors [8]. Let us consider the most significant geographical factors that determine the diversity of stone labyrinths, their location in the landscape, distribution in the regions of Northern Eurasia and the main differences in the drawings.

3.1 Geographical latitude

The points of the ends of the labyrinth spirals form a regular square only at the latitude of Moscow - 55°50'N (Figure 1) because here (and at the same latitude - in the southern hemisphere) the azimuth of sunrise in the days of the solstices is 45° - assuming a straight line of the horizon. To the north of this latitude, the points are shifted to the meridian line, and to the south - to the parallel drawn through the point of installation of the gnomon (Table 1).

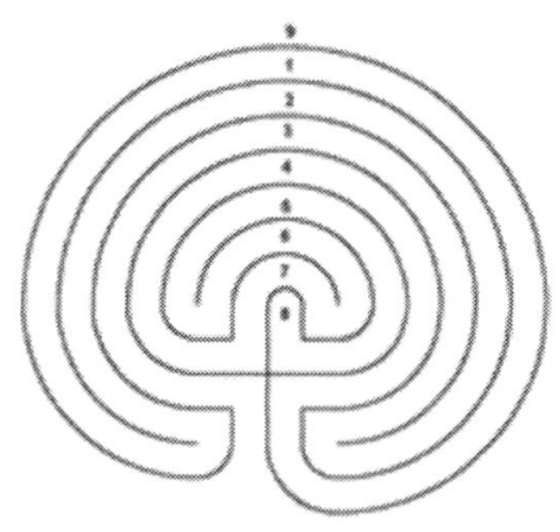

Fig. 1. The points of the ends of the spirals form a regular square only at latitude 55°50'

Table 1. Sunrise azimuths on the days of the solstices in the latitudes of the temperate zone

N (°)	22.12	22.06
66.5	180.00	0
65	160.00	20.03
60	142.86	37.40
55	135.00	45.00
50	128.41	51.82
40	121.29	58.74
30	117.39	62.74
23.5	115.60	64.40

Table 1 also shows that to the north of the parallel of 55°50'N the azimuths of sunrise on the days of solstice change more dynamically than at latitudes south of it. It can be assumed that such differences in the conditions of solar navigation were one of the reasons for drawing the Arctic border at this latitude (56°N - according to Eudoxus).

To the North of the Arctic Circle, the solstice days fall on polar days and, when the Sun does not cross the horizon line (hence, there are no rising/setting phenomena). Under these conditions, the classical bispiral form of the labyrinth is not relevant, and its appearance may indicate borrowing from other regions.

3.2 Changes to the lighting mode in the course of long-period rhythms

In the Arctic region changes in the light regime during the long-period rhythms of the planet (1850-year rhythm of moistening Shnitnikov-Maksimov, 26,000-year equinox precession cycle, 41,000-year Milankovitch cycle, etc.) are most pronounced. Consequently, objects of this region could accumulate the greatest amount of information. The oldest labyrinths are located in the White Sea and are dated approximately 7000 years ago. For this time (climatic optimum of the Holocene), a warmer climate, a decrease in sea ice coverage and a high productivity of the landscape are noted (the territory of the modern tundra was covered with high-quality forests). The main part of labyrinths in Europe belongs to the Middle Ages - this indicates the long-term preservation of the tradition associated with them.

Of great importance for the development of research on labyrinths and other navigation instruments is the doctrine of the rhythm of natural processes by E.V. Maksimov (Figure 2). The graphs show that the structure of the rhythm is the same regardless of the duration. We can see an immutable sequence of periods: warm-dry (WD), warm-moist (WM), cold-moist (CM), cold-dry (CD). In addition, the global climate model reflects several universal rules: 1) the pulsating nature of rhythms reflects the pulsating nature of the matter development; 2) rhythms have a universal manifestation on Earth and in Space; 3) rhythms manifest synchronously in all natural processes; 4) rhythm implies a change in the intensity of natural processes over time; 5) rhythms have a reciprocating character of development: in the active phase "two steps forward – step back", in the decay phase of – "two steps back – one step forward"; 6) no rhythmic process is set by a "zero" background, but always against the background rhythm of a higher rank; 7) the internal structure of asymmetric rhythms reflects an abrupt change in the tempo of natural processes [9].

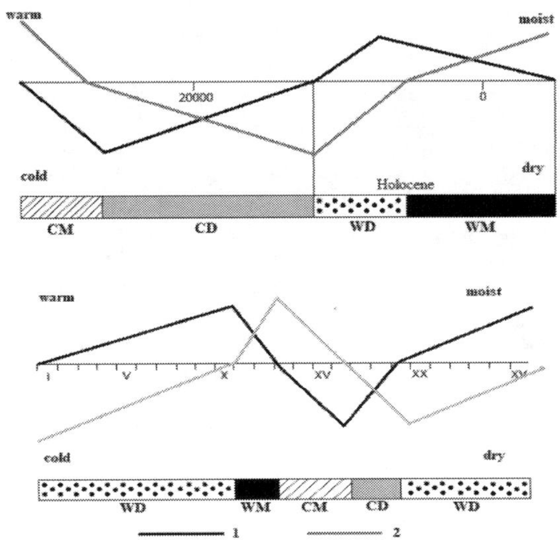

Fig.. 2. The structure of climatic rhythms: on the left- the 41,000-year-old rhythm of the Pleistocene of Milankovitch; on the right - the 1850-year-old rhythm of the Holocene of Shnitnikov-Maksimov [9].

Legend: 1 - temperature, 2 – humidity.

Instrumental objects reflect astronomical conditions at the time of their creation and could retain in their structure information about the features inherent to our planet in the distant past. According to M. Milankovich, the inclination of the Earth's axis changes from 22.1° to 24.5° over 41000 years, which leads to cyclical changes in heat supply and climate. At the same time, the Arctic Circle shifts in the range from 67°54' to 65°30'.

Taking into account the displacement of the Arctic Circle, three zones can be distinguished in the area of the labyrinths of Northern Europe, differing in the light mode: 1. Polar – to the north of the extreme possible position of the Arctic Circle; 2. circumpolar - near the present position of the Arctic Circle (± 1-1.2°); 3. Temperate zone labyrinths, located to the south. In accordance with this division, most labyrinths of the Baltic Sea are located in the temperate zone: the labyrinths of the White Sea are located in the circumpolar zone of the eastern part of the range, only a few objects have survived in the Arctic (according to historical evidence, on the Kola Peninsula, many labyrinths were dismantled in the 19th century).

3.3 Changes in the natural environment and the location of labyrinths in the landscape

The landscape that houses man-made objects has changed over the millennia. Even the most stable component of the landscape - the relief of the earth's surface - has undergone changes. The coastal relief is influenced by several geographic factors: 1) the rate and direction of tectonic movements of the land, 2) natural fluctuations in sea level, 3) the ratio of exogenous abrasion and accumulative processes (erosion/alluvial shores). As a result of the combined action of all these factors, the position of the coastline and the geo-position of the labyrinths changed. Thus, Stone Age labyrinths are often found on high coastal terraces (for example, on B. Zayatsky Island), while younger objects are located near water (Keretsky labyrinth in Krasnaya Luda); some of them are destroyed by waves (labyrinth on Saarema Island, Estonia). Comparison of the age of an objec with its position above sea level makes it possible to estimate the rate and direction of geographic processes.

The age of the relief established by different methods (bioindication, archaeological, geomorphological, limnological) varies greatly. For example, according to the estimates of geomorphologists, the drainage of the peninsula, where the Kandalaksha labyrinth is located, occurred about 1,000 years ago. However, the age of the accompanying archaeological finds reaches 1-2 thousand years BC. Another example: according to the dates of the lakes, the land of Bolshoy Solovetsky Island at an altitude of 13 m is no more than 4000 years old, but on the neighboring Bolshoy Zayatsky Island, the radiocarbon analysis of archaeological finds shows 5000 years. The objective reason for such discrepancies is the different rate of tectonic uplifts (key tectonics, differentiated uplifts), which does not allow extrapolating the obtained results to the neighboring land areas. The peculiarity of bioindication methods is the limited time span, which can be dated. For example, the lichen thallus lives no more than 1000 years, after this period, the process of colonizing the surface begins anew. Perhaps, it is the bioindication methods that rejuvenate the age of most labyrinths of Northern Europe. All this increases the value of new methods based on paleo-astronomical and arche-astronomical data.

4 Labyrinths as navigational elements of water communications

4.1 Toponyms and legends as a source of information about ancient navigation tools

In the work "Historical geography of chronicle Russia" V.I. Paranin (1990) showed that archaic toponyms mark the geographical position of objects (geo-position) in the territorial system. The main landmarks are the directions of the sides of the horizon, determined by the Sun. Toponyms denoting the sides of the horizon directly or indirectly reflect the properties of the Sun in a particular position (high, hot, red; low, cold, white) [10]. For example: the Krasnaya Luda Peninsula is the southern point in the once established spatial system.

Toponyms related to navigation technologies (astronomical instruments or events) can also be added to this group of names. For example, the name of Boshoy Zayatsky island, due to the dense arrangement of astronomical instruments, acquires a new semantic connotation, not associated with hunting for animals (zayats is Russian for hare) or collecting eggs (za-yayitsky literally for eggs. Yayitsa is Russian for eggs). The word "hare" ("janis") in the traditional culture of the peoples of NW Russia, Finland and the Baltic States is used to designate the days of the summer solstice. Comparison of the long-eared hare with the shape of the gnomon's shadow graph on the summer solstice can explain the choice of this image (as a comparison: Seth, brother of Isis and Osiris, the main gods in the cosmogonic mythology of Ancient Egypt, also appears in the image of a hare). The epithet "sunbeam" - a patch of reflected light - is also associated with navigation technologies. It is known that the reflected or directed beam has an independent meaning - as a pointer, which at a certain time illuminates the elements of ornament or objects that play the role of a dial. But the auxiliary role of a reflecting surface (e.g., water) is also great - to increase the total luminous flux, in order to localize the shadow more accurately. In fairness, it should be noted that June is the peak of hare reproduction, which could be an associative basis for fixing the connection between the astronomical cycle and phenological phenomena in nature.

Ancient legends about the "rejuvenating" apples of Freya and the Minotaur are directly connected with the labyrinths, in which the images appear: an apple (the labyrinth looks like an apple in section), a spear (gnomon) and a rope (a flexible tool that can replace a compass, a ruler, a right triangle). Mythopoetic plots with the participation of a bull acquire new content, if we take into account that the bull = an image of time.

4.2 Archaeological and historical evidence of ancient communication networks

The most ancient is the description of the Northern Sea Route, left by the ancient Greek traveler Pytheas, who reached "the last Fule (Thule) by ship". It is generally accepted that the word "thule" is the name of an island. However, it is difficult to imagine several islands with the same name. On the other hand, in the languages of the peoples living in the area of the northern labyrinths, "thule" means "light" (Karelian language) and "instrument" (English).

The first known settlements on the White Sea coast of Karelia date back to the Middle Mesolithic, which coincides with the climatic optimum of the Holocene. The plots of the White Sea petroglyphs show that this time is characterized by a high level of navigation development (images of boats with a capacity of up to 10 people, sea animals catching, etc.). At the end of the 15th - beginning of the 16th century the trade route from Pomorie to Sweden passed here. In 1837, Elias Lönnroth, a researcher of the Karelian epic wrote in his diary that "from mid-March to the end of May, carts pass through Keret to Kandalaksha and from there people go fishing and hunting in the Arctic Ocean".

4.3 Symbolism of signs, derived from technologies for measuring space and time

Galileo Galilei's ingenious idea that Nature speaks to us in the language of geometry, the letters of which are simple figures - circles, lines, triangles - "without them, wandering in a labyrinth is in vain." Comparison of the structure of labyrinths with the graphs of the light and shadow movement of the gnomon allows us to make sure that the primary symbolism of geometric signs and their artistic transformations correlates with the vital processes of nature. The points located to SW and SE from the center of the labyrinth (the ends of the spirals) constrain the graph of the shadow of the longest day, creating the contour of the pyramid. The points at NW and NE from the center of the labyrinth constrain the graph of the shortest winter day, reminiscent of belligerently raised horns. Drawing shadow graphs for a year gives the labrys, a two-sided two-horned axe: according to the legends, the goddess of chaos Neith (Ancient Egypt), the Titanide Rhea and Zeus, the god of light (Ancient Greece).

The two spirals of the classical labyrinth also symbolize the observed natural processes - the spiral trajectories of the rising and falling sun between the solstice days. These spirals are similar in shape, but opposite in direction and meaning. At the same time, they are inseparable in an integral annual cycle, and are one as day and night, light and shadow. The embodiment of these spirals can be considered the images of two snakes: in the labyrinth, in the gnomon-caduceus, as well as the serpent in the hands of the Cretan mother goddess and the snakes "holding the sky" (in the myths of the peoples of the world). Many signs and images are the result of the division and personalization of parts of a single time, hence their ability to reincarnation (for example, Zeus appears in the images of a snake, bull, eagle, etc. like the Sun passing through different constellations of the Zodiac).

5 Conclusion

The northern labyrinths can be viewed as a valuable source of information on the development of nature and culture. To carry out systematic studies, it is necessary to create a database indicating: 1. geographic coordinates of the central points; 2. true (geographic) azimuths; 3. shape of the physical horizon (circular panorama); 4 age of relief elements (coastal terraces) on which labyrinths are located; 5. dating of labyrinths obtained on the basis of paleoastronomical calculations.

Scientific tourism can make a great contribution to the implementation of the project, as a supplier of GPS-data on geo-location and objects and as a new stimulus for the development of ancient communications.

References

1. G.N. Paranina, Northern Labyrinths – gnomon and models of geographical space, *Elsevier Proc. Soc. and Behavior. Sci.*, **19**, pp. 593-601 (2010)
2. A. Paranina and R. Paranin, Primary navigation purpose of the petroglyphs: reconstruction on the basis of the gnomon, *OALib Journal*, **4**, pp. 1-13 (2017)
3. A. Paranina, et al., Cross-disciplinary Research of Objects of Ancient Heritage on the Example of Stone Labyrinths and Petroglyph, *TransNav*, **11(4)**, pp. 729-734 (2017)
4. A.A. Grigoryev, et al., Stone objects of Russian Fennoscandia: potential for recreational use, *IOP Conf. Ser.: Earth Environ. Sci.*, **302(1)**, 012148 (2019)
5. L.S. Marsadolov, et al., Problems of preservation of prehistoric cultural heritage objects in the Arctic, *IOP Conf. Ser.: Earth Environ. Sci.*, **302(1)**, pp. 1-8 (2019)
6. A. Paranina and R. Paranin, Study of Northern Labyrinths as Elements of a Navigation Network, *TransNav*, **10(3)**, pp. 451-456 (2016)
7. L.S. Marsadolov, et al., An integrated approach to the study of megalithic legacy, *Bulletin of Tomsk St. Univer. Hist.*, **22(2)**, pp. 72-75 (2013)
8. V.I. Paranin, *History of the barbarians* (St. Petersburg, Russian Geographical Society, 1998) p. 283
9. O. Pomortsev and E. Kashkarov, Theory of rhythms E V Maximov Geography: Problems of Science and Education, *63rd Herzen Readings, April 22-24, 2010, St. Petersburg*, pp. 8-13 (2010)
10. V.I. Paranin, *Historical geography of chronicle Russia* (Petrozavodsk, Karelia, 1990) p. 152
11. L.V. Larchenko and R.A. Kolesnikov, Regions of the Russian Arctic zone: state and problems at the beginning of the new development stage, I. J. of eng. and technol. (UAE), 7, **14(3)**, pp. 369-375 (2018)
12. L.V. Larchenko, et al., Resources for sustainable development of Russian Arctic territories of raw material orientation, *IOP Conf. Ser.: Earth Environ. Sci.*, **302(1)**, 012121 (2019)
13. L. Larchenko, et al., Industrial Innovation Clusters of Saint Petersburg: Problems and Development Prospects, *SPBPU IDE '19: I. Sci. Conf. on Innovat. in Digital Economy*, **57** (2020)
14. V.D. Sukhorukov, et al., History and Geography of The Russian Arctic in the Domestic Travelogue, *IOP Conf. Ser.: Earth Environ. Sci.*, **625(1)**, 012010 (2021)
15. D.V. Sevastyanov, et al., Recreational environmental management and Arctic tourism as a new trend in the strategy for sustainable development of the Arctic countries, *Ecol., Environ. & Conservation*, **23(4)**, pp. 480-487 (2017)
16. D.V. Sevastyanov, et al., The Barents Sea Region: current status and boundaries of the possible, *IOP Conf. Ser.: Earth Environ. Sci.*, **625(1)**, 012008 (2021)
17. Yu.N. Gladkiy, et al., History and perspectives for the expansion of the Russian Arctic to the Ecumene *IOP Conf. Ser.: Earth Environ. Sci.*, **434(1)**, 012005 (2020)
18. Yu.N. Gladkiy, et al., Geophysical determinants of wind energy development in the Far North of Russia, *IOP Conf. Ser.: Earth Environ. Sci.*, **459(2)**, 022071 (2020)
19. Yu.N. Golubchikov, et al., Arctic tourism: state and prospects for Russia, *Geography, Environ., Sustainab.*, **11(4)**, pp. 5-13 (2018)
20. A.A. Grigoryev, *Prehistoric development of the Arctic: geographical aspects Doistoricheskoe osvoenie Arktiki: Geograficheskie aspekty* (St. Petersburg, Asterion, 2018) p. 226
21. A.A. Grigoriev, et al., Prospects for the rational use of water bodies of the northwestern federal district of Russia in the tourism sector, *IOP Conf. Ser.: Earth Environ. Sci.*, **381(1)**, 012032 (2019)

E3S Web of Conferences **378**, 06022 (2023) https://doi.org/10.1051/e3sconf/202337806022
IIRPCMIA 2021

Opportunities and Limiting Factors for Tourism Development in the Khibiny

Elena Vorobievskaya[1], *Sergey* Kirillov[2,*], *Natalia* Sedova[1] and *Michael* Slipenchuk[3]

[1]Candidate of Geographical Sciences, Senior Researcher, Department of Environmental Management, 119991, Lomonosov Moscow State University, Moscow, Russian Federation
[2]Doctor of Economics, Professor, Department of Environmental Management, 119991, Lomonosov Moscow State University, Moscow, Russian Federation
[3]Doctor of Economics, Department head, Department of Environmental Management, 119991, Lomonosov Moscow State University, Moscow, Russian Federation

Abstract., the Khibiny attracts more and more attention as an interesting and promising object attracting flows of tourists and athletes not only from a number of regions of Russia, but also from abroad. In this regard, significant changes are taking place in Kirovsk in the structure of economic activity, associated, first of all, with the reorganization of recreational activities. However, the presence of mining facilities in this area has caused a large load on natural ecosystems. As a result, an adverse environmental situation is developing in the study area, which limits the development of recreation here. To assess the current state and identify opportunities for the development of recreational activities in the Khibiny, in particular, in the city of Kirovsk and its environs, comprehensive studies were carried out in January 2019, 2020 and in August 2021. The results show that this region is a very attractive recreation center, and vacationers and local residents are interested in the development of tourism in this area. At the same time, the studies revealed a number of problems limiting the development of tourism in the Khibiny, which include, first of all, the deterioration of the quality of the natural environment associated with the emergence of technogenic landscapes as a result apatite-nepheline mining. socio-economic factors that reduce the tourist and recreational attractiveness of the Khibiny include the low level of development of tourist infrastructure and the quality of services.

1 Introduction

Currently, the Khibiny attracts more and more attention as an interesting and promising object attracting flows of tourists and athletes not only from a number of regions of Russia but from abroad. In this regard, significant changes are taking place in Kirovsk, in the structure of economic activity, associated, first of all, with the reorganization of recreational activities [1]. However, the presence of OAO Apatit mining facilities in this area has caused a large load on natural ecosystems (Figure 1).

Such an increased concentration of industrial enterprises and dumps in a relatively small area in one way or another negatively affects the state of both natural geosystems and the environmental situation in the region as a whole. The direct consequences of environmental pollution are: oppression of the vegetation cover in the vicinity of processing plants due to clogging of the stomata of plants with dust and subsequent drying out, changes in the mechanical composition of soils due to enrichment of the sand fraction, eutrophication of lakes, formation of destructive mudflows descending from the slopes of the dumps of open pits, violation of the aesthetics of landscapes, etc. Indirectly, they have an adverse effect on the health of the population: the development of diseases of the skeletal system due to strontium pollution, broncho-pulmonary and allergic diseases due to air pollution, etc. [2]. According to the Kola Science Center, the main air pollutants in the mining area are: dust particles containing strontium, zinc, lead, phosphorus, aluminum salts, magnesium, sodium, calcium, potassium, carbon and nitrogen oxides; for water reservoirs: molybdenum, lead, phosphorus, nitrogen, nickel, copper, lead, sulfates.

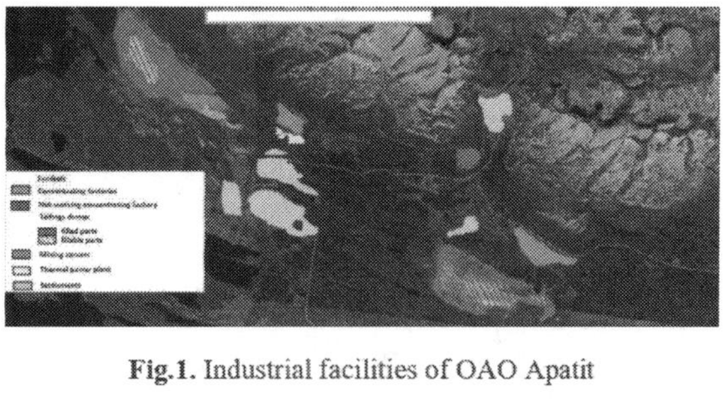

Fig.1. Industrial facilities of OAO Apatit

* Corresponding author: eco-msu@mail.ru

The development of recreational use of natural resources, which, by the nature of its exploitation, has a limited impact on ecosystems in comparison with other types of nature management, can become a serious alternative to industrial nature management [3] and will contribute to improving the environmental situation in the Khibiny region.

Comprehensive studies were carried out in January 2019, 2020 and in August 2021 to assess the current state and identify opportunities for recreational development in the Khibiny [4], in particular, in the city of Kirovsk and its environs, as well as environmental and socio-economic factors limiting the development of recreation in this area.

2 Methodology

A variety of methods have been used to identify opportunities and limiting factors for the development of tourism in the Khibiny. extensive literarture, cartographic, statistical data concerning the natural and socio-economic conditions for the development of this type of nature management in the study area were studied beforehand. By analyzing satellite images and their interpretation, the location of the main objects that pollute the environment was determined. Further, based on the data of environmental studies carried out in the Khibiny in different years, the main environmental problems associated with the extraction and processing of apatite-nepheline ore were identified. Having studied the scientific literature, the main anthropogenic loads resulting from the development of recreational activities were identified. analysis of statistical data made it possible to identify the number of the main flow of tourists and their distribution by seasons of the year.

It also seemed necessary to conduct sociological survey [5] in order to assess the tourism infrastructure and opportunities for local residents in connection with the development of tourism in the Khibiny. To conduct the survey, a questionnaire was prepared for people of different target groups. When processing the information of the survey, the method of serial sampling was used.

3 Results

In the study area, there is an unfavorable environmental situation, which limits the development of recreational activities here due to the presence of mining facilities.

The main environmental problems associated with the extraction and processing of apatite-nepheline ore in the study area are:

- alienation and disturbance of natural landscapes, destruction of soil cover and vegetation, increase in the area occupied by dumps, tailing ponds
- pollution of atmospheric air by dust particles containing strontium, carbon and nitrogen oxides, dioxins
- pollution of water bodies by deposited toxic particles of dust, phosphorus, nitrogen, chemical reagents used in blasting, flotation and industrial processes
- noise pollution (in the areas closest to the blasting operations)
- development of dangerous natural phenomena in the production areas (anthropogenic earthquakes, water and mud flows, avalanches, etc.).

Of special concern is "aesthetic" pollution in the areas of operations of mining and ore processing enterprises, since the aesthetic attractiveness of landscapes is especially important in the development of recreation and tourism. In the Khibiny, one can observe a violation of the landscape aesthetics (destruction, often irrevocable, of natural landscapes): for example, in the places of mining and storage of mining waste, there are mechanical disturbances of slopes, soil and vegetation, which, undoubtedly, significantly reduces the aesthetic value of the territory [6]. Significant and irreparable damage as a result of mining operations has been suffered by: Mounts Aikuayvenchorr, Rasvumchorr, Koashva, Kukis, Nyorkpakhk, Yukspor. Many of OAO Apatit's production areas have created obstacles to popular hiking trails - it is forbidden to stay on their territory and passage is prohibited. For example, now the trail to the Gakman gorge is blocked, there is no passage to Mount Eveslogchorr, etc.

Sociological research [7], which serves as a link between theoretical studies and reality, allowed us to assume the presence of new trends in the development of recreational nature management in the Khibiny, and to identify a number of limiting factors associated with the development of tourism infrastructure.

representatives of various Kirovsk institutions, local residents [8] and tourists took part in the survey.

As a result, data were obtained from a survey of different segments of the population of Kirovsk and its environs. A total of 102 people were interviewed (local residents - 53, tourists - 49). The sample was random and included people of different gender, age, nationality, education and income level. We were interested in the following: 1) how local residents feel about the development of tourism in their region; 2) whether local residents consider it necessary to develop industry and new deposits in the region; 3) for what purpose, mainly, tourists visit the Khibiny; 4) where tourists come from; 4) how people assess the tourist infrastructure of the city and development of recreation in the region; 5) about proposals for improving tourist infrastructure and safety [9]. The results of the survey were compiled into 43 diagrams and one table.

Among tourists 49 people were interviewed, the majority of respondents - 55% - were young people between 17 and 22 years old, 39% were adults between 23 and 55 years old; the overwhelming majority of 49% were students. Most tourists come from cities and villages of the Northwestern Federal District, where the Khibiny is located, and 70% from St. Petersburg. 90% of the interviewed tourists came here for skiing and recreation, 6% to visit, and 4% came for work. 45% arrived by train, 35% by car, and 20% by plane.

Tourists mainly stay in the private sector - rent apartments (47%). 29% stay in hotels, and 24% stay with relatives or friends. tourists themselves explain this by the fact that renting an apartment by a group of people is cheaper than a hotel room. So, the average price for a hotel is 2,190 rubles/day, and the price ranges from 500 rubles/day for a bed in a hostel to 5,000 rubles day for a hotel. You can rent an apartment for an average of 2400 rubles per day.

When asked "Why did you come here?" 46% of respondents answered that they came to a good ski resort, with good slopes and beautiful nature, because they like it here, or at the advice of friends (Figure 2).

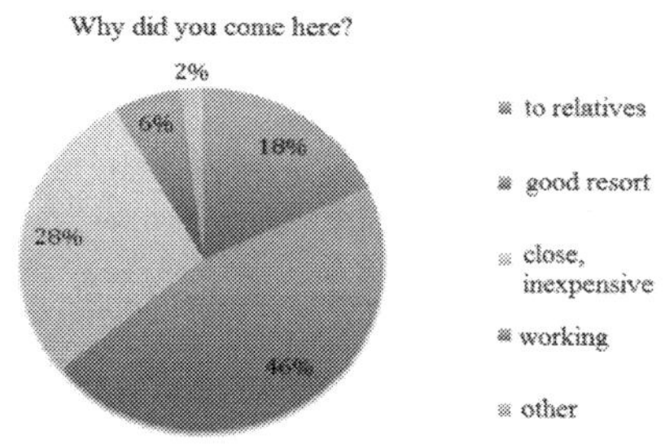

Fig. 2. The reason why tourists come to the Khibiny

Most tourists (61%) learned about the place from relatives, friends and acquaintances. 21% of tourists always knew about this place, 14% learned about it from the Internet, and 4% have been to these places for other reasons.

57% of all tourists came to the Khibiny for vacation for the first time. And of those who have vacationed here before, half of the respondents have been to this place more than 4 times, and they come here regularly - if possible, every year or several times a year.

The infrastructure was rated by tourists on a 5-point scale. The average rating of the accommodation was 4.06. respondents suggested building more hotels and hostels with affordable prices, located near ski resorts, renovating old hotels and improving the appearance of the city. Transportation was rated 3.9. Tourists suggest updating the vehicle fleet and increasing the number of public transport, to make transport accessible (more routes to the ski resorts), to provide transport on time make sure the roads are cleared of snow. Guests also suggested the introduction of more regular and affordable flights from other cities. The quality of service was rated 4.2. The average score for entertainment is 4.06. Here tourists mainly suggest an increase in the number of leisure facilities for youth: Internet cafes, cinemas, bars, clubs, etc., as well as more establishments with good food and places for children's entertainment.

The highest priority goal of the trip for tourists is skiing and snowboarding, as well as free ride for fans of extreme sports (Table 1).

In general, tourists are satisfied with the quality of tourist infrastructure: the place of residence - by 85%; transport - by 73%; technical level of equipment and quality of service - by 94%. Despite this, guests have a number of suggestions for improving the tourist infrastructure. As for accommodation and transport, the suggestions coincide with the previous answers: to build more modern hotels and hostels of different price categories, to adjust the prices at the market of tourist services (price-quality ratio); to increase and renew the fleet of public transport, to expand air accessibility of Kirovsk to Russian and foreign cities. In terms of the technical level of equipment, it is proposed to replace drag lifts with chairlifts and equip more trails. It is also necessary to improve the qualifications of the personnel serving tourists.

Table 1. The purpose of the tourist trip by priority

Type of tourism / Priority	Downhill skiing	Free ride	Ice- fishing	Hiking	Visit to the mine	Cross-country skiing
1	36	10	0	2	0	0
2	7	12	4	12	0	8
3	3	7	2	6	4	14
4	2	6	2	6	6	7
5	0	1	10	7	5	3
6	0	2	10	0	12	3

96% of the interviewed tourists are ready to return to the Khibiny again, because, according to their answers, it is beautiful here, ski resorts have been created, snow lies long, there are slopes of different degree of complexity, designed for any training of vacationers and athletes, close location to the center of European part of Russia, one can engage in other types of tourism and see the northern lights. Some guests would like to come here not only in winter, but also in March-May during the "velvet season", and in summer, to enjoy their holidays in the Khibiny at this time of year.

Tourists in the Khibiny (63%) are ready to do other kinds of recreation besides downhill skiing: cross-country skiing, snowmobiling and ice skating, and in the summer to go hiking, ride ATVs, bicycles, scooters, go mountaineering and diving. However, 37% of respondents want to do only one type of recreation in the Khibiny, because other types of tourism in this area do not attract them.

Tourists assess the level of tourism development in the Khibiny mainly (63%), as "good". 21% of the interviewed tourists assessed it as "excellent" and 16% assessed it as "satisfactory".

Fifty-one people were interviewed among local residents. Those who have lived in Kirovsk for more than 25 years are 63%, and those who are going to live and work here further - 71%. 43% of respondents want their children to stay in Kirovsk, but almost a third of respondents do not want that. When asked "Do you know the history of Kirovsk?", only almost half answered "yes, fine," and 12% answered "no".

The geographical location, size of the city and other factors allow local residents to spend their free time in nature, and 91% of respondents do so. It should be noted that 45% of them stay in nature in the warm season, only 8% more often in the cold season, and for almost half of the respondents, the time of year is not a deterrent, and they go to rest in nature all year round. In the warm season, local residents prefer biking, swimming in lakes, nature walks, hiking, picking mushrooms and berries in the woods, there are also lovers of picnics, fishing, etc. In the cold season, spending time in nature is less diverse, mostly locals spend it on the ski slopes, skiing or skating.

The majority of local residents surveyed (81%) are positive about the development of tourism in their region, while 4% are against it. 65% want improvements in services and recreation, and almost a quarter do not consider it necessary.

Among the desired improvements are: the development of ski resorts and infrastructure in the city, including transportation, accommodations and an increase in the number of leisure facilities. They also noted the need to lower prices for the services of ski resorts and the development of other types of recreation, in addition to skiing, during the cold season. Half of the surveyed locals would not mind to work in the tourism industry. The prospects for tourism development in the region are seen by 90% of the respondents. The majority consider skiing to be the most promising direction (Figure 3).

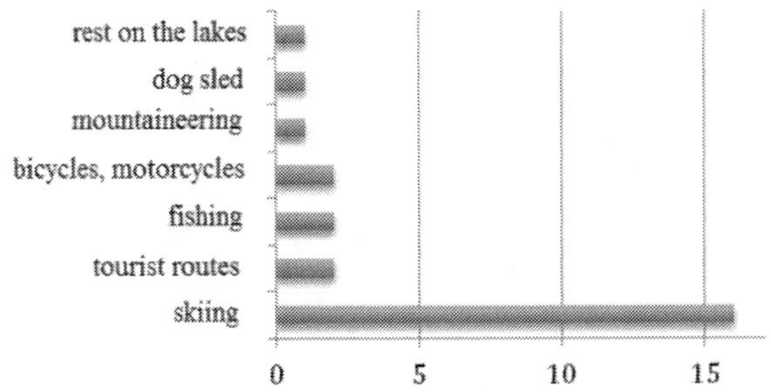

Fig. 3. The most promising type of recreation for the development of tourism in Kirovsk from the point of view of local residents

As for the prospects for the local economy, the majority of local residents (69%) see them in the further development of the mining industry, which provides a large number of jobs. 16% believe it is not necessary, and 15% do not know. About a fifth of all respondents believe tourism will never become a profitable industry in the region. 74% of respondents have a positive attitude to the development of new deposits of apatite ores in the Khibiny (Partomchorr and Oleniy Ruchey), 20% think it is bad, and 6% are neutral. At the same time, almost half of locals are not aware of the creation of a national park in the Khibiny, but the overwhelming majority (96%) express their positive attitude to this.

On the whole, we can say that the opinion of local residents about the prospects for the development of ski tourism is fully justified. Today, more and more tourists from the entire European part of Russia, as well as from the Urals and the Caucasus, come to the Khibiny for recreation at ski resorts. Historically, the region's economy developed thanks to the mining industry, which the locals continue to consider the most important activity. But they also believe that tourism can be a profitable industry in the region, and people are willing to work in this area. relatively good price-quality ratio, close proximity to the center of the European part of Russia, equipped ski slopes, the possibility to ski on them from November to May, all this attracts tourists [10]. The disadvantages are mainly manifested in the underdeveloped recreational infrastructure, lack of inexpensive hostels and hotels, lack of recreational facilities, insufficiently good transport provision, etc. Only one fifth of the tourists rated the level of tourism development as "excellent". This indicates the need to improve conditions for visitors to the region.

4 Conclusion

As a result of the study, the following conclusions were made. The development of recreation in the Khibiny is very promising: this is due to the variety of recreation types, attractive mountain landscapes, and affordability [11]. An overwhelming number of tourists expressed their desire to come to this holiday destination again, and a significant part of the locals would like to work in the tourism sector, despite the fact that it is not considered an alternative to industrial development.

However, there are a number of problems that limit the development of tourism in the vicinity of Kirovsk. An adverse environmental situation is developing in the study area due to the presence of mining facilities [12]. deterioration of quality of the natural environment includes alienation and disruption of natural landscapes, destruction of soil cover and vegetation, air and water pollution, noise pollution, development of dangerous natural processes (anthropogenic earthquakes, avalanches). Special points include the loss of the aesthetic appeal of landscapes and the loss of some valuable tourist sites.

socio-economic factors that reduce the tourist and recreational attractiveness of the region include the low level of development of tourism infrastructure: insufficient number of hotels, boarding houses, tourist centers; discrepancy in the quality of service to the growing needs of vacationers; lack of training in the tourism business (of different categories); insufficient development of health insurance services; unsatisfactory interregional communication of regular transport and specialized transport for tourists; weak advertising activity of tourist services and offers.

The answers of the respondents will serve as a basis for further research and possibly justify the need to take appropriate measures to improve the tourist infrastructure and environmental situation in the Khibiny and adjacent territories, which will contribute to the development of recreational use of natural resources here.

This research was performed according to the development program of the Interdisciplinary Scientific and Educational School of Lomonosov Moscow State University "Future Planet and Global Environmental Change" and within the framework of the state task "Sustainable development of territorial systems of nature management" (project No. 121051100162-6).

References

1. M. Slipenchuk, et al., Multifaceted approach to natural resource management: ethnology, geography, culture, *Journal of the Geographical Institute "Jovan Cvijic" SASA*, **66(3)** (2016)
2. A. Baskov and G. Benker, *Modern sociological theory* (Moscow, 1996)
3. Litvinenko V S et al. 2020 The social and market mechanism of sustainable development of public companies in the mineral resource sector EURASIAN MINING (Business and International Managemen) 1 pp. 36-41
4. M. Slipenchuk, et al., Anthropogenic pollution of the southern part of the Khibiny mountain massif and foothills, *IOP Conference Series: Earth and Environmental Science*, **302**, 012024 (2019)
5. G.V. Osipov Ed. *Sociology: the foundations of a general theory* (Moscow, 1996)
6. M. Huber, et al., *Geoheritage,* **13(3)**, (2021)
7. V.A. Yadov, *Sociological research* (Moscow, 1995)
8. E. Vorobievskaya, et al., Issues of the preservation of historical, cultural and environmental heritage of indigenous peoples of the Arctic, *IOP Conference Series: Earth and Environmental Science*, **302**, 012152 (2019)
9. A. Shnyparkov, et al., Theory and Practice of Individual Snow Avalanche Risk Assessment in the Russian Arctic, *Geography, environment, sustainability*, **5(3)** Available online: https://ges.rgo.ru/jour/article/view/170/168 (2012)
10. D.V. Sevastyanov, et al., *Geography and Natural Resources*, **36**, 4 (2015)
11. S. Zhagina, et al., Development and Prospects of Ecotourism in the Arkhangelsk Region, *International Multidisciplinary Scientific GeoConference SGEM*, **20(5.1)**, pp. 253-258 (2020)
12. G. Buslaev, et al., Ensuring the sustainability of arctic industrial facilities under conditions of global climate change, Resources, 21100808642(10) doi: 10.3390/resources10120128 (2021)

AUTHOR INDEX

Alekseevich Likhomanov Vladimir........26
Alexander Chuikov........39
Alexander Illinsky........182
Alexander Suleymanov........95
Alexandra Chaikovskaya........113
Alexei Chernov........26
Alexei Fadeev........182
Alexey Cherepovitsyn........138
Alexey Grigoryev........160 200
Alexey Shemyakin........85
Alina Cherepovitsyna........124 132
Alina Paranina........160 200
Alisa Mingazheva........90
Alla Trotsenko........152
Anastasia Kornilova........45
Anatoly Scherbina........152
Anatoly Zhirnov........7
Andrey Ingiri........119
Andrey Shchipachev........90
Andrey Volkov........39 45
Anna Kolesnikova........45
Anna Savitskaya........26
Anna Trofimova........60
Anna Yashkina........119
Anton Rybakov........55
Antonovich Likhomanov Vladimir........26
Arina Riadinskaia........132
Arseniy Sukhov........55
Artem Ivanov........171
Artur Kuchumov........107 113 195
Askar Mingazhev........90
Boris Antipov........31
Boris Shifrin........21
Denis Novikov........31
Dmitrii Shabanov........39
Dmitriy Burylov........35
Elena Klyuchnikova........144
Elena Pecheritsa........107 113 195
Elena Vorobievskaya........206
Eugene Istomin........65
Eugene Sundukov........21
Eugene Syroechkovskiy........69
Evgeniya Rutenko........138
Farida Nurbaeva........50
Fyodor Scherbina........152
Galina Artemova........55
Galina Karpova........195
Gleb Klyuchnikov........144
Guizhan Sabyrbayeva........50
Igor Barskii........39
Irma Martyn........65
Ishel Bianco........182
Ivan Svistunov........26
Jean-Paul Markus........102
Konstantin Klokov........69 75
Konstantin Shotidi........31
Lazzat Nurshakhanova........50
Ludmila Dorofeyeva........189
Lyubov Larchenko........160 200
Lyudmila Drachkova........60
Maksim Komarovskiy........12
Margarita Kiseleva........160 200
Marina Kruk........171
Marina Voloshinova........195
Michael Slipenchuk........206
Mikhail Chura........55
Mikhail Shelkov........152
Mikhail Ulchenko........178
Mikhail Vystrchil........55
Nadezhda Kyzyurova........1
Nadezhda Tarabukina........21
Natalia Blazhenkova........107 113
Natalia Ikko........119
Natalia Kirsanova........166
Natalia Roslyakova........189
Natalia Sedova........206
Natalia Yagotinceva........65
Nikita Ponomarev........178
Nikita Volkov........16
Nikolai Kondratov........60
Nina Krupina........26
Olga Kolesnikova........50
Olga Matsko........45
Peter Zvyagintsev........152
Polina Ageeva........119
Polina Patseva........26
Roman Paranin........160 200
Sergey Kirillov........206
Sergey Yakovlev........85
Sergey Yudin........138
Stepan Menshikov........12
Svetlana Yalovitsyna........148
Tatiana Volkova........16
Tatyana Dolgopolova........166
Tatyana Volkova........152
Tatyana Zyabkina........166

Uolan Borisov 102
Vadim Shpenst 35
Valentina Gogoleva 166
Valery Abramov 65
Vasilisa Bobrova 156
Veronica Sundukova 21
Victor Volkov 16
Victoria Solovyova 124
Vladimir Dvoynikov 35
Vladimir Litvin 171
Vladimir Myazin 81
Vladimir Pshenin 12
Vladimir Smirnov 35
Yaroslav Petrov 65
Yuri Gerasimov 69
Yuri Ugrumov 26

EDP Sciences – Web of Conferences
17, Avenue du Hoggar
Parc d'Activité de Courtabouf
BP 112
F-91944 Les Ulis Cedex A
France

ISBN 978-1-7138-7241-2

9 781713 872412

CURRAN ASSOCIATES INC.
proceedings.com